MITOSIS

MITOSIS

Molecules and Mechanisms

Edited by

J. S. HYAMS
University College
London, UK

B. R. BRINKLEY
University of Alabama
Birmingham, USA

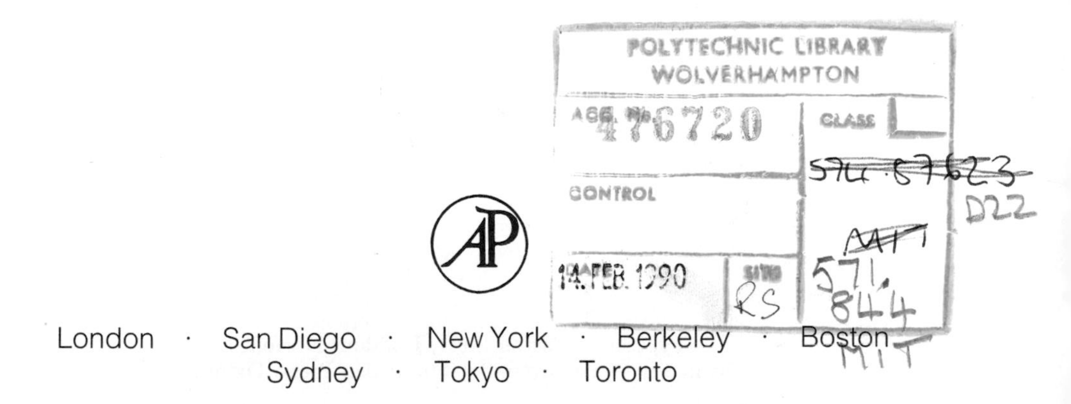

London · San Diego · New York · Berkeley · Boston
Sydney · Tokyo · Toronto

ACADEMIC PRESS LIMITED
24/28 Oval Road
LONDON NW1 7DX

U. S. Edition Published by

ACADEMIC PRESS INC.
San Diego, CA 92101

British Library Cataloguing in Publication Data
Mitosis: molecules and mechanisms
1. Organisms. Cells. Mitosis
I. Hyams, J. S. II. Brinkley, B. R.
574.87′623

ISBN 0-12-363420-2

Typeset by Paston Press, Loddon, Norfolk
Printed in Great Britain by The Alden Press, Oxford

Contributors

R. D. BALCZON Department of Cell Biology & Anatomy, University of Alabama at Birmingham, Basic Health Education Research Building, Birmingham, AL 35294, USA

G. S. BLOOM Department of Cell Biology, University of Texas Health Center at Dallas, 5323 Harry Hines Boulevard, Dallas, TX 75235, USA

G. G. BORISY Laboratory of Molecular Biology, University of Wisconsin, 1525 Linden Drive, Madison, WI 53706, USA

B. R. BRINKLEY Department of Cell Biology & Anatomy, University of Alabama at Birmingham, Basic Health Education Research Building, Birmingham, AL 35294, USA

F. CABRAL Department of Pharmacology, University of Texas Medical School, Houston, TX 77025, USA

W. Z. CANDE Department of Botany, University of California, Berkeley, CA 94720, USA

P. K. HEPLER Department of Botany, University of Massachusetts, Amherst, MA 01003, USA

K. McDONALD Department of Molecular, Cellular and Developmental Biology, University of Colorado, Boulder, CO 80309, USA

J. R. McINTOSH Department of Molecular, Cellular and Developmental Biology, University of Colorado, Box 347, Boulder, CO 80309, USA

E. D. SALMON Department of Biology, CB3280, University of North Carolina, Chapel Hill, NC 27599-3280, USA

S. C. SWEET Department of Anatomy & Cell Biology and Cellular and Molecular Biology Program, The University of Michigan Medical School, Ann Arbor, MI 48109, USA

A. TOUSSON Department of Cell Biology & Anatomy, University of Alabama at Birmingham, Basic Health Education Research Building, Birmingham, AL 35294, USA

M. M. VALDIVIA Department of Cell Biology & Anatomy, University of Alabama at Birmingham, Basic Health Education Research Building, Birmingham, AL 35294, USA

R. B. VALLEE Worcester Foundation for Experimental Biology, 222 Maple Avenue, Shrewsbury, MA 01545, USA

D. D. VANDRE Laboratory of Molecular Biology, University of Wisconsin, 1525 Linden Drive, Madison, WI 53706, USA

M. J. WELSH Department of Anatomy & Cell Biology and Cellular and Molecular Biology Program, The University of Michigan Medical School, Ann Arbor, MI 48109, USA

Preface

There have been a number of excellent books on mitosis over the years. In subtitling this one "Molecules and Mechanisms" we are not trying to suggest that a great deal is known about either. Rather, we feel that mitosis stands at a point where intelligent questions about the molecular basis of spindle formation and the mechanisms of chromosome movement can at least start to be asked. At such a time it is useful to take stock, to try to assess what is known and to define those areas where progress will emerge over the next few years. For many of us, one of the most appealing features of mitosis is that it brings together researchers from a number of diverse backgrounds, such as cell biology, genetics, biochemistry, molecular biology and physiology. In planning this volume we have tried to reflect this breadth of interest. This approach is neither historical nor nostalgic but, rather, reflects our belief that true understanding will only be achieved if we continue to attack these problems on a broad front.

We have been told that this book represents an important milestone, namely, that it will be the last book on mitosis to be published before everything is understood. Perhaps this will indeed prove to be the case. However, the optimist should be cautioned that the history of mitosis has seen many false dawns. Almost every dogma has at some stage been revised or rejected, almost every model of chromosome movement abandoned as yet another unsuspected property of the spindle is revealed. Our own view is that the mechanisms which underlay the formation of the spindle, the rapid erratic behaviour of the chromosomes at prophase and their orderly, rather sedate segregation at anaphase, will continue to test the imagination and experimental ingenuity of those working in this area for some time to come. We hope that this volume may contribute in some small part to this exciting exercise. We hope too that it might stimulate the next generation to take up the challenge.

J. S. Hyams
B. R. Brinkley

Contents

CHAPTER 1

Mitotic Spindle Ultrastructure and Design

KENT McDONALD

Department of Molecular, Cellular and Developmental Biology,
University of Colorado, Boulder, Colorado, USA

I. Introduction

A model application of electron microscopy (EM) to biological problem solving can be seen in the history of T4 phage research, where a combination of genetics, biochemistry, EM and *in vitro* reconstitution experiments was employed in a coordinated fashion to understand the molecular biology of T4 bacteriophage (Wood, 1979). In cell motility research, similar logic is being used in the molecular analysis of flagellar beat (Luck, 1984). EM is used to define the macromolecular structure of the flagellum whilst biochemical techniques establish a catalogue of protein components. Mutants are then selected that show abnormal function and their ultrastructural and biochemical patterns are compared with wild type. Missing parts of the machinery can be correlated with missing proteins on gels and a correspondence between gene products, their structural location and/or function is established. Final corroboration can be achieved by add-back experiments *in vitro* or by genetic means. The challenge to the mitosis researcher is to characterize the parts of the mitotic machinery in the same detail, if not in the same way. In comparison with ciliary (and muscle) research we are still in the early descriptive stages of mitotic ultrastructure and biochemistry. Unlike muscle and cilia, mitotic spindles are fragile structures that are all too easily destroyed by the methods used to study their structure and chemistry. Also, they lack the almost-crystalline arrangement of parts that characterizes these other motility machines. Nevertheless, the importance of mitosis as a biological research problem requires that we surmount these difficulties and gain an understanding of spindle structure and function at a molecular level.

MITOSIS: Molecules and Mechanisms
ISBN 0-12-363420-2

The first electron-microscope images of dividing cells were published over 35 years ago (Rozsa and Wyckoff, 1950) and were less informative than many light micrographs. Since that time, thousands of micrographs have been published illustrating the ultrastructure of mitosis. Perhaps the main results of all these efforts is an even greater appreciation for the difficult nature of the problem and the realization of the need to break down the events of mitosis into manageable subsets, to ask more specific questions and to interpret the experimental results in a narrower and more meaningful context. In this chapter we will identify those instances in which EM has succeeded in clarifying the mitotic picture, where it has failed, and where current technological innovations seem to be leading us. We will stress the new information and ideas that have emerged since the last review of mitotic ultrastructure (Heath, 1981) and some of the areas that have received little attention in the past, such as the ultrastructure of prometaphase chromosome movements. Following the lead of Heath (1981) we will give the methodology of EM more consideration than is usual, because it is becoming obvious that conventional methods of EM preparation are inadequate for the task at hand, i.e. the direct visualization of the molecules responsible for moving chromosomes. Finally, preference will be given to those studies that have used EM in conjunction with other methods such as micromanipulation, microbeam irradiation or cinematography, or which use serial section analysis to produce high-resolution, 3D information on the structure of the mitotic apparatus.

II. Brief History of Mitotic Ultrastructure Studies

A. Types of morphological papers

Most of the papers concerning mitotic ultrastructure fall more or less into one of three categories:

1. survey studies, which include nearly all the papers from the pre-glutaraldehyde era and most of the papers published between 1964 and 1975;
2. analytical or quantitative studies, in which EM is combined with experimental manipulations and/or morphometric techniques to analyse images; and
3. incidental studies, which may use an electron micrograph in an otherwise non-morphological paper, or a morphological study that is essentially non-mitotic but includes a micrograph of a cell in mitosis.

The difference between types 1 and 2 can be illustrated by the problem of chromosomal (kinetochore) fibre structure. In a survey paper, the fact that microtubules are continuous between pole and kinetochore is illustrated by a longitudinal section that includes both structures and one to several microtubules running continuously between them (e.g., see references in Luykx 1970, p. 103). In analytical studies, all the microtubules between kinetochore and pole are traced in serial sections and their lengths and end distributions are displayed in two or even three dimensions (see references in Rieder 1982). Both kinds of information have been useful in shaping our ideas about spindle structure, although in recent years there has been more emphasis on analytical studies. An example of an incidental study is the paper by Ledbetter and Porter (1963) on cell-wall formation in root tips, which also included a micrograph of a dividing cell. This micrograph is of historical interest because it is the first illustration of a glutaraldehyde-fixed spindle.

B. Pre-glutaraldehyde studies (1950–1963)

After the first application of electron microscopy to biological specimens in the early 1940s, it took nearly twenty years to develop the specimen-preparation techniques we take for granted today. The development of commercially available microscopes and microtomes, embedding resins, stains and the techniques of thin sectioning were all major milestones in allowing us to visualize the fine structure of cells. Accounts of these early years can be found in the articles by Pease and Porter (1981), Porter (1980), and in the personal reminiscences of Irene Manton (1978).

Porter (1980) has referred to the years 1953–1963 as the "decade of discovery" because many features of cell fine structure were then being seen for the first time. This was in spite of the fact that buffered osmium and/or potassium permanganate were the fixatives of choice. The most prominent structural features of the mitotic spindle were also being described during this period, although very crudely resolved by today's standards. Porter (1955) was one of the first to study the changes in spindle fine structure during mitosis. Earlier studies (Rozsa and Wyckoff, 1950, 1951; Beams *et al.*, 1950; Selby, 1953) are of historical interest because they give the first EM images of dividing cells, but they are of little scientific value. The papers of Bernhard and deHarven (1956; deHarven and Bernhard, 1956) described centriole replication, pericentriolar material, kinetochores and the tubular nature of the spindle fibre. The membranous components of the spindle were characterized by Porter and Machado (1960) in root-tip cells fixed in permanganate, the preferred fixative for plant cells at the time. By 1962, reasonable descriptions of spindle fibres were available (Roth and Daniels, 1962; Harris, 1962) but the need for better fixation was well recognized (Harris and Mazia, 1962). As is now well known, the discovery by Sabatini *et al.* (1963) of glutaraldehyde fixation provided the improvement and the stage was set for an explosion of ultrastructure papers in the next 10 years or so. Following Porter's (1980) terminology, we might think of these years as the "decade of description".

C. Post-glutaraldehyde studies: survey papers

As mentioned earlier, the first micrograph of a glutaraldehyde-fixed spindle appeared in Ledbetter and Porter's (1963) paper on "microtubules" (Slautterback, 1963; but see also, Brachet, 1957, p. 65) and plant cell-wall formation. Ledbetter and Porter recognized that spindle fibres, flagellar fibres and cytoplasmic microtubules were probably homologous structures. The first glutaraldehyde-fixed material studied specifically for mitosis was the Ascites tumour cells studied by Robbins and Gonatas (1964a,b) and the alga *Prymnesium* by Manton (1964). Within a few years, good ultrastructural studies were available for many cell types, mostly the so-called lower eukaryotes, which exhibit considerable structural diversity. These survey studies continued well into the late 1970s and, among lower eukaryotes especially, were used to study phylogenetic relationships and/or the evolution of the mitotic apparatus (see references and reviews cited in Heath, 1980). By the time of Heath's (1980) review, the mitotic spindles of over 280 different species of algae, fungi and protozoa had been examined by electron microscopy.

D. Post-glutaraldehyde studies: analytical papers

The technique of serial-section analysis is an old one in morphological studies and was used extensively by light microscopists to work out the details of embryology and other aspects of morphology and anatomy. The advantages of the method were also recognized in the early years of electron microscopy (Porter and Blum, 1953) but it had to await the development of reliable automatic microtomes before it became generally useful. One of the earliest applications of EM serial section analysis to mitosis research was by Paweletz (1967) when he addressed the question of whether the light microscopist's "continuous fibres" were truly continuous from pole to pole or whether they were composed of overlapping half-spindle microtubules. By counting microtubules on either side and in the middle of HeLa cell "stem bodies", he showed that microtubules in the spindle midzone overlap and that few, if any, microtubules are really continuous. Similar conclusions were reached by McIntosh and Landis (1971) and Brinkley and Cartwright (1971) in their studies of mammalian spindles and Manton *et al.* (1969a,b; 1970a,b) in their analysis of diatom central spindles. All of the above studies sampled microtubule number along the spindle pole-to-pole axis and conclusions about microtubule length were by inference. The first study that actually mapped the lengths of individual microtubules and the positions of their ends was by Heath (1974), who took advantage of the fact that spindle microtubules in the fungus *Thraustotheca* are few in number, widely spaced, and could be reconstructed from relatively few serial sections. Other serial-section studies of this type form the basis of this chapter and will be discussed in later sections.

Another way to use serial sections is to superimpose selected structures from each section and build up a picture of the relative positions of these spindle components. We can call this type of imaging a two-dimensional reconstruction. Two-dimensional reconstructions make no attempt to establish continuity of structures from section to section and quantitative comparisons are only approximate. An early example of this type of reconstruction is the paper by Mole-Bajer and Bajer (1968) in which cells that had been filmed during mitosis were then fixed and prepared for EM; selected regions were reconstructed and compared with the film data. The paper by Kubai and Ris (1969), in which they used serial section information to construct a three-dimensional physical model of a dinoflagellate spindle, is another variation on this type of reconstruction. Recently, Nicklas and his co-workers (Nicklas *et al.*, 1982; Nicklas and Kubai, 1985; Nicklas and Gordon, 1985) have used two- and three-dimensional reconstructions in conjunction with micromanipulation experiments to study the relationship between microtubule distributions and chromosome movements at various stages of mitosis.

High-voltage EM (HVEM) can give information similar to that of 2D reconstructions because a half-micrometre section is the equivalent of 6–7 superimposed thin sections. Early HVEM images of spindles were published by McIntosh *et al.* (1975a,b) and Coss and Pickett-Heaps (1974). HVEM has been used to advantage with very small spindles (Peterson and Ris, 1976) and very large spindles (McIntosh *et al.*, 1979b; Sluder and Rieder, 1985; see also Fig. 5 in Chapter 4). An excellent review of the advantages of HVEM and serial sections can be found in the article by Rieder (1985).

A final quantitative method to consider is the use of computers in analysing serial sections. This is probably the only way the vast amount of information available in serial sections can be adequately managed. McIntosh *et al.* (1975a,b) first applied computer morphometric techniques to the study of the microtubules in the mammalian midbody, and later (McIntosh *et al.*, 1979a) to the 3D reconstruction of diatom central spindles (Fig. 1.1) and the near-neighbour analysis of antiparallel microtubules in the diatom spindle overlap zone (McDonald *et al.*, 1979). Moens and Moens (1981) developed a computer program for converting serial section data into 3D reconstructions, which has been used with good success by Heath (Heath *et al.*, 1984) (Fig. 1.2) and by Nicklas and co-workers (Nicklas *et al.*, 1982; Nicklas and Kubai, 1985). Because of the rapid advances in computer capabilities, we can expect greater application of these techniques in future quantitative studies of mitosis.

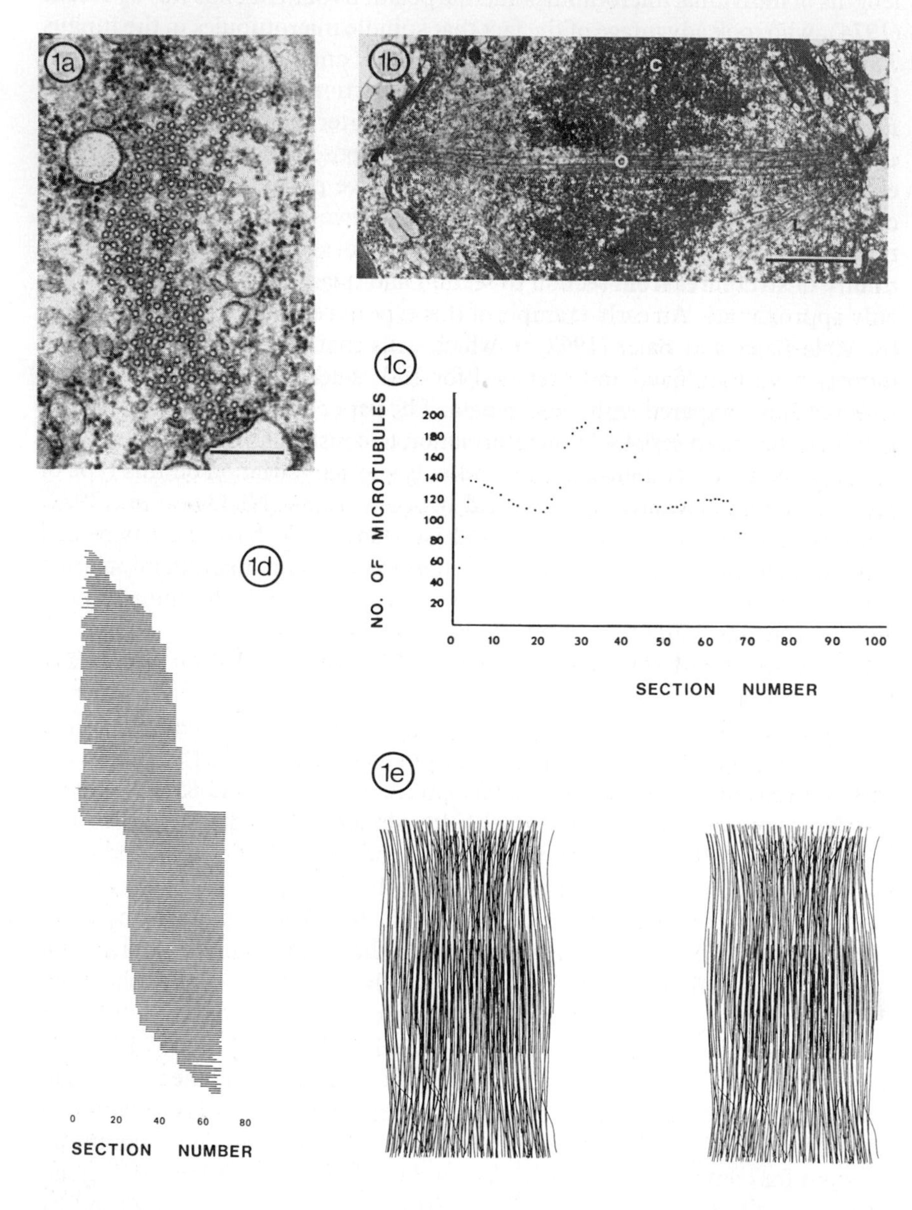
1a
1b
C
O
L
1c
NO. OF MICROTUBULES
200
180
160
140
120
100
80
60
40
20
0
10
20
30
40
50
60
70
80
90
100
SECTION NUMBER
1d
0
20
40
60
80
SECTION NUMBER
1e

Figure 1.1.

III. Basic Spindle Design and Variations

A. Terminology

Light-microscope descriptions of spindle structure (Schrader, 1953) recognized three types of spindle fibres: (1) astral rays, (2) chromosomal fibres, and (3) continuous fibres. Serial-section reconstructions have allowed electron microscopists to define spindle fibres more precisely by classifying microtubules according to where they begin and end relative to other spindle components. For this chapter, we will use the terminology of Heath (1981), which is a hybrid of the classifications of Bajer and Mole-Bajer (1972), Heath (1974) and McIntosh *et al.* (1975b). In addition, we will add the aster microtubule, which was not included in Heath's classification.

All spindle microtubules that insert into the kinetochore region of the chromatid are called kinetochore microtubules (kMTs). The poleward end of a kMT may end on the pole itself, it may end short of the pole by some variable distance, or it may be laterally associated with the other class of spindle

---

Figure 1.1. Different ways of representing the microtubule components of spindles, in this case, the central spindle of the diatom *Diatoma vulgare.* (a) A section perpendicular to the spindle long axis showing central spindle microtubules in cross-section. When spindle microtubules are tightly bunched, this is the preferred orientation for making microtubule counts such as in (c) and for making reconstructions such as those shown in (d) to (e) and Fig. 7(c). (b) Longitudinal thin section showing parallel microtubule arrays and some density in the spindle midregion. It is difficult to know whether the midregion density is due to overlapping microtubules, or just matrix material, or both. C = chromatin. O = presumed zone of overlap, arrows = poles, L = lateral spindle microtubules. (c) Microtubule distribution profile derived from serial cross-sections, where the number of microtubules in each section is plotted against position along the spindle pole-to-pole axis. The increase in microtubule number in the midregion suggests overlapping half-spindle microtubules but is inconclusive. Increased numbers of free microtubules in this region would give the same result. (d) From the same data set used to construct (c) but this time each individual microtubule was tracked through the serial cross-sections then arranged according to its point of origin and termination along the spindle axis. From this type of representation it is clear that the increase in microtubule number in the midregion is due to overlapping half spindle microtubules. (e) From the same data set used to make (c) and (d) but this time all microtubule positional coordinates were entered into a computer and computer graphics were used to generate stereo images. (Figs. 1(b) and (d) from McDonald *et al.* (1977). Fig. 1(e) from McIntosh *et al.* (1979a). Fig. 1(a), bar = 0.2 μm. Fig. 1(b), bar = 1 μm.

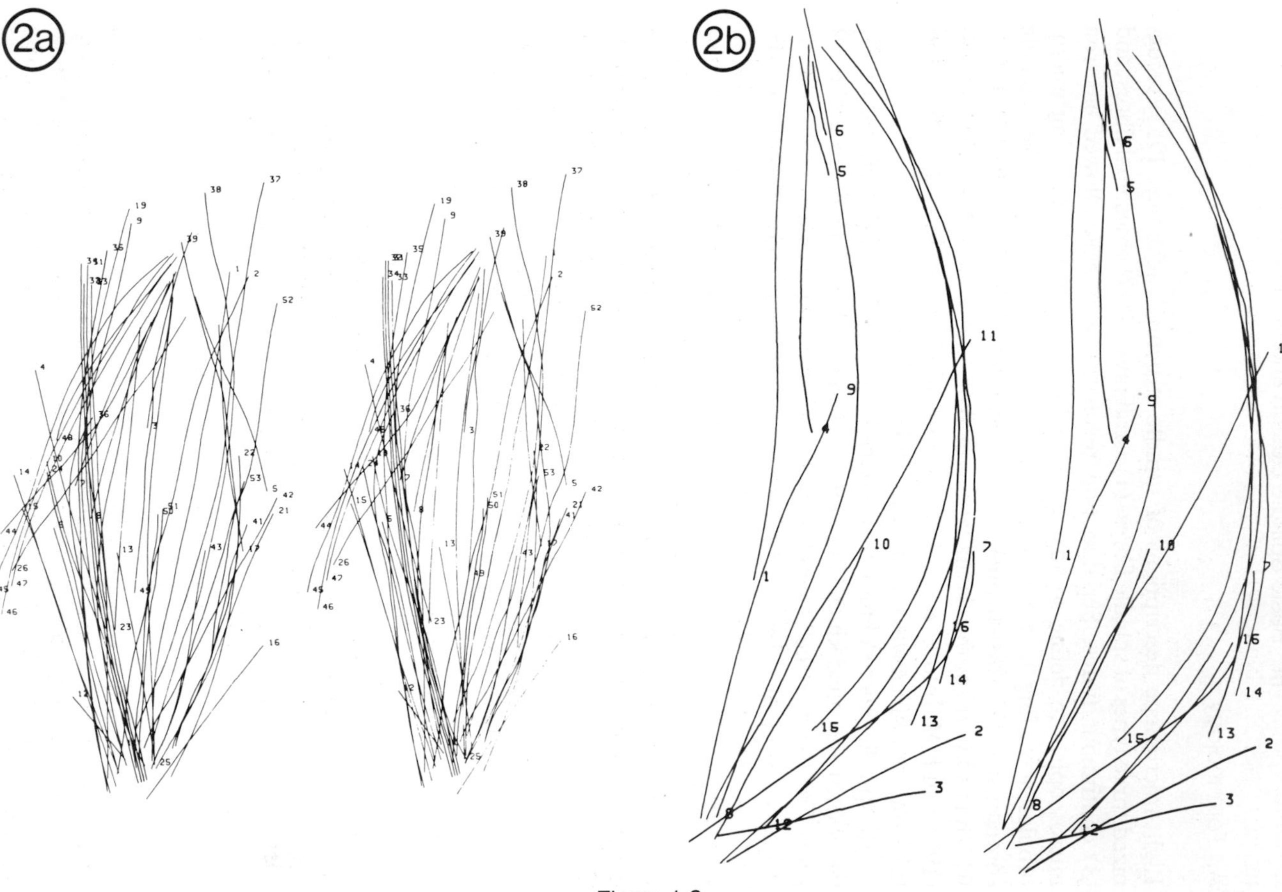

Figure 1.2.

microtubules, which are referred to collectively as non-kinetochore microtubules (nkMTs). Non-kinetochore microtubules (MT) are of five types:

1. polar MTs—one end at the pole, the other free in the spindle;
2. aster MTs—one end at the pole, the other ending outside the spindle;
3. interdigitating MTs—one end at the pole, the other laterally associated with a microtubule from the opposite pole;
4. continuous MTs—an end at each pole;
5. free MTs—neither end associated with a pole or chromosome.

Polar and interdigitating microtubules are probably the most common type of nkMT; aster MTs are the next most frequent; and, except in special cases (Heath and Heath, 1976), relatively few spindles have significant numbers of continuous or free MTs.

Because relatively few spindles have been reconstructed from serial sections, the above terminology has limited application. For general discussion of spindle structure it is useful to classify spindle microtubules by function. In this regard, Heath (1980) distinguishes two classes of microtubules in the spindle: chromosomal microtubules and framework microtubules. Framework microtubules connect the spindle poles, establish the bipolar spindle axis and determine the direction of chromosome movement. Framework microtubules may include continuous microtubules, interdigitating microtubules or even free microtubules if they are laterally associated. Chromosomal microtubules have one end associated with chromatin and the other end either at the pole or associated with the framework microtubules. This definition is more inclusive than the term kMT because it includes all microtubules inserted into the chromatin, not just those at the kinetochore.

B. Variation in the arrangement of framework microtubules

Although functionally equivalent, the arrangement of framework microtubules shows considerable variability, especially in lower eukaryotes. These variations are reviewed in detail by Kubai (1975), Heath (1980) and Fuge (1974, 1977, 1978) and need only be summarized here. Many lower eukaryote mitoses occur within an intact nuclear envelope (so-called "closed" division). The framework microtubules can be outside the envelope (extranuclear

---

Figure 1.2. Stereo projections of non-kinetochore microtubules from (a) early and (b) late anaphase spindles of *Saprolegnia*, reconstructed from serial sections. In this type of reconstruction the spatial relationships among individual microtubules is easily visualized as well as the overall difference in spindle structure at the two mitotic stages. (From Heath *et al.*, 1984.)

spindle) or inside (intranuclear spindle). Among extranuclear spindles there are three different arrangements of microtubules:

- as a sheath around the nucleus (Jenkins, 1967, 1977);
- as several bundles that form channels through the nucleus (Kubai and Ris, 1969); or
- as a single bundle through the nucleus (Ris and Kubai, 1974; Hollande, 1974)

Among intranuclear spindles there are many different arrangements of framework microtubules:

- as a sheath just inside the nuclear envelope (LaFountain and Davidson, 1979, 1980);
- as an elliptical bundle just inside the nuclear envelope (Tucker *et al.*, 1980);
- as an interdigitated but widely spaced array (Heath, 1974);
- as an interdigitated but closely spaced array (Tippit *et al.*, 1984; McIntosh *et al.*, 1985); or
- as a bundle of pole-to-pole microtubules (Heath and Heath, 1976).

Among those organisms in which the nuclear envelope breaks down during mitosis (open division), there is less diversity in the arrangement of framework microtubules. They may be arranged in a single large bundle of interdigitating microtubules (Pickett-Heaps *et al.*, 1975), as several bundles of microtubules (Manton *et al.*, 1969a,b, 1970a,b), or as interdigitating microtubules that are in many small bundles (of as few as two microtubules) (Tippit *et al.*, 1983). Most higher plants and animals are probably of the last category but, because spindle microtubules have not been tracked by serial sectioning it cannot be said for certain that the microtubules are not continuous. Limited reconstructions of mammalian cells in late division stages (McDonald and Euteneuer, 1983; Paweletz, 1967) and microtubule distribution profiles (Brinkley and Cartwright, 1971; McIntosh and Landis, 1971) indicate that the spindle framework microtubules are mostly interdigitating microtubules.

C. Variations in the arrangement of chromosomal microtubules

In some cases, chromatin interacts with the sides of microtubules, and if this lateral interaction helps move the chromosome, as some believe (Fuge, 1974), then we should include these as a class of chromosomal microtubules. It may also be that the interactions are non-specific, since these chromosomes also have distinct kinetochores. A clearer variation of the chromosomal microtubules is the case in which microtubules end in the chromatin but not at a localized site. These chromosomes are said to have diffuse kinetochores (see, for example Fig. 4 in Chapter 3) and the fact that kinetochore activity is distributed all along the chromatin was shown by breaking it into fragments, all of which moved to the pole normally during anaphase (references in Schrader, 1953; Brinkley *et al.*, this volume). In the cases above, it is not known whether the poleward ends of the microtubules end on the pole or interact with framework microtubules.

In spindles with discrete kinetochores there are two basic variations, depending on how the poleward end of the chromosomal microtubule interacts with the rest of the spindle. If the chromosomal microtubules interact primarily with the framework microtubules, we have what the light microscopists (Schrader, 1953) called an indirect spindle. The spindles of the *Tetrahymena* micronucleus (LaFountain and Davidson, 1980) and diatoms (Pickett-Heaps and Tippit, 1978) are of this type although in the former case the framework microtubules form a peripheral sheath just inside the intact nuclear envelope whilst in the latter they usually consist of a single bundle of microtubules (the central spindle) in an open division. In most higher plants and animals, the chromosomal microtubules have one end on the kinetochore and the other at or near the pole (Rieder, 1982). In the light-microscope literature (Schrader, 1953), this type is known as a direct spindle. Electron microscopy (Jensen, 1982a,b; Bajer and Mole-Bajer, 1973) shows that some of these microtubules with direct connection to the poles are also laterally associated with framework microtubules.

D. The molecular polarity of spindle microtubules

One of the most important contributions of electron microscopy to mitosis research in recent years has been the identification of the molecular polarity of the different classes of spindle microtubules. Heidemann and co-workers (Heidemann and McIntosh, 1980; Heidemann *et al.*, 1980) discovered that neurotubulin in 0.5 M PIPES buffer containing DMSO, will add to the side walls of the native microtubules as curved sheets that look like "hooks" in cross section. By correlating the direction of hook curvature with microtubules of known polarity, it was possible to show that when the hooks curve counter-clockwise then you are looking toward the plus end of the microtubule—that which adds subunits fastest when the microtubule is growing (Bergen and Borisy, 1980; Borisy, 1978). Euteneuer and McIntosh (1980, 1981; McIntosh and Euteneuer, 1984) used this method to show that all spindle microtubules have their plus ends distal to the spindle poles (Fig. 1.3). The same conclusion was reached by Telzer and Haimo (1981) using the binding of exogenous axonemal dynein from *Tetrahymena* or *Chlamydomonas* to native microtubules as a molecular polarity probe. This method has the advantage that polarity can be determined in longitudinal sections as well as cross sections.

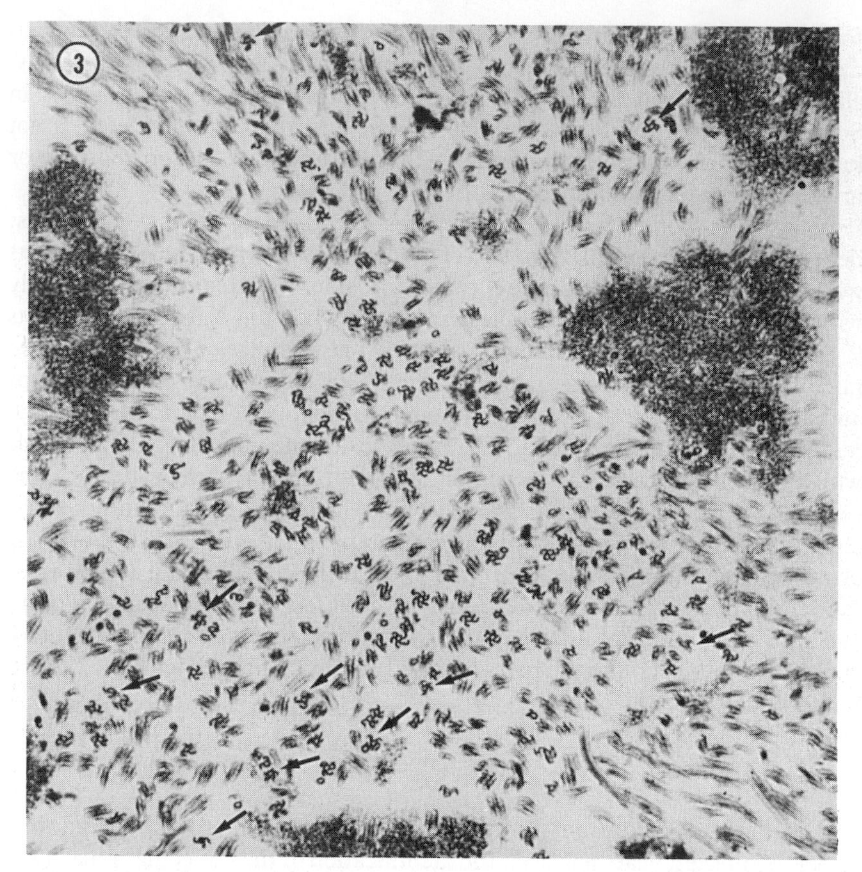

Figure 1.3. "Hook" decoration of spindle microtubules to indicate molecular polarity. This is a cross-section through an early anaphase PtK_1 cell in the region between the kinetochores and the poles and near the kinetochore. Most microtubules seen in cross-section are decorated with hooks and over 90% of the hooks are curved counterclockwise, which means we are looking toward the plus end of the microtubule. These kinds of analyses have shown that all microtubules in a given half-spindle have the same molecular polarity with the fast-growing microtubule end (the plus end) distal to the pole. (From Euteneuer and McIntosh (1981). Bar = 0.5 μm.)

IV. Spindle Formation

While there has been some recent progress in our understanding of the structural basis of spindle formation, the main questions remain unanswered.

1. What structures are responsible for mitotic centrosome (pole) separation?

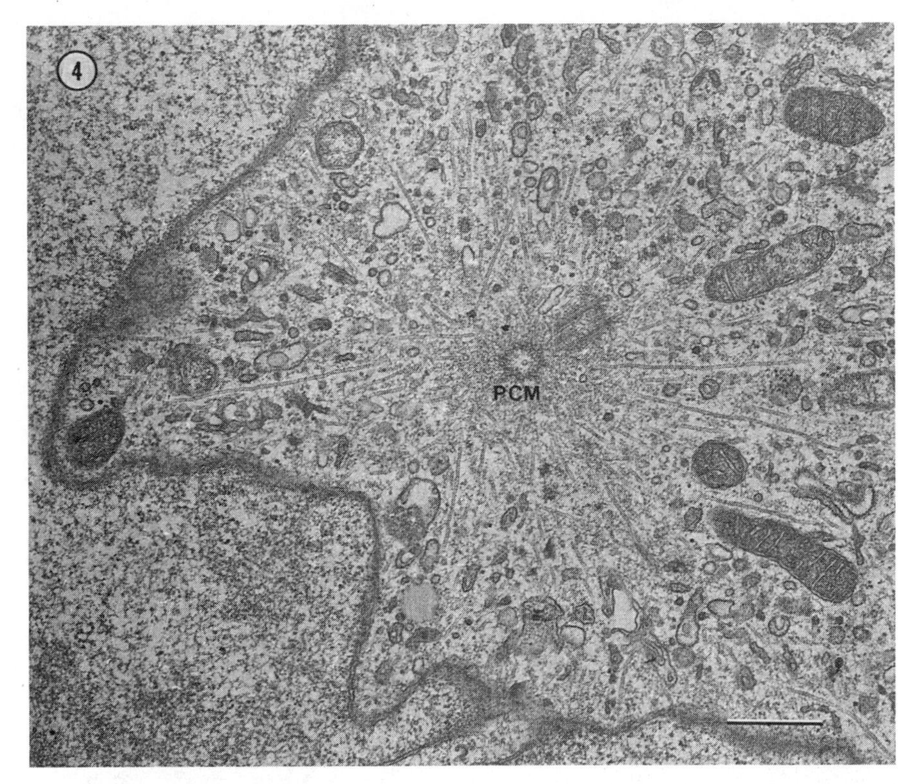

Figure 1.4. A typical mammalian centrosome containing orthogonally arranged centrioles with pericentriolar material (PCM) predominantly around the parent centriole. This is one pole of a spindle forming in prophase. The microtubules are radially arranged around the centrosome and some seem to make indentations in the nuclear envelope. (Bar = 0.5 μm.)

2. What is the structural relationship between the pericentriolar material (osmiophilic "fuzz") and the microtubules that emanate from it?
3. How do antiparallel framework microtubules interact to form the bipolar spindle axis.

A fourth question, about how the chromosomes/kinetochores attach to the spindle, will be considered in the next section on prometaphase.

The behaviour of centrosomes during the cell cycle has been an important area of study for over a century (Mazia, 1984). Electron microscopy has contributed to this body of knowledge by revealing the structural diversity of mitotic centrosomes. In addition to the typical animal centrosome with its paired, orthogonally arranged centrioles (Fig. 1.4; Vandré and Borisy, this volume) there exists a wide variety of plaques, rods, and spheres, especially among lower eukaryotes (reviewed in Heath, 1980), whilst higher plant cells typically lack any defined structures at the spindle poles. Most of the recent information about mitotic centrosomes has come from studies of animal cells and the discussion that follows will concentrate on the centriole-containing centrosome. A discussion of non-centriole structural variations can be found in Heath (1980). Electron microscopy has revealed that centrioles *per se* are not the structures responsible for microtubule nucleation and anchoring. Instead, it is the osmiophilic pericentriolar material (PCM) which appears to perform this function (Gould and Borisy, 1977). In tissue culture cells, the PCM is preferentially associated with the parent centriole and increases to maximum volume in prometaphase and metaphase (Rieder and Borisy, 1982; Robbins *et al.*, 1968). In dividing sea-urchin zygotes the form of the PCM is even more plastic. Paweletz *et al.* (1984) took advantage of the fact that in cells treated with β-mercaptoethanol the chromosome cycle is arrested at metaphase but the centrosome cycle continues. They were able to show that the centrosome is a compact sphere early in mitosis but that it changes to a flat disk in later stages. The flat disk then divides, giving rise to two compact spheres that serve as the spindle poles for the next division. During this cycle, the shape of the spindle poles changes from pointed to broad and back to pointed, corresponding to the changing shape of the centrosomes. In lower organisms with non-centriole centrosomes the shape is usually less plastic and microtubules are frequently nucleated from one side only (Fig. 1.5). During spindle formation in these organisms, a continuous or overlapping set of microtubules connect the

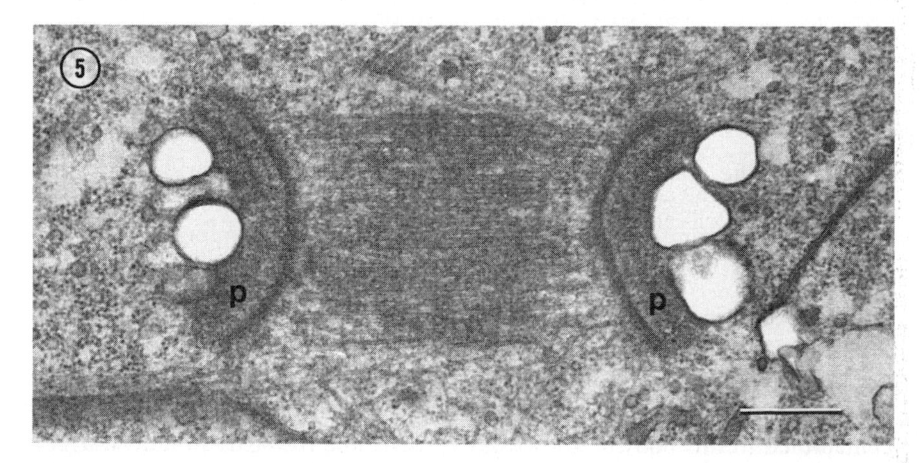

Figure 1.5. Spindle formation between two asymmetric mitotic centrosomes in a diatom (*Stephanopyxis turris*). All microtubules are nucleated from one side of the plates (p), which serve as the mitotic poles. (Bar = 0.5 μm.)

separating centrosomes and it is likely that the mechanism of pole separation is microtubule growth and/or microtubule sliding (Tippit *et al.*, 1978; McDonald *et al.*, 1979). In animal cells, the structural relationship of the poles to each other and the cytoskeleton is less clear. However, Paweletz *et al.*'s (1984) analysis of the centrosome cycle in sea-urchin zygotes found microtubules intimately associated with the osmiophilic material as it changed shape. Likewise, Rattner and Berns (1976) in their serial section analysis of microtubule distribution during centriole separation in PtK_2 cells found that microtubules from opposite poles displayed sufficient interaction to provide the force necessary for pole separation. And while drug studies (reviewed in McIntosh, 1982) confirm that microtubules are required for pole migration, a "pulling" mechanism based on interaction of astral microtubules with the cytoplasm has not been ruled out. A fuller discussion on the structure, replication and separation of centriole-containing centrosomes may be found in the following chapter.

V. Prometaphase

Prometaphase is one of the most dynamic phases of mitosis and one of the most demanding to characterize ultrastructurally. In cells with open spindles, this phase begins when the nuclear envelope breaks down and the chromosomes attach to the spindle. Nuclear envelope breakdown always occurs in the vicinity of the centrosomes (Roos, 1973a) and, although it is typical to see microtubules creating invaginations in the envelope (Fig. 1.4), it is evident from the results of Rieder and Borisy (1981) that microtubules are not strictly required for membrane breakdown. In this study, PtK_1 cells were chilled to 6°C in late prophase and the nuclear envelope disintegrated in the absence of centrosomal microtubules. After the envelope is lost, the chromosomes nearest the centrosomes become attached to invading microtubules at their kinetochores (Roos, 1976). Before nuclear envelope breakdown, kinetochores in mammalian cells are recognizable as small "balls" (less than a micrometre in diameter) in an electron-dense "cup" at the primary constriction of the chromosome (Rieder, 1982; Chapter 3, this volume). At early prometaphase, kinetochores nearest the centrosome differentiate first from the "ball and cup" form to a single fibrillar plate surrounded by lightly staining fibrillar material referred to as a "corona" (Jokelainen, 1967; Roos, 1973b). Later in prometaphase, kinetochores begin to differentiate into the familiar trilaminar form characteristic of metaphase and anaphase (reviewed in Rieder, 1982; Brinkley *et al.*, Chapter 3).

Historically, there have been three explanations of chromosome–spindle attachment (Schrader, 1953):

1. that chromosomes attach to pre-existing polar microtubules;
2. that chromosomes nucleate microtubules at the kinetochore and these microtubules become anchored at the pole; and
3. that interaction between polar microtubules and kinetochore-generated microtubules forms a stable kMT fibre.

According to Schrader (1953), "The first of these hypotheses—that the pole originates the fibres and sends them to the chromosomes—is perhaps the most firmly ensconced in the biological mind. It is also the oldest hypothesis . . ." But, with the advent of electron microscopy and the popularization of the microtubule-organizing centre (MTOC) concept (Porter, 1966; Pickett-Heaps, 1969), more emphasis was given to the idea that kinetochores served as MTOCs (Pickett-Heaps, 1974). So, until recently, it was perhaps more generally believed that kMT bundles were nucleated at the kinetochores, an idea that received support from the fact that isolated chromosomes would form kMT bundles *in vitro* (Telzer *et al.*, 1975; McGill and Brinkley, 1975). It was this climate of thinking that led Tippit *et al.* (1980b) to offer a "new" proposal for kinetochore function during prometaphase, namely that kinetochores attached to pre-existing polar MTs.

In the past five years there have been a number of good analytical EM studies on the structural relationship between spindle microtubules and chromosomes during prometaphase (Nicklas *et al.*, 1979; Tippit *et al.*, 1980a; Witt *et al.*, 1980; Rieder and Borisy, 1981; Ris and Witt, 1981; Church and Lin, 1982, 1985; Jensen, 1982a; Steffen and Fuge, 1982; Nicklas and Kubai, 1985). They have all employed serial-section reconstruction techniques to map the exact distribution of microtubules relative to chromosomes and other spindle components, and in some cases (Church and Lin, 1985; Nicklas and Kubai, 1985) the behaviour of selected chromosomes was recorded up to the moment of fixation so that the direction of movement could be correlated with microtubule distributions. An extensive discussion of how these and earlier results relate to the models of kinetochore fibre formation is beyond the scope of this chapter, but interested readers should consult the chapters by Brinkley *et al.* and Salmon in this volume and the earlier review of Rieder (1982). There is now a general acceptance that the initial interaction of chromosomes and microtubules is capture of polar microtubules by kinetochores. This model most easily explains the fact that chromosomes nearest the pole are the first to develop kMT fibres while sister kinetochores on the same chromosomes lack microtubules (Jokelainen, 1965; Roos, 1973a, 1976; Mole-Bajer *et al.*, 1975; Rieder and Borisy, 1981; Rieder, 1982). Attempts to explain these facts by gradients of tubulin (Roos, 1976) or factors affecting microtubule polymerization (DeBrabander *et al.*, 1979, 1980) are less convincing because of their speculative nature and do not really explain why kinetochores near poles but facing away from them do not develop microtubules. Recent light-microscope observations by Mitchison and Kirschner (1985a,b) also provide evidence in support of kinetochore capture. In these experiments, microtubules were preformed from microtubule seeds, centrosomes or axonemes, and then isolated chromosomes without microtubules were added. The chromosomes captured the microtubules specifically at the kinetochore region and dilution experiments showed that these were more stable than uncaptured microtubules. The incubation conditions were such that nucleation of new microtubules from the kinetochores was ruled out. The experiments are discussed more fully in Chapters 3 and 4.

Evidence from experiments on chromosomes *in vitro* (Telzer *et al.*, 1975; Gould and Borisy, 1978), in lysed cells (Snyder and McIntosh, 1975; McGill and Brinkley, 1975) and *in vivo* (Witt *et al.*, 1980; Ris and Witt, 1981; DeBrabander *et al.*, 1979, 1980, 1981) indicate that nucleation of microtubules from the kinetochore is possible. The results of those studies with virgin kinetochores *in vivo* (Witt *et al.*, 1980; Ris and Witt, 1981; DeBrabander *et al.*, 1980) show that microtubules do not simply arise from the outer layer of the kinetochore plate. Instead, they seem to form first in the fibrous corona material and at various angles to the kinetochore. Ris and Witt (1981) point out that the kinetochore corona and pericentriolar material have a similar morphology and may "contain similar factors which enhance microtubule nucleation". Whatever the mechanism of nucleation, it remains to be shown whether nucleation is required for the normal development and functioning of kinetochore fibres.

Once chromosomes become attached to the spindle they begin to move. The movement is rapid at first (up to 7 times faster than anaphase velocity) then slows down to about anaphase speeds at the end of prometaphase (Church and Lin, 1985). As speed diminishes, the number of microtubules per kinetochore increases, suggesting that microtubules function as governors of chromosome movement (Nicklas, 1975). In later prometaphase, kMTs show increased interaction with bundles of nkMTs (Jensen, 1982), which may also contribute to the restriction of movement. Cinematic studies combined with EM (Church and Lin, 1985; Nicklas and Kubai, 1985) show that a single microtubule is sufficient to move a chromosome. Nicklas and Kubai (1985) were able to study the initial interaction of chromosomes and microtubules under carefully controlled conditions by using micromanipulation to remove a chromosome from the spindle, then fix it at the moment it began to move back. We can summarize their findings as follows:

1. The junction between kinetochores and kMTs is weak. EM reconstructions of detached chromosomes show that all microtubules are left behind in the spindle and that they depolymerize shortly after detachment. These results

are consistent with the observation that assembly of kMTs is at the plus (kinetochore) end (McIntosh and Euteneuer, 1984; Mitchison and Kirschner, 1985a,b).

2. Movement requires microtubules. Detached chromosomes fixed prior to movement back toward the spindle had no microtubules or, at most, one. After movement began, microtubules were invariably found at the kinetochore nearest the spindle (Fig. 1.6).
3. The direction of movement was always parallel to the long axis of the kinetochore microtubules.
4. The kMTs did not have to extend all the way to a pole for movement to take place. Only one half-bivalent in 14 was found to have a kMT with its kinetochore-distal end at the pole. This result is difficult to reconcile with the model based on capture of polar microtubules unless those microtubules become detached somehow from their poleward end. In this regard, it is worth remembering that there are forces that act to push particles and chromosomes away from the poles (reviewed in Pickett-Heaps *et al.*, 1982) and it is possible that polar microtubules may be uprooted and translocated also.
5. Initial attachments are mostly inappropriate and reorientation movements continue until appropriate bipolar orientation is achieved. In general, reorientation movements involve new microtubule attachments and loss or rearrangement of existing microtubules. "Dynamic instability" of microtubules (Mitchison and Kirschner, 1984, 1985a,b) can explain the appearance of new microtubules and the disappearance of others, but it remains unclear how existing microtubules might be moved around.

The forces acting on chromosomes at prometaphase are probably more complicated than simply competition between oppositely directed poleward forces. Observations on monopolar spindles (Bajer, 1983; Mazia *et al.*, 1981; Rieder *et al.*, 1986) show that chromosomes will move towards and away from a single pole as if both pulling and pushing forces were operating. Pickett-Heaps *et al.* (1982) suggest that movement away from the pole is due to dynein-like activity on the part of the pole-distal kinetochore. However, the observations of Rieder *et al.* (1986) argue against this idea and favour a mechanism acting at the proximal kinetochore alone. The evidence from Mitchison and Kirschner (1985a,b) suggests that growth of kMTs by insertion of new subunits at the proximal kinetochore could account for movement away from the pole. Depolymerization of kMTs due to dynamic instability (Mitchison and Kirschner, 1984) would permit poleward movement as a result of pole-directed forces acting on kMTs. Recent evidence for kinesin-like molecules in the mitotic apparatus (Scholey *et al.*, 1985) raise the possibility that kinesin is responsible for this movement. Some of these points are

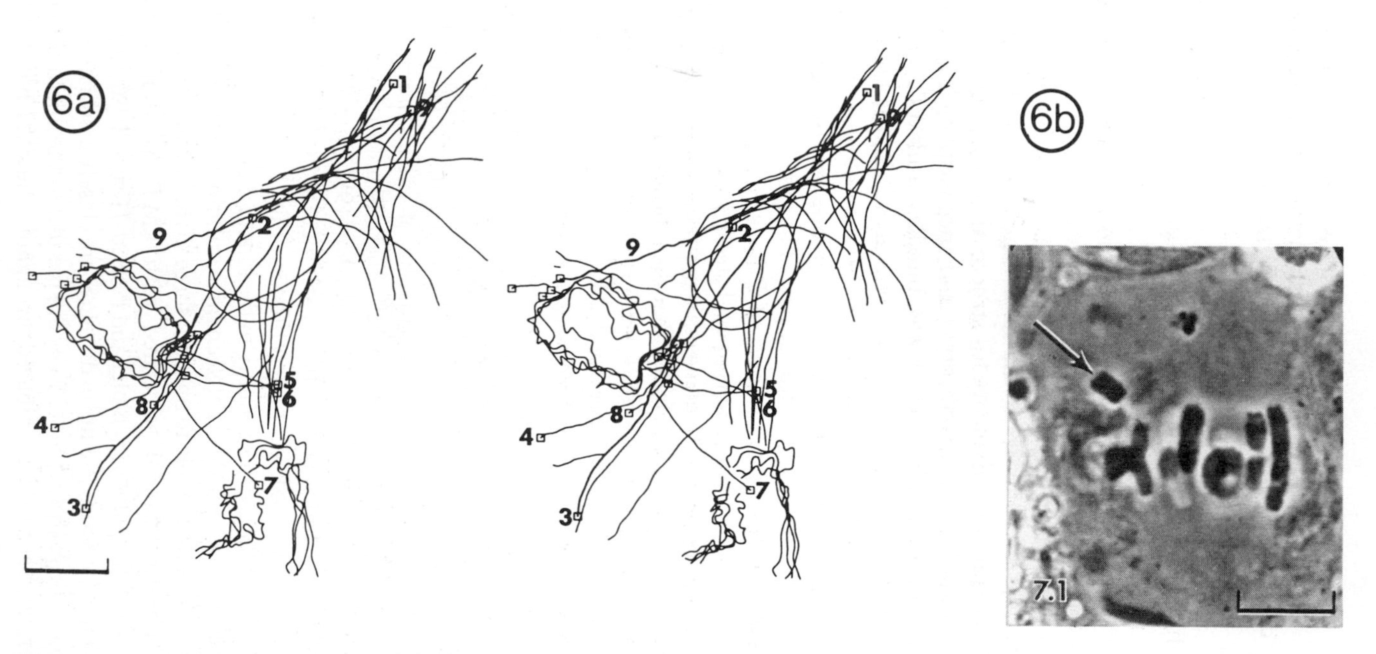

Figure 1.6. (a) Three-dimensional (stereopair) reconstruction of a grasshopper (*Melanoplus differentialis*) spermatocyte bivalent that had been removed from a prometphase spindle by micromanipulation, filmed, then fixed just as the outer end began rotation toward a pole. All of the kinetochore microtubules of the manipulated bivalent are marked with a box and several are numbered. From the inner end of the bivalent, kMTs extend in a variety of directions: from the outer end, a single long kMT (number 9) extends toward the upper pole, in the direction the bivalent was rotating. (Bar = 2 μm.) (b) Print from the cine record showing the position of the bivalent at the time the cell was fixed. (Bar = 10 μm.) (From Nicklas and Kubai, 1985.)

considered further in the chapter by Salmon later in this volume. Suffice to say that ultrastructural evidence for any molecule responsible for prometaphase movements in any direction is, at present, completely lacking.

VI. Chromosome-to-pole Movement (Anaphase A)

The structural basis for the forces that move chromosomes to the poles remain one of the least well understood aspects of mitosis. We know that the average length of kinetochore microtubules decreases during anaphase but we do not know whether the loss of subunits is the force-generating mechanism or whether it is simply permissive. Evidence for structural associations between kMTs and other components is fragmentary and inconsistent. The most frequent observation is that kMTs interact with bundles of nkMTs (Jensen, 1982; review in Heath, 1981). In *Haemanthus* (Jensen, 1982), the bundle of kMTs at early anaphase (average number about 70) may be surrounded by bundles of 10–30 nkMTs that extend into the interzone and become intermingled with the kMTs about 1 μm toward the pole. The same arrangement of microtubules is found in kinetochore fibres of PtK_1 cells (McDonald and Cande, 1980), although in reconstructions of cold stable kinetochore fibres in PtK_1 cells (Rieder, 1981; Fig. 6 in Chapter 4) this particular class of nkMTs is absent. Likewise, Nicklas *et al.* (1982) found that the few nkMTs penetrating the kinetochore fibre near the kinetochore were easily displaced when the chromosome was moved by micromanipulation. Thus, the proximity of kMTs and nkMTs may be due to the chance distribution of fibres in a limited space rather than an indication of mechanochemical coupling. But Nicklas *et al.* (1982) did find that there were strong mechanical associations among microtubules near the poles. They discount the possibility that short (e.g., dynein-like) cross-bridges connect the microtubules and argue that a gel-like net (McIntosh, 1981) is a more reasonable explanation of their observations. Unfortunately, even in these well-preserved cells there is no ultrastructural evidence for such a "net" near the poles. Both Rieder (1981) and Witt *et al.* (1981) show that within the kinetochore bundle itself there is an electron-dense matrix material that connects adjacent microtubules (Fig. 1.7). They reason that this material may confer greater stability to the kinetochore fibre and Witt *et al.* (1981) suggest that cross-bridges between microtubules make a mechanically coherent fibre that is better able to transmit force to the chromosome. The best example of fibrous material associated with kMTs is seen in the chromosomal fibres of *Oedogonium* (Schibler and Pickett-Heaps, 1980). Microfilaments (not shown to be actin) 5–8 nm in diameter are closely associated with kMTs from prophase through anaphase. They are parallel to the microtubules and in cross-sections are seen to be distributed throughout the kMT bundle. In later stages of mitosis, the microfilaments are found mostly near the

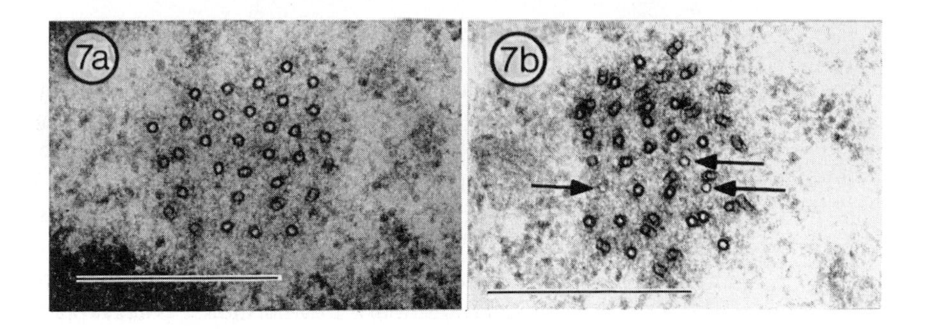

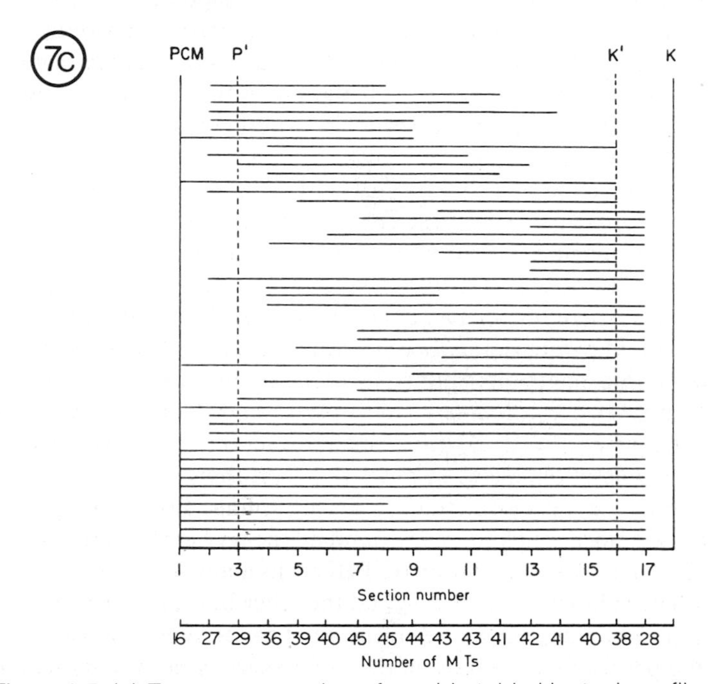

Figure 1.7. (a) Transverse section of a cold-stable kinetochore fibre taken within 1.0 μm of the kinetochore. The fibre microtubules are embedded in an amorphous material that is slightly more electron-opaque than the surrrounding spindle matrix. (b) Section 9 from the fibre reconstructed in (c). The arrows indicate polar microtubules that terminate in this section. (Bars in (a) and (b) = 0.5 μm.) (c) Reconstruction of the kinetochore fibre pictured in (b). The number of each serial section through the fibre, beginning at the pole and extending to the kinetochore, is indicated on the horizontal axis. Each individual microtubule of the fibre is tracked through these sections and is represented by a longitudinal line. Together these lines show the end points of all the microtubules comprising this cold-stable kinetochore fibre. This method of reconstruction provides no information about the position of individual microtubules in the fibre relative to each other. The number of microtubules in each section is indicated below each section number. K–K′ is the area of the kinetochore. PCM–P′ is the area of the pole. (From Rieder, 1981.)

kinetochore and may not extend the length of the entire fibre. The function of these filaments may be to stabilize the kinetochore fibre, or, as the authors suggest, "they could be involved in chromosome movement during both metakinesis and anaphase" (Schibler and Pickett-Heaps, 1980, p. 694).

The number of kinetochore microtubules per chromosome may be as few as one (Heath, 1980) or as many as 100 or more (Jensen and Bajer, 1973), but Moens (1979) and Lin *et al.* (1981) have shown that there is no correlation between chromosome size and kMT number. Furthermore, Nicklas and Kubai (1985) demonstrated that one microtubule is sufficient to move chromosomes in grasshopper spermatocytes (see above). Most spindles probably have a great deal of redundancy in the amount of force generation capability per chromosome (Nicklas, 1983, 1984). Jensen (1982) has good evidence that the number of microtubules per kinetochore changes with the stage of mitosis in *Haemanthus*. In other cell types, however, the numbers seem to remain constant (reviewed in Heath, 1981).

Micromanipulation experiments (Nicklas and Stahely, 1967) show that chromosomes are firmly anchored at the poles, presumably because of microtubules that span the distance between these two structures. Early descriptive EM studies (reviewed in Luykx, 1970) showed longitudinal sections through chromosomal fibres where one or two microtubules were apparently continuous between pole and chromosome. Recent analytical EM studies give a more accurate account of the length distribution of kMTs in the chromosomal fibre. In cases in which there are only a few kMTs per chromosomee (reviewed in Heath, 1981; Tippit *et al.*, 1984) these span the entire distance between chromosome and pole at all stages of mitosis. When there are more microtubules per kinetochore, only a fraction of the true kMTs (those with one end inserted into the kinetochore plate) may extend all the way to the pole. Rieder (1981) found that in chromosomal fibres of PtK_1 cells about 50% of the kMTs were continuous from kinetochore to pole and the other 50% ended well before the pole. In addition, from 1/3 to 1/8 of the total microtubules in the fibre originated at the pole but ended short of the kinetochore and a small fraction were free microtubules with neither end at a pole or kinetochore. Results from Cande's laboratory (McDonald and Cande, unpublished data) on untreated PtK_1 cells are in general agreement with Rieder's figures. The smaller chromosomal fibres of CHO cells tend to have a greater percentage of continuous microtubules between the chromosomes and the poles but some polar microtubules are also present and some kMTs end short of the pole (Witt *et al.*, 1981). Fuge (1984) reported that relatively few kMTs extend between pole and chromosome in *Pales* spermatocyte spindles and that the chromosomal fibre is composed mainly of free microtubules. The results of Scarello *et al.* (1985) for the same cells are contradictory in that they find long kMTs, many of which reach the poles. The discrepancy may well be due to the difference in methods of reconstruction, a point which we will discuss in a later section.

There are three popular ideas of how chromosomes might move to the poles:

1. by dynamic equilibrium; i.e. conditions favouring kMT disassembly cause them to get shorter and in doing so they pull the chromosomes poleward. The poles do not move toward the chromosomes because the poles are held apart by the framework microtubules (reviewed in Inoué, 1981);
2. by kMT–nkMT interactions mediated by cross-bridges that propel the kMTs and attached chromosomes poleward by dynein-like bridges (McIntosh *et al.*, 1969), by "zipping" (Bajer and Mole-Bajer, 1973), or by treadmilling (Margolis *et al.*, 1978);
3. by interactions between kMTs and a non-microtubule spindle component that produces a pole-directed force.

Several reconstruction studies argue against model (2) because they show that the spacing between kMTs and nkMTs is too great for much direct interaction (reviewed in Heath, 1981). The evidence for model (3) is scarce at best, and suggestions that the microfilaments of *Oedogonium* (Schibler and Pickett-Heaps, 1980) or the collar material of diatoms (Tippit and Pickett-Heaps, 1977; Pickett-Heaps *et al.*, 1978, 1982) might be the force-generating source are only speculative. So, in the absence of any contradictory evidence, the dynamic equilibrium model is in the best shape (see also Salmon, this volume). However, as we will discuss in a later section, there is good reason to believe that the quality of fixation is limiting and that, if improved, evidence for model (3) could be forthcoming.

VII. Spindle Elongation (Anaphase B)

In many cells, the movement of chromosomes to the poles is accompanied by further separation of the chromosomes/poles owing to spindle elongation. The extent of spindle elongation ranges from a few per cent of the metaphase spindle length to a 12-fold increase (Tucker *et al.*, 1980). The fact that anaphase may include two different types of movement has been known for a long time (literature reviewed in Inoué and Ritter, 1978) but only in recent years, with the advent of quantitative studies (either EM or polarization optics), has there been much appreciation of the fact that two different mitotic motors are probably involved. Once identified as a separate research problem, anaphase B has received a lot of attention and we now suspect that at least three different mechanisms can contribute to the separation of spindle poles: (1) microtubule sliding, (2) microtubule growth, and (3) cytoplasmic forces pulling on the poles.

The earliest analytical EM studies concentrated on the question of whether "continuous fibres" (also called interpolar fibres), as defined by the light microscopists (Schrader, 1953) were in fact continuous from pole to pole, or whether they were interdigitated half-spindle microtubules. In every case (Paweletz, 1967; Manton *et al.*, 1969, 1970; Brinkley and Cartwright, 1971; McIntosh and Landis, 1971) it appeared that interdigitating microtubules were most common. These results supported the idea that spindle elongation took place by the sliding apart of half-spindles (McIntosh *et al.*, 1969). More convincing evidence was provided by mapping individual framework microtubules during metaphase, anaphase and telophase in the diatom central spindle (McDonald *et al.*, 1977). The pattern of redistribution is exactly what one would expect if half-spindle microtubules were sliding over each other. Analysis of these same spindles in cross section also showed that microtubules of one half-spindle were preferentially associated with microtubules from the other half-spindle at a spacing that was consistent with the involvement of a dynein-like cross-bridge (McDonald *et al.*, 1979). Even in some spindles where there is clear evidence of microtubule growth during spindle elongation (Heath and Heath, 1976; Tippit *et al.*, 1984; McIntosh *et al.*, 1985), the spacing between interdigitated microtubules is sufficiently close that a sliding mechanism is still possible. Physiological studies (see Cande, this volume) also support a dynein-like mechanochemistry for spindle elongation.

An alternative model for spindle elongation is supported by the laser microbeaming studies of Aist and Berns (1981) and the structural studies of King *et al.* (1982) and Heath *et al.* (1984). Aist and Berns (1981) used a microbeam to sever the elongating spindle in the fungus *Fusarium* and, after laser ablation, the rate of spindle elongation increased up to 3-fold. This observation led them to suggest that the force for spindle elongation was generated by a cytoplasmic microfilament system pulling on the astral microtubules and that the central spindle microtubules act as a "brake" or governor to regulate the rate of elongation. In *Saccharomyces* (King *et al.*, 1982) the final 4 μm of spindle elongation take place when there is only one continuous microtubule connecting the poles. This would seem to rule out interaction of antiparallel microtubules through cross-bridges unless a very short microtubule near one pole was acting on the long microtubule. King *et al.* (1982) suggest instead that the single microtubule may be pushing the poles apart as the result of microtubule assembly, or that cytoplasmic forces are pulling on the spindle poles. In either case, the central spindle microtubule acts to regulate the rate of elongation. Finally, Heath *et al.* (1984) show that *Saprolegnia* spindles have very few closely spaced microtubules in early anaphase; and at later stages the number of near neighbours increases but the rate of elongation slows down. But, as Heath *et al.* (1984) point out, while this evidence suggests a regulatory role for the central spindle, it is not mutually exclusive of a model based on force-generating cross-bridges. It simply means that, if the rate of elongation from cross-bridge activity is slower than the rate of pulling on the poles, the cross-bridges act as a regulator. If there is no pulling force, then the cross-bridges could be solely responsible for elongation. It is worth noting that all the evidence supporting a regulatory role for central spindle microtubules comes from fungi, in which migration of nuclei in the cytoplasm is fairly common (Girbardt, 1968). It may be that these organisms have adapted the machinery for nuclear migration to spindle elongation (Heath *et al.*, 1984).

VIII. Non-microtubule Spindle Components

We know from immunofluorescence and biochemical studies (McIntosh, 1984; Vallee and Bloom, this volume) that mitotic spindles contain many proteins besides tubulin, and while tubulin is the dominant species, it may be as little as 10% of the total protein (Forer *et al.*, 1980). But at the ultrastructural level, only three types of non-microtubule structures (other than spindle pole bodies and kinetochores) appear with any consistency. These are the pericentriolar material, the matrix material between interzonal microtubules and membranes. Other structures that have been reported less consistently include microfilaments, intermediate filaments, and bridges between microtubules.

A. Electron-dense material at the poles and interzone

In most animal cells there is a halo of osmiophilic, filamentous or "fuzzy" material surrounding the centrioles and a similar material between microtubules in the interzone. In the interzone, this material is best visualized during metazoan midbody formation (McIntosh and Landis, 1971; Mullins and Biesele, 1977) or in diatoms during the later stages of spindle elongation (McDonald *et al.*, 1977). This midbody matrix material is seen only where half-spindle microtubules overlap, and as the zone of overlap shortens the staining intensity increases. In general, it is found where microtubules are tightly bunched and near microtubule ends. At the poles it is very likely that this pericentriolar matrix material has a role in nucleating microtubule growth (Gould and Borisy, 1977). Given what we know about microtubule polarity (McIntosh and Euteneuer, 1984) and the likelihood that microtubules preferentially grow from one end, it is unlikely that the osmiophilic material in the interzone performs a nucleating function. In other ways, however, they may be similar. Gould and Borisy (1977) have suggested that the pericentriolar material anchors microtubules at the pole and it seems feasible that the interzonal material might have the same anchoring function. In the interzone of both prophase and telophase spindles, the electron-dense material may be the imperfectly preserved remains of a contractile network that serves to bring antiparallel microtubules together. Because metazoan microtubules begin to bunch together in mid anaphase, the contraction of the cleavage furrow cannot be the mechanism of microtubule bunching. Some kind of lateral interaction between microtubules similar to the "zipping" of microtubules proposed by Bajer (1973) is another possible mechanism.

B. Structures connecting adjacent microtubules

There is considerable evidence, both direct and indirect, that there are structural connections (bridges) between mitotic spindle microtubules. Direct visualization of bridges was reviewed by McIntosh (1974), and since then several studies (McDonald *et al.*, 1979; Oakley and Heath, 1978; Ritter *et al.*, 1978) have shown connections between microtubules. Indirect evidence for microtubule–microtubule interactions includes near-neighbour analyses (McDonald *et al.*, 1979; Roos, 1981; Tippit *et al.*, 1983; McIntosh *et al.*, 1985) and computer-enhanced image analysis (Jensen, 1982b; 1986a,b; Fig. 1.8). None of these studies has demonstrated that these connections are functionally significant. Nevertheless, the occurrence of bridges can be used to support the idea that microtubules slide over one another by means of a dynein-like cross-bridge (McIntosh *et al.*, 1969; Nicklas, 1971). A "cytoplasmic" dynein may be present in some spindles (Pratt *et al.*, 1980) although its current status is somewhat controversial (Vallee and Bloom, this volume). Alternatively, if one believes that central spindle microtubule cross-bridges have no mechanochemical function but act like a "brake" during spindle elongation (Aist and Berns, 1981; King, 1983) then the cross-bridges equally could be the morphological expression of this "brake".

C. Membranes

For a detailed account of the contribution of membranes to spindle structure and their role in spindle function, see Chapter 7 by Hepler in this volume.

Figure 1.8. (a) Electron micrograph of a section through a kinetochore of a *Dictyostelium discoideum* metaphase mitotic spindle . Arrows mark the area scanned to produce the densitometer record in (b). K = kinetochore, KMT = kinetochore microtubule. (Bar = 0.2 μm.) (Original micrograph courtesy of U.-P. Roos.) (b) to (d) Illustration of techniques used in microdensitometer–computer correlation analysis. (b) Initial densitometer record obtained from scanning the region indicated in (a). (c) Effect of applying standard image-filtering techniques to the record in (b). (d) The best-fit correspondence between the filtered densitometer record in (b) and a model that has projections arranged according to a symmetrical helical superlattice that repeats after 12 tubulin dimers (see, e.g., articles by Jensen and Smaill, 1986a,b). (Figures courtesy of Dr. Cynthia Jensen.)

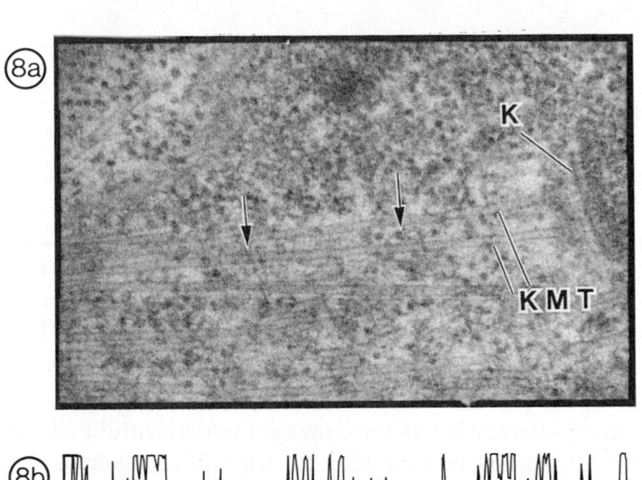

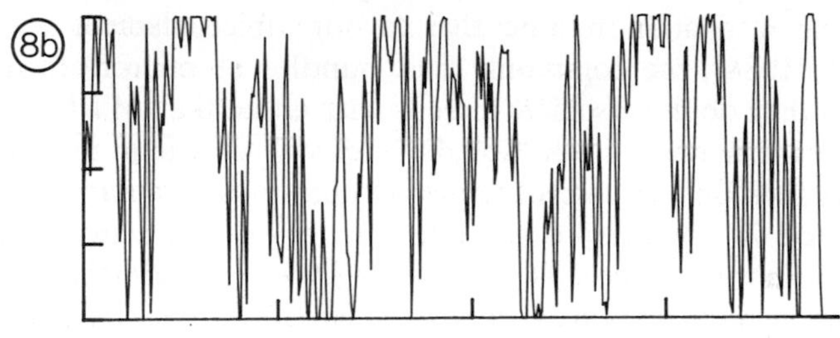

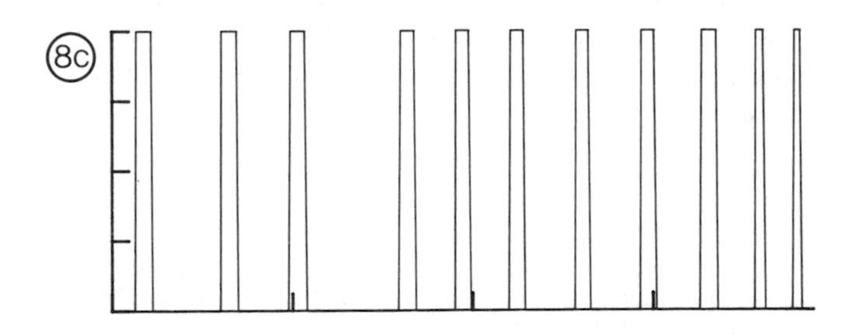

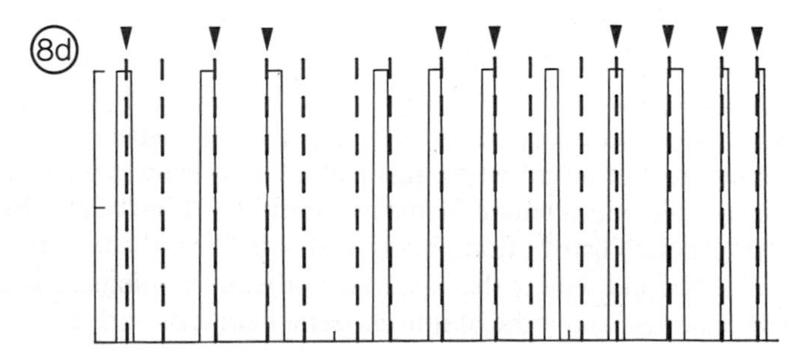

Figure 1.8.

D. Other structures

1. *Microfilaments*

The current thinking on actin in the spindle seems to be that actin is present but is non-functional as far as concerns moving chromosomes. The evidence showing that actin is present is primarily morphological, either EM or immunofluorescence (reviewed in Heath, 1981; Aubin, 1981). The evidence against actin as a functional motor is mostly pharmacological. In the presence of cytochalasin, chromosome movement continues normally, while, in the same cells, actomyosin activity in the cleavage furrow is inhibited (Cande *et al.*, 1981). Using a fixation regime that favours microfilament preservation (McDonald, 1984), we commonly find bundles of microfilaments in the spindle, but they do not associate directly with either the kMT bundles or the chromosomes. Likewise, in the spindle interzone they tend to be associated with membranes but not microtubules. Other kinds of microfilaments have been reported in spindles; e.g., individual 5 nm filaments that connect microtubules in the spindles of *Cryptomonas* (Oakley and Heath, 1978) and the 5–8 nm filaments in the bundle of *Oedogonium* kMTs (Schibler and Pickett-Heaps, 1980). While both these examples show specific microfilament–microtubule associations, the functional implications are unclear. Perhaps the clearest evidence against the involvement of actin in chromosome movement comes from genetic studies in *Saccharomyces*, where mitosis proceeds normally in cells carrying temperature-sensitive mutations in the single actin gene (Novick and Botstein, 1985).

2. *Intermediate filaments*

There is some evidence from both immunofluorescence (Aubin *et al.*, 1980; Blose, 1981; Blose and Meltzer, 1981; Blose and Bushnell, 1982; Gordon *et al.*, 1978; Hynes and Destree, 1978; Aubin, 1981) and EM studies (Zieve *et al.*, 1980; Blose, 1981; Starger *et al.*, 1978) that intermediate filaments are regularly associated with spindle poles during mitosis. Some of these authors suggest that intermediate filaments might act as a scaffold, serving to orient the spindle relative to the rest of the cell during division. This is a difficult hypothesis to check because of the lack of agents which specifically interfere with the deployment of intermediate filaments. Other evidence argues against a significant role for intermediate filaments in mitosis. First of all, they have a limited phylogenetic distribution, being restricted mostly to differentiated metazoan cells. Second, where they are found, the morphological associations are quite general, e.g., a bundle of intermediate filaments with a spindle pole, as opposed to one-to-one correspondence between microtubules and intermediate filaments. Third, disruption of keratin filaments by microinjection of anti-keratin antibodies has no discernible effect on mitosis (Klymkowsky *et al.*, 1983). Given the prevalence of intermediate filaments in the tissue culture cells in which most mitotic studies are done, it might be more surprising if some intermediate filaments were not associated with the spindle.

IX. Methods and Materials

A. The limitations of conventional EM processing

Nearly all the information we have about the ultrastructure of mitotic spindles comes from material that has been fixed in glutaraldehyde, post-fixed in OsO_4, dehydrated, embedded in resin, and thin sectioned. There are several lines of evidence that suggest that this kind of treatment alters or destroys the molecular architecture of the spindle, at least at the level of microtubules and smaller structures.

Monitoring birefringence changes during glutaraldehyde fixation is a way of detecting rearrangements or loss of certain spindle components such as microtubules and other linear elements arranged in parallel arrays. There is some debate as to whether microtubules are responsible for all the birefringence seen in spindles (summarized in Heath, 1981), but that is less important than the fact that during glutaraldehyde fixation something is lost or rearranged. Early studies by Inoué and Sato (1967) showed a 50% loss of birefringence after fixation of sea-urchin eggs. In smaller cells, where one would expect better fixation due to faster penetration of the fixative, the loss of birefringence is less. McIntosh *et al.* (1975) reported 0–50% loss in mammalian tissue culture cells; Nicklas *et al.* (1979) reported losses of 10–30% in insect spermatocytes; and LaFountain (1974, 1976) found a different response depending on the stage of mitosis. In cranefly spermatocytes in metaphase there was a 50% loss, but in anaphase there was no reduction in birefringence. None of these studies followed birefringence changes through later stages of processing such as post-fixation with OsO_4 or dehydration, where it is likely that further losses might take place.

A number of studies have been done which use serial-section reconstruction to measure the effects of different fixative conditions on the number and length of microtubules. This method is much more accurate than use of polarization optics because it allows one to measure actual numbers of microtubules as well as their lengths. The composition of the primary fixative solution, including choice of buffer, added electrolytes and even the type of glutaraldehyde used, can make a substantial difference to microtubule preservation. When Nicklas *et al.* (1982) used glutaraldehyde that had been mixed with a slurry of agar to fix insect spermatocyte spindles, they found a 50% increase in the number of preserved microtubules over material fixed with untreated glutaraldehyde. However, the same agar-treated glutaraldehyde gave no improvement in the preservation of PtK_2 spindle microtubules (Nicklas *et al.*, 1982). In our own laboratory we have found that a practical grade glutaraldehyde (TAAB Industries, Reading, England), used at the suggestion of Dr. Beth Burnside (University of California at Berkeley) gives better preservation of spindle ultrastructure than more highly purified stocks (McDonald, 1983). Luftig *et al.* (1977) and Seagull and Heath (1980) found that microtubule stabilizing buffers preserved more and longer cytoplasmic microtubules than phosphate or cacodylate buffers. Shigenaka *et al.* (1974) have shown that Mg^{2+} added to the fixative will stabilize microtubules but Ca^{+}, on the other hand, is known to disrupt microtubule patterns (Schliwa, 1976). It is worth noting that most of the results on microtubule preservation have been from studies of cytoplasmic microtubules. Various lines of evidence (references in Saxton *et al.*, 1984; Salmon *et al.*, 1984) indicate that mitotic microtubules are more labile than cytoplasmic microtubules, and thus the possibilities of fixation-related artefacts are even greater.

Quantitative studies on the same organism that produce significantly different results as to the number and arrangement of microtubules also suggest that the methods of specimen preparation are perhaps more crucial than we have learned to believe. Both Forer and Brinkley (1977) and LaFountain (1976) have counted numbers of microtubules in the spindle of *Pales* spindles, but LaFountain (1976) found considerably more microtubules. This may have been due to differences in the fixative solution or to the physiological condition of the cells prior to fixation.

So far we have only been concerned with how the composition of the primary fixative solution can affect the preservation of spindle microtubules. But the initial fixation is only the first in a long series of steps between the living cell and the image we see in the negative or print. Of course, if the primary fixation fails then the subsequent problems are irrelevant. Assuming that the fine details of structure can be preserved by glutaraldehyde, there are still the problems of post-fixation with osmium and dehydration in acetone or ethanol. It is well known that OsO_4 will destroy filamentous actin that has been preserved by glutaraldehyde (Maupin-Szamier and Pollard, 1978; Pollard and Maupin, 1982; Small, 1981). In some cell types the destructive effects of OsO_4 can be reduced by adding potassium ferricyanide to the osmium step (McDonald, 1984). It is also reported that tannic acid and saponin added to the primary fixative will help preserve F-actin (Maupin and Pollard, 1983). Actin preservation has received a lot of attention because of its importance in cell motility events (Pollard and Maupin, 1982) and it has the advantage of being a molecule that can readily be tested *in vitro* or assayed *in vivo* by immunocytochemistry. Thus, from immunofluorescence evidence (reviewed in Aubin, 1981; Traas *et al.*, 1987) we know that mitotic spindles can contain some F-actin, but this actin does not appear in electron micrographs. Presumably, other molecules that show up by immunofluorescence, such as MAPs, cytoplasmic dynein, and myosin (reviewed in McIntosh, 1984), but not in thin sections, must suffer a similar fate.

Using F-actin again as a test subject, Small (1981) has shown that, even if one succeeds in preserving long, straight actin filaments through OsO_4 (by low concentrations and short incubation times), dehydration will cause them to collapse. Studies with striated muscle actin (reviewed in Forer, 1978) show that acetone is less destructive than ethanol. With non-muscle actins, however, both dehydration fluids cause filament collapse (Small, 1981). Shrinking during dehydration is also thought to be responsible for the wavy microtubules seen in many spindle micrographs (Bajer and Mole-Bajer, 1972).

In summary, we have seen that the accurate preservation of spindle fine structure can be thwarted at many steps along the way to the electron microscope. Some of these problems can be corrected, e.g. by careful attention to the composition of the primary fixative, but others are more problematical, like the destruction of fine detail by osmium and dehydration. Is there any way around these inherent difficulties? Some believe that rapid freezing methods may be the answer. In the next section we will consider the advantages (and disadvantages) of cryotechnology as it applies to the ultrastructure of dividing cells.

B. The promise of rapid-freezing technology

The chief advantage of rapid-freezing fixation is the rate at which cells/tissues are fixed relative to chemical fixation. Gilkey and Staehelin (1986) estimate that the increase is about three orders of magnitude. So while it may take seconds to minutes to partially fix cells chemically, freezing immobilizes all cell components in milliseconds. After the material is frozen, it can be processed in a number of different ways; fractured, fractured and etched, freeze-substituted, or in some cases, viewed in the hydrated state. Most mitotic studies have used freeze-substitution methodology, although there are a few freeze-fracture images (Moor, 1967; LaFountain and Thomas, 1975; Mullins, 1984) and one paper (Hirokawa *et al.*, 1985) on rapidly-frozen, deep-etched spindles.

Heath and his co-workers (Heath and Rethoret, 1982; Heath *et al.*, 1984; McKerracher and Heath, 1985) have done some careful quantitative comparisons of chemically fixed and freeze-substituted fungal cells, including some in mitosis. In *Saprolegnia* (Heath *et al.*, 1984) the two methods gave identical results with regard to number, length and distribution of microtubules. In

Zygorhynchus (Heath and Rethoret, 1982) some interphase microtubules were preserved by freezing but not chemical fixation; and in *Basidiobolus* (McKerracher and Heath, 1985) freeze substitution preserved significantly more cytoplasmic microtubules. Hoch and Staples (1980) showed that freeze substitution gave better preservation of microfilaments as well as microtubules. These results suggest that freeze substitution is the method of choice for fixing mitotic fungal cells. Qualitative comparisons of freezing and chemical fixation for cytoplasmic preservation in higher plants (Lancelle *et al.*, 1986) and animals (Porter and Anderson, 1982; Bridgeman and Reese, 1984) also show freezing to be superior. As yet, there are no freeze-substitution studies of mitosis in higher plants or animals. It is important to remember that freeze-substitution methods do not eliminate chemical fixation, it merely takes place at low temperature, and that dehydration and embedding are still necessary. Thus, the artefacts associated with these phases of the processing are still potential problems. In addition, new freezing-related distortions are possible such as ice-crystal damage from poor freezing or recrystallization.

Freeze-fracture is a way to avoid the problems of chemical artefacts but, unfortunately, the images of freeze-fracture replicas are not very informative about the fibrillar components of the mitotic apparatus. Most of the detail beyond the level of microtubules is obscured by the background cytoplasm. Even if these preparations were deep-etched it is likely that the granular-appearing soluble proteins would mask the details of the fibrous components. This was the experience of Hirokawa and Heuser (1981) in their study of intestinal brush border cytoskeleton. One way to improve the visualization of the cytoskeleton is to remove soluble proteins by extraction. Hirokawa *et al.* (1985) used this approach in the only published work on rapidly frozen deep-etched mitotic spindles. Unfortunately, the spindles are no longer functional after treatment with 1 M glycerol, 10% DMSO and 1% detergent, and so it is difficult to know whether the structures seen bear any resemblance to the living state. The cross-bridges they see could be artefacts of deep-etching (cf., Schnapp and Reese, 1982) and the "buttons" may be microtubule-associated proteins (MAPs) that coil up after exposure to glutaraldehyde (Langford, 1983).

The studies mentioned above are the pioneering efforts in what promises to be an exciting new era in the ultrastructure of mitosis. Many technical problems remain to be solved, but they will be solved sooner or later because the logic of rapid-freeze fixation and the results from some other motility systems (Goodenough and Heuser, 1984) are too compelling to ignore. Both quick-freeze deep-etch techniques and the observation of frozen-hydrated specimens (Stewart and Vigers, 1986) offer the most potential for visualizing the "true" structure of the mitotic spindle.

Acknowledgements

Thanks are due to W. Z. Cande for thoughtful discussions and criticisms and to the authors who provided illustrations for this chapter. This work was supported in part by NIH grant GM-23238 and National Science Foundation grant PCM-8408594 to W. Z. Cande.

REFERENCES

Aist, J. R. and Berns, M. W. (1981). *J. Cell Biol.* **91**, 446–458.

Aubin, J. E. (1981). In *Mitosis/Cytokinesis* (ed. A. M. Zimmerman and A. Forer), pp. 211–244. Academic Press, New York.

Aubin, J. E., Osborn, M., Franke, W. W. and Weber, K. (1980). *Exp. Cell Res.* **129**, 149–165.

Bajer, A. (1973). *J. Cell Biol.* **8**, 139–160.

Bajer, A. (1983). *J. Cell Biol.* **93**, 33–48.

Bajer, A. and Mole-Bajer, J. (1972). *Int. Rev. Cytol. Suppl.* **3**, 1–271.

Bajer, A. and Mole-Bajer, J. (1973). *Cytobios* **8**, 139–160.

Beams, H. W., Evans, T. C., Baker, W. W. and von Breeman, V. (1950). *Anat. Rec.* **107**, 329–345.

Bergen, L. G. and Borisy, G. G. (1980). *J. Cell Biol.* **84**, 141–150.

Bernhard, W. and deHarven, E. (1956). *C. R. Acad. Sci., Paris* **242** (Ser. D), 288–290.

Bernhard, W. and deHarven, E. (1958). *IV Int. Conf. Electron Microscopy, Berlin* basnd II, pp. 217–227.

Blose, S. H. (1981). *Cell Motility* **1**, 417–431.

Blose, S. H. and Bushnell, A. (1982). *Exp. Cell Res.* **142**, 57–62.

Blose, S. H. and Meltzer, D. I. (1981). *Exp. Cell Res.* **135**, 299–309.

Borisy, G. G. (1978). *J. Mol. Biol.* **124**, 565–570.

Brachet, J. (1957). *Biochemical Cytology*, p. 535. Academic Press, New York.

Bridgeman, P. C. and Reese, T. S. (1984). *J. Cell Biol.* **99**, 1655–1668.

Brinkley, B. R. and Cartwright, J. Jr., (1971). *J. Cell Biol.* **50**, 416–431.

Cande, W. Z., McDonald, K. and Meeusen, R. L. (1981). *J. Cell Biol.* **88**, 618–629.

Church, K. and Lin, H.-P. P. (1982). *J. Cell Biol.* **93**, 365–373.

Church, K. and Lin, H.-P. P. (1985). *Chromosoma* **92**, 273–282.

Coss, R. A. and Pickett-Heaps, J. D. (1974). *J. Cell Biol.* **63**, 84–98.

DeBrabander, M., Geuens, G., DeMey, J. and Jonaiau, M. (1979). *Biol. Cell.* **34**, 213–226.

DeBrabander, M., Geuens, G., Nuydens, R., Willebrords, R. and DeMey, J. (1980). In *Microtubules and Microtubule Inhibitors 1980* (ed. M. Debrabander and J. DeMey), pp. 255–268. Elsevier, Amsterdam.

DeBrabander, M., Geuens, G., Nuydens, R., Willebrords, K. and DeMey, J. (1981). *Cell Biol. Int. Rep.* **5**, 913–920.

deHarven, E. and Bernhard, W. (1956). *Z. Zellforsch. Mikrosk. Anat.* **45**, 387–398.

Euteneuer, U. and McIntosh, J. R. (1980). *J. Cell Biol.* **87**, 509–515.

Euteneuer, U. and McIntosh, J. R. (1981). *J. Cell Biol.* **89**, 338–345.

Forer, A. (1978). In *Principles and Techniques of Electron Microscopy, Biological Applications* (ed. M. A. Hyat), Vol. 9, pp. 126–174. Van Nostrand, New York.

Forer, A. and Brinkley, B. R. (1977). *Can. J. Genet. Cytol.* **19**, 503–519.

Forer, A., Larson, D. E. and Zimmerman, A. M. (1980). *Can. J. Biochem.* **58**, 1277–1285.
Fuge, H. (1974). *Protoplasma* **82**, 289–320.
Fuge, H. (1977). In *Mitosis: Facts and Questions* (ed. M. Little, N. Paweletz, C. Petzelt, H. Ponstingl, D. Schroeter and H.-P. Zimmerman), pp. 51–77. Springer-Verlag, Berlin.
Fuge, H. (1978). *Int. Rev. Cytol. Suppl.* **6**, 1–58.
Fuge, H. (1984). *Chromosoma* **90**, 323–331.
Gilkey, J. and Staehelin, A. (1986). *J. Electron Microsc. Tech.* **3**, 177–210.
Girbardt, M. (1968). In *Aspects of Cell Motility*, XXII Symp. Soc. Exptl. Biol., pp. 249–260. Cambridge University Press, Cambridge.
Goodenough, U. and Heuser, J. E. (1984). *J. Mol. Biol.* **180**, 1083–1118.
Gordon, W. E., Bushnell, A. and Burridge, K. (1978). *Cell* **13**, 249–261.
Gould, R. R. and Borisy, G. G. (1977). *J. Cell Biol.* **73**, 601–615.
Gould, R. R. and Borisy, G. G. (1978). *Exp. Cell Res.* **113**, 369–374.
Harris, P. (1962). *J. Cell Biol.* **14**, 475–488.
Harris, P. and Mazia, D. (1962). In *The Interpretation of Ultrastructure* (ed. R. J. C. Harris), Symp. Int. Soc. Cell Biol. **1**, 279–305. Academic Press, New York.
Heath, I. B. (1974). *J. Cell Biol.* **60**, 204–220.
Heath, I. B. (1980). *Int. Rev. Cytol.* **64**, 1–80.
Heath, I. B. (1981). In *Mitosis/Cytokinesis* (ed. A. M. Zimmerman and A. Forer), pp. 245–275, Academic Press, New York.
Heath, I. B. and Heath, M. C. (1976). *J. Cell Biol.* **70**, 592–607.
Heath, I. B. and Rethoret, K. (1982). *Eur. J. Cell Biol.* **28**, 180–189.
Heath, I. B., Rhethoret, K. and Moens, P. B. (1984). *Eur. J. Cell Biol.* **35**, 284–295.
Heidemann, S. R. and McIntosh, J. R. (1980). *Nature* **286**, 517–519.
Heidemann, S. R., Zieve, G. W. and McIntosh, J. R. (1980) *J. Cell Biol.* **87**, 152–159.
Hirokawa, N. and Heuser, J. E. (1981). *J. Cell Biol.* **91**, 399–409.
Hirokawa, N., Takemura, R. and Hisanaga, S.-I. (1985). *J. Cell Biol.* **101**, 1858–1870.
Hoch, H. C. and Staples, R. C. (1983). *Mycologia* **75**, 795–824.
Hollande, A. (1974). *Protistologica* **10**, 413–451.
Hollenbeck, P. J., Suprynowicz, F. and Cande, W. Z. (1984). *J. Cell Biol.* **99**, 1251–1258.
Howard, R. J. and Aist, J. R. (1979). *J. Ultrastruct. Res.* **66**, 224–234.
Hynes, R. D. and Destree, A. T. (1978). *Cell* **13**, 151–163.
Inoué, S. (1981). *Discovery in Cell Biology: J. Cell Biol.* (*Suppl.*) **91**, 131s–147s.
Inoué, S. and Ritter, Jr., H. (1978). *J. Cell Biol.* **77**, 655–684.
Inoué, S. and Sato, H. (1967). *J. Gen. Physiol.* **50**, 259–292.
Jenkins, R. A. (1967). *J. Cell Biol.* **34**, 463–488.
Jenkins, R. A. (1977). *J. Protozool.* **24**, 264–275.
Jensen, C. G. (1982a). *J. Cell Biol.* **92**, 540–558.
Jensen, C. G. (1982b). *J. Cell Biol.* **95**, 335a.
Jensen, C. G. and Bajer, A. (1973). *Chromosoma* **44**, 73–89.
Jensen, C. G. and Smaill, B. H. (1986a). *Ann. N.Y. Acad. Sci.* **466**, 417–419.
Jensen, C. G. and Smaill, B. H. (1986b). *J. Cell Biol.* **103**, 559–569.
Jokelainen, P. T. (1965). *J. Cell Biol.* **27**, 48a. [abstr.]
Jokelainen, P. T. (1967). *J. Ultrastruct. Res.* **19**, 19–44.
King, S. M. (1983). *J. Theor. Biol.* **102**, 501–510.
King, S. M., Hyams, J. S. and Luba, A. (1982). *J. Cell Biol.* **94**, 341–349.
Klymkowsky, M. W., Miller, R. H. and Lane, E. B. (1983). *J. Cell Biol.* **96**, 494–509.

Kubai, D. (1975). *Int. Rev. Cytol.* **43**, 167–227.
Kubai, D. and Ris, H. (1969). *J. Cell Biol.* **40**, 508–528.
LaFountain, J. R., Jr. (1974). *J. Ultrastruct. Res.* **46**, 268–278.
LaFountain, J. R., Jr. (1976). *J. Ultrastruct. Res.* **54**, 333–346.
LaFountain, J. R., Jr. and Davidson, L. A. (1979). *Chromosoma* **75**, 293–308.
LaFountain, J. R., Jr. and Davidson, L. A. (1980). *Cell Motility* **1**, 41–61.
LaFountain, J. R., Jr. and Thomas, H. R. (1975). *Ultrastruct. Res.* **51**, 340–347.
Lancelle, S. A., Callaham, D. A. and Hepler, P. K. (1986). *Protoplasma* **131**, 153–165.
Langford, G. M. (1983). *J. Ultrastruct. Res.* **85**, 1–10.
Ledbetter, M. C. and Porter, K. R. (1963). *J. Cell Biol.* **19**, 239–250.
Lin, H.-P. P., Ault, J. G. and Church, K. (1981). *Chromosoma* **83**, 507–521.
Luck, D. J. L. (1984). *J. Cell Biol.* **98**, 789–794.
Luftig, R. B., McMillan, P. N., Weatherbee, J. A. and Weihing, R. R. (1977). *J. Histochem. Cytochem.* **25**, 175–187.
Luykx, P. (1970). *Int. Rev. Cytol. Suppl.* **2**, 1–173.
Manton, I. (1964). *J. Roy. Microsc. Soc.* **83**, 317–325.
Manton, I. (1978). *Proc. Roy. Microsc. Soc.* **13**, 45–57.
Manton, I., Kowallik, K. and von Stosch, H. A. (1969a). *J. Microsc.* (*Oxford*) **89**, 295–320.
Manton, I., Kowallik, K. and von Stosch, H. A. (1969b). *J. Cell Sci.* **5**, 271–298.
Manton, I., Kowallik, K. and von Stosch, H. A. (1970a). *J. Cell Sci.* **6**, 131–157.
Manton, I., Kowallik, K. and von Stosch, H. A. (1970b). *J. Cell Sci.* **7**, 407–444.
Margolis, R. L., Wilson, L. and Kiefer, B. I. (1978). *Nature* **272**, 450–452.
Maupin, P. and Pollard, T. D. (1983). *J. Cell Biol.* **96**, 51–62.
Maupin-Szamier, I. P. and Pollard, T. D. (1978). *J. Cell Biol.* **77**, 837–852.
Mazia, D. (1984). *Exp. Cell Res.* **153**, 1–15.
Mazia, D., Paweletz, N., Sluder, G. and Finze, E.-M. (1981). *Proc. Natl. Acad. Sci. USA* **78**, 377–381.
McDonald, K. (1983). In *41st Ann. Meeting Electron Microscopy Soc. Amer., Proc.* (ed. G. W. Bailey), pp. 544–547.
McDonald, K. (1984). *J. Ultrastruct. Res.* **86**, 107–118.
McDonald, K. and Cande, W. Z. (1980). *J. Cell Biol.* **87**, 236a.
McDonald, K. and Euteneuer, U. (1983). *J. Cell Biol.* **97**, 88a.
McDonald, K., Pickett-Heaps, J. D., McIntosh, J. R. and Tippit, D. H. (1977). *J. Cell Biol.* **74**, 377–388.
McDonald, K. L., Edwards, M. K. and McIntosh, J. R. (1979). *J. Cell Biol.* **83**, 443–461.
McGill, M. and Brinkley, B. R. (1975). *J. Cell Biol.* **67**, 189–199.
McIntosh, J. R. (1974). *J. Cell Biol.* **61**, 166–187.
McIntosh, J. R. (1981). In *International Cell Biology 1980–1981* (ed. H. G. Schweiger), pp. 359–368. Springer-Verlag, Berlin.
McIntosh, J. R. (1982). In *Developmental Order: Its Origin and Regulation* (ed. P. W. Green and S. Subtelny), pp. 77–115. Alan R. Liss, Inc., New York.
McIntosh, J. R. (1984). *Trends Biochem. Sci.* April, 195–198.
McIntosh, J. R. and Landis, S. C. (1971). *J. Cell Biol.* **49**, 468–497.
McIntosh, J. R., Hepler, P. K. and VanWie, D. G. (1969). *Nature* **224**, 659–663.
McIntosh, J. R., Cande, W. Z., Snyder, J. A. and Vanderslice, K. (1975a). *Ann. N.Y. Acad. Sci.* **253**, 407–427.
McIntosh, J. R., Cande, W. Z. and Snyder, J. A. (1975b). In *Molecules and Cell Movement* (ed. S. Inoué and R. E. Stephens), pp. 31–76. Raven Press, New York.

McIntosh, J. R., McDonald, K. L., Edwards, M. K. and Ross, B. M. (1979a). *J. Cell Biol.* **83**, 428–442.
McIntosh, J. R., Sisken, J. E. and Chu, L. K. (1979b). *J. Ultrastruct. Res.* **66**, 40–52.
McIntosh, J. R. and Euteneuer, U. (1984). *J. Cell Biol.* **98**, 525–533.
McIntosh, J. R., Roos, U.-P., Neighbors, B. and McDonald, K. L. (1985). *J. Cell Sci.* **75**, 93–129.
McKerracher, L. J. and Heath, I. B. (1985). *Protoplasma* **125**, 162–172.
Mitchison, T. and Kirschner, M. (1984). *Nature* **312**, 237–242.
Mitchison, T. and Kirschner, M. (1985a). *J. Cell Biol.* **101**, 755–765.
Mitchison, T. and Kirschner, M. (1985b). *J. Cell Biol.* **101**, 766–777.
Moens, P. B. (1979). *J. Cell Biol.* **83**, 556–561.
Moens, P. B. and Moens, T. (1981). *J. Ultrastruct. Res.* **75**, 131–141.
Mole-Bajer, J. and Bajer, A. (1968). *Cellule* **67**, 257–265.
Mole-Bajer, J., Bajer, A. and Owczarzak, A. (1975). *Cytobios.* **13**, 45–65.
Moor, H. (1967). *Protoplasma* **64**, 89–103.
Mullins, J. M. (1984). *Cell Biol. Int. Rep.* **8**, 107–115.
Mullins, J. M. and Biesele, J. J. (1977). *J. Cell Biol.* **73**, 672–684.
Nicklas, R. B. (1971). In *Advances in Cell Biology* **2**, 225–294.
Nicklas, R. B. (1975). In *Molecules and Cell Movement* (ed. S. Inoué and R. E. Stephens), pp. 97–117. Raven Press, New York.
Nicklas, R. B. (1983). *J. Cell Biol.* **97**, 542–548.
Nicklas, R. B. (1984). *Cell Motility* **4**, 1–5.
Nicklas, R. B. and Gordon, G. W. (1985). *J. Cell Biol.* **100**, 1–7.
Nicklas, R. B. and Kubai, D. F. (1985). *Chromosoma* **92**, 313–324.
Nicklas, R. B. and Staehly, C. A. (1967). *Chromosoma* **21**, 1–16.
Nicklas, R. B., Brinkley, B. R., Pepper, D. A., Kubai, D. F. and Rickards, G. K. (1979). *J. Cell Sci.* **35**, 87–104.
Nicklas, R. B., Kubai, D. F. and Hays, T. S. (1982) *J. Cell Biol.* **95**, 91–104.
Novick, P. and Botstein, D. (1985). *Cell* **40**, 405–416.
Oakley, B. R. and Heath, I. B. (1978). *J. Cell Sci.* **31**, 53–70.
Paweletz, N. (1967). *Naturwissenschaften* **54**, 533–535.
Paweletz, N., Mazia, D. and Finze, E.-M. (1984). *Exp. Cell Res.* **152**, 47–65.
Pease, D. C. and Porter, K. R. (1981). *Discovery in Cell Biology: J. Cell Biol.* (*Suppl.*) **91**, 287s–292s.
Peterson, J. B. and Ris, H. (1976). *J. Cell Sci.* **22**, 219–242.
Pickett-Heaps, J. D. (1969). *Cytobios* **1**, 257–280.
Pickett-Heaps, J. D. (1974). *BioSystems* **6**, 37–48.
Pickett-Heaps, J. D. and Tippit, D. H. (1978). *Cell* **14**, 455–467.
Pickett-Heaps, J. D., McDonald, K. L. and Tippit, D. H. (1975). *Protoplasma* **86**, 205–242.
Pickett-Heaps, J. D., Tippit, D. H. and Andreozzi, J. A. (1978). *Biol. Cell.* **33**, 79–84.
Pickett-Heaps, J. D., Tippit, D. H. and Porter, K. R. (1982). *Cell* **29**, 729–744.
Pollard, T. D. and Maupin, P. (1982). In *Electron Microscopy in Biology* (ed. J. D. Griffith), vol. 2, pp. 1–53. Wiley, New York.
Porter, K. R. (1955). *Symp. on Fine Structure of Cells, Leiden, 1954*, pp. 236–250. Noordhoff Ltd., Groningen.
Porter, K. R. (1966). In *Principles of Biomolecular Organization* (ed. G. E. W. Wolstenholme and M. O.'Connor), pp. 308–356. Little Brown, Boston.
Porter, K. R. (1980). In *Microtubules and Microtubule Inhibitors 1980* (ed. M. DeBrabander and J. DeMey), pp. 555–568. Elsevier, Amsterdam.

Porter, K. R. and Anderson, K. L. (1982). *Eur. J. Cell Biol.* **29**, 83–96.
Porter, K. R. and Blum, J. (1953). *Anat. Rec.* **117**, 685–710.
Porter, K. R. and Machado, R. D. (1960). *J. Biophys. Biochem. Cytol.* **7**, 167–180 + plates 81–96.
Pratt, M. M., Otter, T. and Salmon, E. D. (1980). *J. Cell Biol.* **86**, 738–745.
Rattner, J. B. and Berns, M. W. (1976). *Cytobios* **15**, 37–43.
Rieder, C. L. (1981). *Chromosoma* **84**, 145–158.
Rieder, C. L. (1982). *Int. Rev. Cytol.* **79**, 1–58.
Rieder, C. L. (1985). *J. Electron Microsc. Tech.* **2**, 11–28.
Rieder, C. L. and Borisy, G. G. (1981). *Chromosoma* **82**, 693–716.
Rieder, C. L. and Borisy, G. G. (1982). *Biol. Cell.* **44**, 117–1322.
Rieder, C. L., Davison, E. A., Jensen, L. C. W., Cassimeris, L. and Salmon, E. D. (1986). *J. Cell Biol.* **103**, 581–591.
Ris, H. and Kubai, D. (1974). *J. Cell Biol.* **60**, 702–720.
Ris, H. and Witt, P. L. (1981). *Chromosoma* **82**, 153–170.
Ritter, H., Inoué, S. and Kubai, D. F. (1978). *J. Cell Biol.* **77**, 638–654.
Robbins, E. and Gonatas, N. K. (1964a). *J. Cell Biol.* **20**, 356–359.
Robbins, E. and Gonatas, N. K. (1964b). *J. Cell Biol.* **21**, 429–463.
Robbins, E., Jentzsch, G. and Micali, A. (1968). *J. Cell Biol.* **36**, 329–339.
Roos, U.-P. (1973a). *Chromosoma* **40**, 43–82.
Roos, U.-P. (1973b). *Chromosoma* **41**, 195–220.
Roos, U.-P. (1976). *Chromosoma* **54**, 363–385.
Roos, U.-P. (1981). Quantitative structure analysis of the mitotic spindle. In *International Cell Biology 1980–1981* (ed. H. Schweiger), pp. 369–381. Springer-Verlag, Berlin.
Roth, L. E. and Daniels, E. W. (1962). *J. Cell Biol.* **12**, 57–78.
Rozsa, G. and Wyckoff, R. W. G. (1950) *Biochim. Biophys. Acta* **6**, 334–339.
Rozsa, G. and Wyckoff, R. W. G. (1951) *Exp. Cell Res.* **2**, 630–641.
Sabatini, D. D., Bensch, K. G. and Barnett, R. J. (1963). *J. Cell Biol.* **17**, 19–58.
Salmon, E. D., Leslie, R. J., Saxton, W. M., Karow, M. C. and McIntosh, J. R. (1984). *J. Cell Biol.* **99**, 2165–2174.
Saxton, W. M., Stemple, D. L., Leslie, R. J., Salmon, E. D., Zavortink, M. and McIntosh, J. R. (1984). *J. Cell Biol.* **99**, 2175–2186.
Scarello, L. A., Janicke, M. A. and LaFountain, J. R., Jr. (1985). *J. Cell Biol.* **101**, 148a.
Schibler, M. J. and Pickett-Heaps, J. D. (1980). *Eur. J. Cell Biol.* **22**, 687–698.
Schliwa, M. (1976). *J. Cell Biol.* **70**, 527–540.
Schnapp, B. J. and Reese, T. S. (1982). *J. Cell Biol.* **94**, 667–679.
Scholey, J. M., Porter, M. E., Grissom, P. M. and McIntosh, J. R. (1985). *Nature* **318**, 483–486.
Schrader, F. (1953). *Mitosis*, 2nd edn. Columbia University Press, New York.
Seagull, R. W. and Heath, I. B. (1980). *Protoplasma* **103**, 205–229.
Selby, C. C. (1953). *Exp. Cell Res.* **5**, 386–393.
Shigenaka, Y., Watanabe, K. and Kaneda, M. (1974). *Exp. Cell Res.* **85**, 391–398.
Slautterback, D. B. (1963). *J. Cell Biol.* **18**, 367–388.
Sluder, G. and Rieder, C. L. (1985). *J. Cell Biol.* **100**, 887–896.
Small, J. V. (1981). *J. Cell Biol.* **91**, 695–705.
Snyder, J. A. and McIntosh, J. R. (1975). *J. Cell Biol.* **67**, 744–760.
Starger, J. M., Brown, W. E., Goldman, A. E. and Goldman, R. D. (1978). *J. Cell Biol.* **78**, 93–109.

Steffen, W. and Fuge, H. (1982). *Chromosoma* **87**, 363–371
Stewart, M. and Vigers, G. (1986). *Nature* **319**, 631–636.
Telzer, B. R. and Haimo, L. T. (1981). *J. Cell Biol.* **89**, 373–378.
Telzer, B. R., Moses, M. J. and Rosenbaum, J. L. (1975). *Proc. Natl. Acad. Sci. USA* **72**, 4023–4027.
Tippit, D. H. and Pickett-Heaps, J. D. (1977). *J. Cell Biol.* **73**, 705–727.
Tippit, D. H., Schulz, D. and Pickett-Heaps, J. D. (1978). *J. Cell Biol.* **79**, 737–763.
Tippit, D. H., Pillus, L. and Pickett-Heaps, J. D. (1980a). *J. Cell Biol.* **87**, 531–545.
Tippit, D. H., Pickett-Heaps, J. D. and Leslie, R. (1980b). *J. Cell Biol.* **86**, 402–416.
Tippit, D. H., Pillus, L. and Pickett-Heaps, J. D. (1983). *Eur. J. Cell Biol.* **30**, 9–17.
Tippit, D. H., Fields, C. T., O'Donnell, K. L., Pickett-Heaps, J. D. and McLaughlin, D. J. (1984). *Eur. J. Cell Biol.* **34**, 34–444.
Traas, J. A., Doonan, J. H., Rawlins, D. J., Shaw, P. J., Watts, J. and Lloyd, C. W. (1987). *J. Cell Biol.* **105**, 387–395.
Tucker, J. B., Beisson, J., Roche, D. L. J. and Cachon, J. (1980). *J. Cell Sci.* **44**, 135–151.
Witt, P. L., Ris, H. and Borisy, G. G. (1980). *Chromosoma* **81**, 483–505.
Witt, P. L., Ris, H. and Borisy, G. G. (1981). *Chromosoma* **83**, 523–540.
Wood, W. B. (1979). *Harvey Lectures* **73**, 203–224.
Zieve, G. W., Heidemann, S. R. and McIntosh, J. R. (1980). *J. Cell Biol.* **87**, 160–169.

CHAPTER 2

The Centrosome Cycle in Animal Cells

DALE D. VANDRÉ and GARY G. BORISY

Laboratory of Molecular Biology, University of Wisconsin,
Madison, Wisconsin, USA

I. Introduction

The function of the centrosome has posed a challenging problem to cell biologists since the first light-microscopic descriptions of the polar corpuscles by van Beneden (1876). (For a more complete historical perspective of centrosome biology see Wheatley, 1982.) E. B. Wilson's (1911) 2nd edition of the classic treatise, *The Cell in Development and Heredity*, presents the hypothesis of van Beneden and Boveri describing the centrosome as a "permanent cell-organ which forms the dynamic centre of the cell and multiplies by division to form the centre of the daughter cells". By the appearance of the 3rd edition of Wilson's book in 1925, the pre-eminent role for the centrosome in cell division had been modified, primarily because observations on plant cells had failed to detect the presence of typical centriole containing centrosomes like those found in animal cells. Wilson suggested that "van Beneden's hypothesis must be abandoned in favour of the view that the centrosome is but a subordinate part of the general apparatus of mitosis, and one which may be entirely dispensed with."

The different opinions expressed on the role centrosomes play in cell division were due, in part, to difficulties in defining the functional and structural components present at the mitotic poles. Unfortunately, the same dilemma faces modern cell biologists. The centrosome has been defined in both structural and functional terms, but these are not necessarily synonymous. With the advent of the electron microscope, the stereotypical animal centro-

MITOSIS: Molecules and Mechanisms
ISBN 0-12-363420-2

some was ultrastructurally defined as consisting of two major constituents, the centrioles and associated pericentriolar material (deHarven and Bernhard, 1956; Szöllösi, 1964; Stubblefield and Brinkley, 1967; Brinkley and Stubblefield, 1970). A centriole is a cylindrical structure comprised of 9 triplet tubules running in slightly twisted bundles to form the outer wall of the cylinder. When viewed in cross-section the tubules are arrayed in a pinwheel pattern. Surrounding the centriole is a fine electron-dense matrix referred to as the pericentriolar material. The centrosome serves as a focal point for microtubules in the interphase cell and was therefore termed a microtubule organizing centre (MTOC) (Porter, 1966; Pickett-Heaps, 1969). Mitotic cells contain two centrosomes, which serve as the poles of the spindle apparatus. In most animal cells, the centrosome may conveniently be defined as a microtubule organizing centre containing centrioles and associated pericentriolar material.

A number of cell types, however, do not contain centrioles at their mitotic spindle poles. This is most notably observed in plant cells and many lower eukaryotes. As it is beyond the scope of this article to describe these different mitotic systems in detail, the reader is referred to a number of earlier reviews (Cleveland, 1963; Pickett-Heaps, 1969; Luykx, 1970; Fuge, 1977). Even in animal cells in which centrioles are normal components of centrosomes, they have been shown to be dispensable. Unlike the case of most animal eggs, centrioles do not appear at the spindle poles in mouse eggs until after the third division cycle (Schatten *et al.*, 1985). This may be typical for all mammalian eggs, and demonstrates that centrioles are not obligatory components of the spindle pole. A second example of a cell that is capable of division and yet apparently lacks a centriole is the mutant *Drosophila melanogaster* cell line 1182-4 (Debec *et al.*, 1982). Under some experimental conditions, functional spindles may also be induced to form with one acentriolar pole. Flattening of tipulid spermatocytes prevents the normal migration of one of the centrosomes around the nucleus in some cells during diakinesis. This results in the formation of a spindle with one centriole-free pole (Steffen *et al.*, 1986). Prolonged colcemid exposure promotes the formation of multipolar spindles, and a large percentage of these functional spindles have also been shown to contain at least one acentriolar pole (Keryer *et al.*, 1984). As a final example, laser ablation experiments using cells in prophase have demonstrated that the centriole is not required for normal bipolar spindle function (Berns and Richardson, 1977).

The centriole exhibits other cellullar functions that are not directly involved with centrosomal function. A number of studies have shown the flagellar basal bodies to be structurally analogous to the centriole (Wheatley, 1982). Indeed, centrioles and basal bodies are generally considered as "manifestations of the same organelle" on the basis of structural identity and the ability of a centriole to become a basal body or of a basal body to detach and become a centriole in both function and location (Avers, 1976). For example, in *Chlamydomonas,* at the onset of mitosis the two flagella are resorbed and the basal bodies organize the mitotic spindle (Coss, 1974). After mitosis is completed, the centrioles serve as basal bodies for flagellar regrowth (Johnson and Porter, 1968; Fulton, 1971; Wolfe, 1972; Heidemann and Kirschner, 1975). The basal body is directly connected to the outer microtubule doublets of the axoneme, and functions in a different context from the centriole within the centrosome. In a similar manner, the centriole can serve as the site for growth of the primary cilium (Sorokin, 1968; Albrecht-Buehler and Bushnell, 1980; Wheatley, 1982). It appears, then, that the centriole (basal body) is involved with cellular movement, and it has been suggested that the centriole may also be involved in sensory transduction (Albrecht-Buehler, 1977). A specific centriolar role in the mitotic process has not been unequivocally determined. As discussed below, the centriole may serve to focus the pericentriolar material at the mitotic pole in animal cells, but this may be a dispensable part of centrosomal function at mitosis.

The centrosome is clearly a dynamic organelle, changing throughout the cell cycle; so what then is the most appropriate definition of the centrosome? A definition based purely on the structure of centrosomes in animal cells, a centriole and its surrounding pericentriolar material, would be insufficient, since most functional centrosomes do not contain centrioles, as discussed above. It would also be inaccurate simply to define the centrosome as a microtubule organizing centre, since other structures within the cell have distinct microtubule organizing capacity, i.e. kinetochores (Rieder, 1982), midbodies, basal bodies (Snell *et al.*, 1974), and the nuclear envelope in both plant cells (Wick *et al.*, 1981; Bajer and Molé-Bajer, 1981) and myotubes (Tassin *et al.*, 1985). The biochemical relationship between these organizing centres is at present not known; however, these different microtubule organizing centres may share some common protein components. The interphase cytoplasmic microtubule array of many different tissue culture cell lines radiates from a perinuclear region out to the cell periphery. The absence of a well defined or focused origin for many cytoplasmic microtubules, such as those observed in Madin–Darby canine kidney cells (Bré *et al.*, 1987), suggests that there are other microtubule organizing sites (or initiating sites) lying within the perinuclear region of interphase cells. Similarly, microtubules reorganize and emanate from a centralized region in centrosome-free cytoplasmic fragments of teleost melanophores (McNiven and Porter, 1988). Therefore, other cytoplasmic factors, such as additional microtubule nucleating sites, microtubule-associated proteins and local tubulin concentration, regulate the nucleation, distribution, orientation, and stability of cytoplasmic microtubules.

It is clear, however, that the centrosome serves as a focal point for microtubule nucleation at the mitotic poles and has unique cellular functions that are not common to other mitotic microtubule organizing centres. This has led to a definition equating the centrosome with the mitotic spindle pole irrespective of the different forms it may assume (McIntosh, 1983; Mazia, 1984). It implies no common unit at the ultrastructural level, but does imply functional equivalence. In each case, no matter the form, the centrosome is required to establish a bipolar spindle and organize microtubules in that spindle. Although this definition is satisfactory for mitotic cells, it does not directly address the centrosome of interphase cells.

Interphase microtubules are not generally organized about a focused point source, as discussed above. However, upon recovery of microtubules depolymerized by drug or cold treatments, or upon regrowth of microtubules from exogenous tubulin added to lysed cells, the newly initiated microtubules arise from a focal point(s) in the cytoplasm (see the recent review by Brinkley, 1985). Therefore, the centrosome of interphase cells can clearly be distinguished as the preferred site of microtubule nucleation. Similar results have been obtained with mitotic cells. We will simply define the centrosome here as the *preferred* site of microtubule nucleation within the cell. This definition holds regardless of cell cycle stage or ultrastructural form of the centrosome. Changes in the nucleating capacity, in part, allow for the progression or maturation of the interphase to the mitotic centrosome. Implicit in this view of the centrosome is its indispensable role in cell reproduction, as Mazia suggests, "If mitotic poles are embodied in centrosomes (whatever forms they assume) the reproduction of centrosomes is a necessary part of the reproduction of cells. A cell inherits a centrosome." (Mazia, 1984).

We will restrict the following discussion primarily to animal cells, which normally have a centriole present at the centrosome. Although much of the above discussion revolved around the importance, or lack thereof, of the centriole to centrosomal function, where centrioles do occur they can be utilized as markers for the centrosome. Also, in most of the systems to be discussed, centriole replication is linked to centrosomal replication and so may reflect centrosomal behaviour generally.

Defining the centrosome as the preferred site of microtubule nucleation allows us to address the following questions. What structures make up the centrosome, and what is their composition? How are they formed between interphase and mitosis, and how are they replicated? How do centrosomes function throughout mitosis and the remainder of the cell cycle? How are centrosomal functions regulated within the cell? Obviously, the answers to these questions are not known completely, and are the topics for a number of recent reviews (Wheatley, 1982; McIntosh, 1983; Bornens and Karsenti, 1984; Brinkley, 1985; Mazia, 1987; Vorobjev and Nodezhdina, 1987). We review here the most recent literature addressing these questions, and where possible emphasize the relationship between the centrosome, microtubule, cell and mitotic cycles.

II. The Microtubule Cycle: A Reflection of Centrosome Dynamics

As a cell completes the mitotic process, each daughter cell inherits a single centrosome. This centrosome served as one of the two functional spindle poles in the parent, but as the two progeny cells enter interphase the mitosis-specific functions of the centrosome are no longer required. At least one functional characteristic is retained by both mitotic and interphase centrosomes, however, and that is their capacity to serve as organizing sites for microtubules. This has been demonstrated clearly by a number of investigators using indirect immunofluorescence with antibodies against tubulin (Osborn and Weber, 1976; Brinkley *et al.*, 1976). Microtubules form a complex cytoplasmic array that in part radiates out to the cell periphery from a focal region adjacent to the nucleus (Osborn and Weber, 1976; Brinkley *et al.*, 1976; Frankel, 1976). Upon recovery from cold or drug treatments that induce microtubule depolymerization, the microtubules also initiate growth from a localized centre near the nucleus. In interphase, the focal point of microtubule growth serves to define the position of the centrosome. The changes that occur in the microtubule network as the cell progresses through the cell cycle partially reflect changes in centrosome dynamics. We briefly review the microtubule cycle as observed in PtK_1 cells, and will use these observations as a basis for further discussions of centrosome activity.

The overall appearance of the cytoplasmic microtubule network (as described above) changes very little throughout interphase, (see Fig. 2.1(a)), although recent evidence suggests that individual microtubules within this array are highly dynamic (Soltys and Borisy, 1985; Schulz and Kirschner, 1986; Salmon, this volume). Shortly after the onset of chromosome condensation and the beginning of prophase, the microtubule pattern begins a significant rearrangement, progressively changing from the generalized cytoplasmic array to a discretely focused astral array. This is initially established with a single focus, but as chromosome condensation becomes more pronounced two foci appear (Fig. 2.1(b)). In terms of the centrosome, two major events have occurred at prophase. First, the microtubule nucleating capacity of the centrosome has increased, as evidenced by the development of highly focused astral arrays of microtubules, and second, the centrosome has split into two entities, each a focal point of microtubules, and separated. The replace-

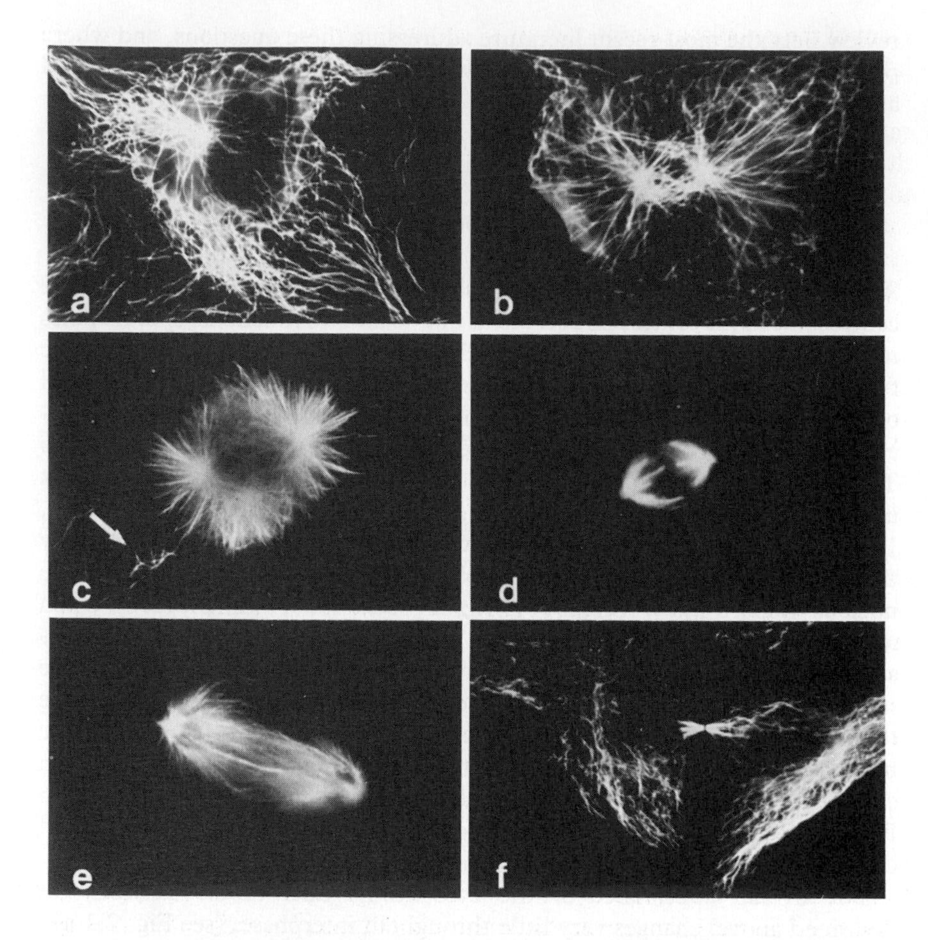

Figure 2.1. Indirect immunofluorescence staining of tubulin during the cell cycle of PtK_1 cells. (a) Interphase, the cytoplasmic microtubule array is only partially focused at the centrosome. (b) Late prophase, two arrays of microtubules are present focused at each of the two centrosomes. (c) Prometaphase, astral arrays of microtubules are focused at the two well-separated centrosomes. The cytoplasmic microtubules breakdown after nuclear envelope dissolution, residual cytoplasmic microtubules are still apparent at early prometaphase (arrow). (d) Metaphase, all microtubules are associated with the spindle. (e) Anaphase, microtubules are still focused at the separated spindle poles, but are in the process of regrowing into the cytoplasm. (f) Telophase, the interphase cytoplasmic array is re-established in the daughter cells. The midbody microtubules are still present, but the strong focus of microtubules at the centrosome is lost.

ment of the cytoplasmic microtubule network during prophase may reflect the inherent dynamic properties of individual microtubules coupled with the activation of microtubule nucleating sites at the centrosome. Incorporation rates for microinjected tubulin derivatized with dichlorotriazinylamino-fluorescein, and fluorescence redistribution following photobleaching, suggest total interphase microtubule turnover rates of the order of 30 minutes in mammalian tissue culture cells (Saxton *et al.*, 1984; Sammak *et al.*, 1987). If we assume that centrosomal microtubule nucleating sites must be occupied, i.e. that they dominate all other possible microtubule nucleating activities within the cell, all microtubule regrowth during prophase would be restricted to newly activated centrosomal nucleating sites. Therefore, as individual fibres of the interphase microtubule network disassemble, all replacement fibre growth would be nucleated off of the centrosome with the resultant development of the prophase astral microtubule array. This transition would occur at the same rate as for normal turnover of the interphase microtubule network. This model is consistent with the transit time for tissue culture cells through prophase.

At prometaphase (nuclear envelope breakdown), the cytoplasmic array of microtubules breaks down abruptly and the length of the astral microtubules becomes shorter. The cell in Fig. 2.1(c) was fixed immediately after nuclear envelope breakdown and the transition in microtubule pattern is clearly evident. Fragments of microtubules are apparent in the cytoplasm, reflecting the remnants of the cytoplasmic array. At prometaphase the centrosome clearly continues to support microtubules in an overall cytoplasmic environment that is void of microtubules. At metaphase (Fig. 2.1(d)) microtubules are found only as part of the mitotic spindle and are focused at the two spindle poles and the kinetochores. Spindle formation may in part result from the stabilization of microtubules within the kinetochore fibre by capping of their free ends at the kinetochore (Kirschner and Mitchison, 1986; Salmon, this volume). Similarly, interzonal microtubules may be partially stabilized by cross-linking of their free ends.

With the onset of anaphase (Fig. 2.1(e)), microtubules appear to have grown into the cytoplasm, suggesting a change in the cytoplasmic environment favouring microtubule assembly. In addition to being the focal points for these regrowing microtubules, the centrosomes are also the foci to which the chromatids are being drawn. Thus, the chromosomes are in a localized region for efficient packaging during nuclear envelope reassembly. With the formation of the midbody, and completion of the mitotic process, the interphase microtubule pattern is re-established (Fig. 2.1(f)). One centrosome has been efficiently delivered to each daughter cell. The activity of the centrosome, with regard to its microtubule organizing activity, has returned to interphase levels even at this early point in G_1.

III. Ultrastructure of the Centrosome

It has been known from early electron-microscopic observations that a centriolar pair and a surrounding cloud of electron-dense, osmiophilic material constituted the ultrastructural entity located at the focal point of the cellular microtubule array (deHarven and Bernhard, 1956; Szöllösi, 1964; Stubblefield and Brinkley, 1967; Brinkley and Stubblefield, 1970). The ultrastructural features of the centriole have been reviewed by Wheatley (1982) and recently the centriole structure of isolated centrosomes has been described (Bornens *et al.*, 1987). Only in recent years have detailed ultrastructural studies been completed that document changes in centrosome structure with progression through the cell cycle (Zeligs and Wollman, 1979; Rieder and Borisy, 1982; Vorobjev and Chentsov, 1982). As just discussed, the microtubule arrays organized by the centrosome change dramatically as the cell prepares for entry into mitosis and during progression through mitosis. These differences in the organization of microtubules have been correlated with ultrastructural changes that occur at the centrosome.

In general, interphase cells contain a single centrosome that contains a pair of centrioles in an orthogonal relationship. In centrosomes isolated from the human lymphoblastic cell line KE37, the centrioles have an average length of 0.4 μm, and appear to be linked by a network of thin filaments surrounding each centriole (Bornens *et al.*, 1987). The lumen of the centriole was filled with an uncharacterized material to over half of its length. Thus, the proximal centriole end is defined as the end having an empty lumen. The pericentriolar material extended the length of the centriole, forming a sheath 40 nm thick covering the proximal half and extending with radial symmetry 0.1 μm at the distal end. The two centrioles can be differentiated from one another, however, in sections of whole cells (Zeligs and Wollman, 1979; Rieder and Borisy, 1982; Vorobjev and Chentsov, 1982). The parent centriole has the majority of the associated pericentriolar material, distributed 100–150 nm in diameter around the centriole. The primary cilium is also associated with the parent centriole, while the daughter centriole is non-ciliated. In addition to fibrilar osmiophilic material, the pericentriolar material also consists of compact satellite structures, approximately 70 nm in diameter, and centriolar appendages or feet that project 0.1–0.2 μm off the centriolar wall. The satellite material is primarily at the periphery of the pericentriolar cloud, which extends on either side of the parent centriole (Fig. 2.2(a)). The two centrioles split and replicate as the cell progresses towards prophase, a process that will be discussed in further detail below. The replicated centrosomes are structurally equivalent, each comprising a parent and daughter centriole pair. The parent centriole of each pair has acquired an associated complement of pericentriolar satellites.

As the mitotic cycle continues towards prometaphase the pericentriolar material undergoes a number of alterations (Zeligs and Wollman, 1979; Rieder and Borisy, 1982; Vorobjev and Chentsov, 1982). The primary cilium is resorbed at prophase and does not reappear until telophase. An extensive cloud of pericentriolar material is accumulated throughout prophase with a concomitant reduction in the pericentriolar satellites. By metaphase the pericentriolar cloud has reached a diameter of 0.75 μm in PtK_2 cells (Fig. 2.2(b)). The pericentriolar cloud persists until telophase, when satellite material reappears. This expansion of the pericentriolar cloud correlates with an increased microtubule nucleating capacity of the centrosome during mitosis. The association of the satellite structures with the periphery of the cloud material, and their disappearance and reappearance with the respective formation and degradation of the pericentriolar cloud, led Rieder and Borisy to suggest that the satellite structures may be an aggregated and less active transitional form of the pericentriolar cloud (Rieder and Borisy, 1982). Dynamic changes in the structure of the centrosome through mitosis, especially with regard to the pericentriolar material, can be correlated with the changes in the microtubule patterns observed by indirect immunofluorescence microscopy. For example, the expansion of the pericentriolar cloud throughout prophase correlates with the development of strongly focused astral arrays of microtubules at this same time.

IV. Centrosome Composition

A detailed analysis of centrosome composition has been hampered because techniques for the preparation of centrosomal samples in sufficient quantity and biochemical purity have only recently been reported (Bornens *et al.*, 1987). A number of methods exist for the isolation of the whole mitotic apparatus (Sakai and Kuriyama, 1974; Forer and Zimmerman, 1974; Chu and Sisken, 1977; Zieve and Solomon, 1982; Kuriyama, 1982; Kuriyama *et al.*, 1984) and crude centrosome preparations (Blackburn *et al.*, 1978; Nadezhdina *et al.*, 1978), but identification of specific centrosomal components from the contaminating cytosolic proteins and other spindle components has been hampered. Recently, however, Mitchison and Kirschner have reported the 3000-fold purification of interphase centrosomes from N115 and CHO cells (Mitchison and Kirschner, 1984a). From their initial compositional studies, tubulin comprises 3–5% of the total protein in the preparation, approximating the centriolar content of the samples. Using a modification of this isolation procedure, Bornens *et al.* (1987) have prepared purified centrosomes from the human lymphoblastic cell line KE37, leading to a more direct compositional

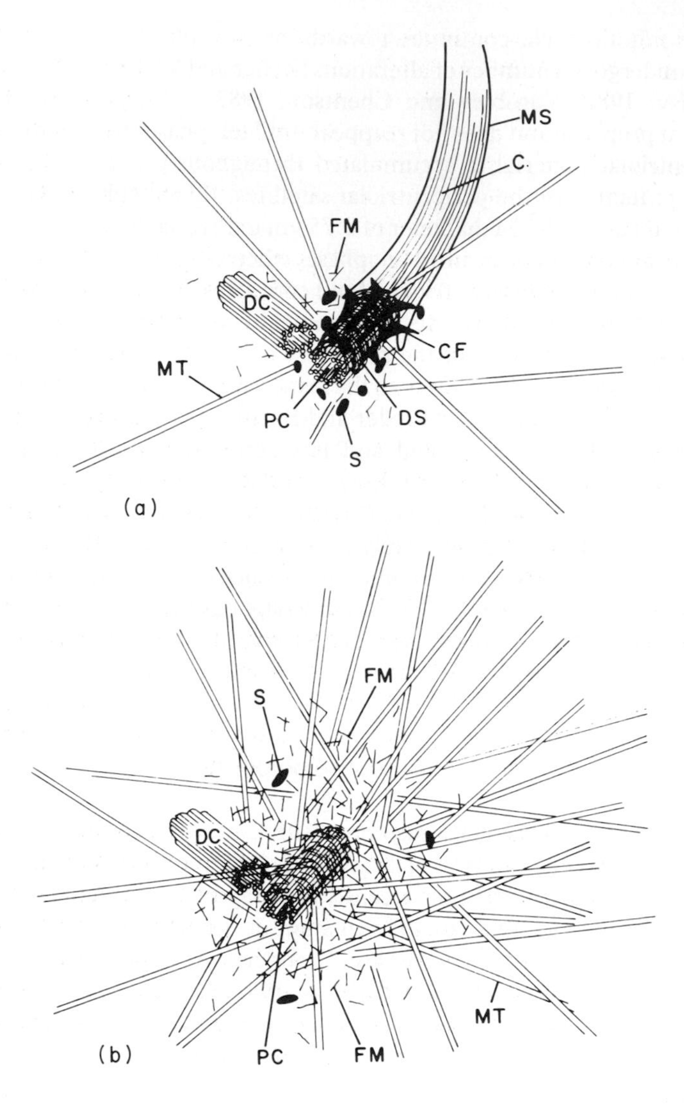

Figure 2.2. A diagram of the centrosome of (a) interphase and (b) mitotic animal tissue culture cells incorporating ultrastructural observations from several reports (Rieder and Borisy, 1982; Vorobjev and Chentsov, 1982). (a) Interphase centrosome containing an orthogonally orientated pair of centrioles, primary cilium, pericentriolar structures and microtubules. PC, parent centriole; DC, daughter centriole; C, primary cilium; MS, membrane sheath around cilium; DS, dense sheath surrounding parent centriole; CF, centriolar feet; S, centriolar satellite; FM, fibrous pericentriolar material. The pericentriolar material is primarily associated with the parent centriole. Microtubules arise from the pericentriolar material,

analysis. These isolated centrosomes are functionally active, as determined by their ability to nucleate microtubules. The centrosomal purity is indicated by the relative amount of protein in the centrosomal fraction compared to the whole cell. The protein concentration of an individual centrosome ranges between 2 and 3×10^{-2} pg. The protein composition is complex and consists of groups of proteins around 300 and 180, and between 50 and 65 kDa, as well as major bands of 130, 85 and 45 kDa.

Information regarding centrosomal composition has also been obtained by more indirect approaches. Although the presence of both DNA and RNA at the centrosome has been suggested (Berns and Richardson, 1977; Rieder, 1979), we do not believe that sufficient evidence currently exists to conclude that either nucleic acid is a specific component of the centrosome. The evidence for a DNA component is based on cytological staining (Randall and Disbrey, 1965; McDonald and Weijer, 1966). A definite conclusion cannot be drawn from these results owing to the possibility of non-specific staining. While the evidence for an RNA component is more compelling, consisting of cytological staining (Hartman *et al.*, 1974; Rieder, 1979), RNase digestion (Stubblefield and Brinkley, 1967; Brinkley and Stubblefield, 1970; Heidemann *et al.*, 1977; Rieder, 1979), and sensitization of the centrosome to laser irradiation after acridine orange or psoralen exposure (Berns *et al.*, 1977; Debec *et al.*, 1982), a specific RNA has never been characterized (Pennell *et al.*, 1988). These results may also reflect the presence of virus-like particles observed by electron microscopy at some centrosomes. For a more complete discussion of nucleic acids as a centrosomal component, the reader is referred to Wheatley (1982).

A number of proteins have been localized to the centrosome by indirect immunofluorescence. These include cAMP-dependent protein kinase (Nigg *et al.*, 1985), transferrin receptor (Willingham and Pastan, 1985), purine nucleotide phosphorylase (Oliver *et al.*, 1981), kinesin, a microtubule-activated Mg ATPase functionally involved in the movement of microtubule attached vesicles to the plus end of the microtubule (Neighbors *et al.*, 1988), and microtubule-associated protein 1 (MAP-1) (see references in Table 2.1).

Figure 2.2 continued
occasionally growing off of satellite structures or the centriole. (b) Mitotic centrosome containing an orthogonally orientated pair of centrioles, associated material and microtubules. PC, parent centriole; DC, daughter centriole; S, centriole satellite; FM, fibrous pericentriolar material. The pericentriolar material surrounding the parent centriole expands during mitosis with concomitant loss of the centriolar feet and the majority of the satellite structures. The primary cilium is resorbed and the dense sheath is less prominent. The microtubule-initiating capacity increases approximately 10-fold when compared to the interphase centrosome.

Table 2.1. Antibodies against MAP-1 that recognize centrosomal components

Type of antibody	Antibody class	Antigen prepared against	Proteins identified by immunoblot	Staining pattern	References
Polyclonal		Rat brain MAP-1		Centrosomes Mitotic spindle	(*a*)
Monoclonal	IgM	Bovine brain MAPs	Brain MAP-1	Nuclear spots Interphase fibres Mitotic cytoplasm Mitotic centrosomes	(*b*)
Monoclonal	IgG(MPM-2) IgM(MPM-1)	HeLa mitotic cell extracts	Brain MAP-1 CHO mitotic cell lysates many bands including 415 kDa band	Nuclear patches Mitotic spindle Mitotic centrosomes Kinetochores, midbody	(*c*)
Monoclonal		Bovine brain MAP-1B	Brain MAP-1B	Centrosomes Primary cilium Microtubules	(*d*)
Monoclonal	IgM	Rat brain microtubules	Brain MAP-1 CHO mitotic cell lysates 415 kDa band	Nuclear patches Mitotic spindle Mitotic centrosomes Kinetochores, midbody	(*e*)
Monoclonal	IgG	Rat brain MAPs	Brain MAP-1 cell lysates 280 kDa band	Punctate fibre pattern Nuclear patches Interphase centrosome Mitotic centrosomes	(*f*)

[a] Sherline and Mascaro (1982).
[b] Sato *et al.* (1983, 1984, 1985).
[c] Vandré *et al.* (1984, 1986).
[d] Vallee *et al.* (1984).
[e] DeBrabander *et al.* (1986), DeMey *et al.* (1987).
[f] Bonifacino *et al.* (1985).

Several different antibodies generated against MAP-1 show staining at the centrosome, but in each case other cellular structures were also stained. The antibodies that recognize MAP-1 and stain the centrosome by indirect immunofluorescence microscopy are summarized in Table 2.1. It is not known whether the proteins identified by these different antibodies at the centrosome are actual MAP-1 molecules, degradation products of MAP-1, or proteins sharing some common epitopes with MAP-1, such as phosphorylated domains (Vandré *et al.*, 1986; Luca *et al.*, 1986; DeMey *et al.*, 1987). Since many of the antibodies recognize proteins of different molecular weight from MAP-1, it seems more likely that these may be proteins sharing an immunologically cross-reactive domain. The specific function of MAP-1 or MAP-1 cross-reactive proteins at the centrosome is not known. However, brain MAP-1 has microtubule assembly-promoting activity *in vitro* (Vallee and Davis, 1983), and a reasonable possibility is that molecules having a similar function to MAP-1 could be located at the centrosome.

Several autoimmune or non-immune sera obtained from humans and rabbits also stain centrosomes when examined by indirect immunofluorescence microscopy (see references in Table 2.2). It has been reported that nearly 10% of non-immune rabbit sera stain centrioles and basal bodies (Connolly and Kalnins, 1978). The majority of the centriole-staining non-immune sera have not been fully characterized. One particular serum, however, has been localized to "peribasal body" material that may be similar to pericentriolar material (Fung and Kasamatsu, 1985). This antibody recognizes proteins of 14 and 17 kDa on immunoblots (Lin *et al.*, 1981), and the antigenic expression has been linked to varying growth conditions of TC7 African green monkey kidney cells (Segarini *et al.*, 1983).

A second non-immune rabbit serum, designated 0013, has also been reported to stain centrosomes, but with a specificity for monkey and human cells (Gosti-Testu *et al.*, 1986). The 0013 serum identified a family of proteins in the 180–250 kDa range, a 130 kDa band, and a doublet of proteins between 60 and 65 kDa by immunoblot analysis of enriched centrosome preparations obtained from the human T-lymphoblast cell line KE37 (Gosti-Testu *et al.*, 1986). Antibody staining was localized to the pericentriolar material, but antibody pretreatment did not block microtubule nucleation off isolated centrosomes (see section VII). This antibody was also reported not to react with brain microtubule-associated proteins such as MAP-1, a potential centrosome component (discussed above). Further studies have shown that at least one of the antigenic determinants recognized by serum 0013 is shared between a 36 kDa subunit of human lactate dehydrogenase and the centrosomal components (Gosti-Testu *et al.*, 1987).

Recently, a human autoantibody (5051) from a scleroderma patient has been characterized that stains the pericentriolar material (Calarco-Gillman *et al.*,

Table 2.2. Autoimmune and non-immune sera that recognize centrosomal components

Sera	Source	Specificity	Characteristics	References
Z-3, Z-7	Non-immune rabbit sera	Mammalian and avian cells	Stains centrioles and basal bodies	(*a*)
CT-2, HMW-3	Non-immune rabbit sera	Widespread	Stains centrioles and basal bodies, 50 kDa protein detected on blots	(*b*)
Pooled serum	Human scleroderma sera		Stains centrosomes (microtubule organizing centres)	(*c*)
0013	Non-immune rabbit sera	Human and monkey cell lines	Stains pericentriolar material, react with proteins ranging in molecular weight from 180 to 250 kDa, 130 kDa and a 60–65 kDa doublet	(*d*)
POPA	Human sera, CREST variant of sclero-derma	Tissue culture cells	Stains spindle poles, absent from interphase, stains multiple cyto-plasmic foci in prophase, recog-nizes 110/115 kDa protein specific for spindle pole	(*e*)
Anti-Cag	Non-immune rabbit sera	Widespread	Stains centriolar and basal body regions, stains proteins of 14 and 17 kDa, present during G_1 phase of cell cycle	(*f*)
5051	Human scleroderma serum	Widespread	Stains pericentriolar material, reacts with acentriolar spindle poles	(*g*)

[a] Connolly and Kalnins (1978).
[b] Wang *et al.* (1979), Turksen *et al* (1982).
[c] Brenner and Brinkley (1982).
[d] Gosti-Testu *et al.* (1986), Gosti *et al.* (1987).
[e] Sager *et al.* (1986).
[f] Lin *et al.* (1981), Segarini *et al.* (1983), Fung and Kasamatsu (1985).
[g] Calarco-Gillman *et al.* (1983), Tuffanelli *et al.* (1983), Maro *et al.* (1985), Clayton *et al.* (1985).

1983; Tuffanelli *et al.*, 1983). The 5051 serum has been used to identify the presence of pericentriolar material at the acentriolar barrel-shaped spindle poles of mouse oocytes (Calarco-Gillman *et al.*, 1983; Maro *et al.*, 1985) and higher plant cells (Clayton *et al.*, 1985). Similarly, the 5051 autoimmune serum localizes to each mitotic pole in colcemid-induced multipolar spindles, many of which also do not contain centrioles (Sellitto and Kuriyama, 1988). The monoclonal antibody MPM-2, which recognizes a set of mitotic phosphoproteins, also stains centrosomes or related microtubule organizing centres in diverse cell types (Vandré *et al.*, 1984). In each case, the antiserum 5051 and monoclonal antibody MPM-2 recognize what must be highly conserved antigenic determinants on centrosomal protein components, but whose presence is not necessarily dependent upon the presence of a centriole. The immunoreactive components recognized by the 5051 antiserum have not been determined, since this serum does not identify reactive protein by immunoblot analysis. The MPM-2 antibody, however, recognizes several high-molecular-weight phosphoproteins in taxol-isolated CHO spindles (Vandré and Borisy, in preparation), but it is not clear which of these are specifically associated with the centrosome. Other than differences in phosphorylation (see section VII), compositional changes within the centrosome have not been correlated to progresssion through the cell cycle. Unlike the rabbit non-immune serum 0013 (see above), the MPM-2 antibody does block centrosome nucleated microtubule regrowth in lysed cell model systems (Centonze *et al.*, 1986). These results suggest that phosphorylated centrosomal components are involved in centrosomal microtubule nucleating activity.

A human autoantibody present in a CREST serum has been reported to stain cytoplasmic foci in prophase cells that appear to aggregate at the spindle pole (Sager *et al.*, 1986). The POPA autoantibody (prophase-originating polar antigen), reacts with 115 and 110 kDa proteins on immunoblots that appear to be centrosomal components. The aggregation of POPA coincides with the activation of centrosomal nucleating sites and microtubule assembly during prophase. Whether POPA accumulation at the pole functions directly in microtubule organization and nucleation, or is dependent upon microtubule assembly, is not known. It is tempting to speculate that POPA accumulation, may in part, reflect the expansion of the pericentriolar material during prophase.

V. The Centrosome Replication Cycle

A. Echinoderm eggs

The centrosome replication cycle is the sequence of events by which one centrosome or pole duplicates at mitosis. Mazia and coworkers (1960) provided the first experimental evidence enabling the dissection of the centrosome replication process into its component stages. A number of more recent reports, based primarily on extensions of the technique utilized in the initial experiments have added to these observations (Sluder, 1978; Paweletz *et al.*, 1984; Sluder and Rieder, 1985; Sluder and Begg, 1985). Therefore, it is worth examining the findings of Mazia *et al.* (1960) in some detail before discussing the more recent observations.

In sea-urchin and sand-dollar eggs, treatment with β-mercaptoethanol was shown to block division in a reversible fashion. If the block was applied in early metaphase, and was not removed until control eggs had entered their second division, the treated eggs were observed to divide from one cell directly into four. The result of this quadripartition was that the mitotic centres normally destined for two cells were distributed among four cells. These cells, after quadripartition, formed monopolar spindles during the next mitotic cycle and did not divide. Upon entering the next mitosis, the second after β-mercapto-ethanol release, the cells formed a normal bipolar spindle and cleaved from four to eight cells.

Several conclusions were drawn from these results (Mazia *et al.*, 1960). First, if a mitotic centre (spindle pole) is defined in terms of potential pole-forming units, it is normally duplex. Under the experimental conditions used, the two mitotic centres split into four functional units, two from each centre. The full potential pole-forming units of each mitotic centre were then only expressed after the mitotic centre had split, yielding the quadripartition event. Normally, cells enter mitosis with four potential poles, but, because of their duplex nature, only two poles form. Each daughter cell inherits a single mitotic centre with two potential poles. By second division the daughter cell once again has four potential pole-forming units. Therefore, a duplication of the pole-forming units occurs between first and second cleavage. It is this duplication event that is blocked by β-mercaptoethanol, the splitting and separation of the pole-forming units occurring normally.

By timing the addition and release of β-mercaptoethanol, the splitting and duplication events were determined to occur during the same period of the cell cycle. The duplication event was possibly dependent upon splitting of the pole-forming units, but not vice versa. To reiterate, the normal mitotic centre is duplex in nature and goes through several distinct stages during its reproduction.

1. *Splitting*. The original potential units within the mitotic centre that were segregated to the daughter cell become separable.
2. *Duplication*. The original units each replicate (this stage could possibly be further subdivided into initiation and growth phase).

3. *Separation*. The units in the process of duplication move apart from one another forming the two spindle poles.

The interpretations of the β-mercaptoethanol experiments were based strictly upon functional consequences of the treatment. Ultrastructural analysis of eggs treated in this way has only recently been completed (Paweletz *et al.*, 1984; Sluder and Rieder, 1985). Together with other functional assays that have been completed since Mazia *et al.*'s (1960) original report, most if not all of the conclusions inferred from the first β-mercaptoethanol experiments have been confirmed. The monopolar spindles formed after the quadripartition experiments have been shown to be functional half-spindles (Sluder, 1978; Mazia *et al.*, 1981; Sluder and Begg, 1985). A small percentage of the β-mercaptoethanol-treated eggs divide from one to three cells when two poles of the tetrapole do not move far enough apart at anaphase. In such cases, two monopolar spindles and one bipolar spindle form at the next mitosis. Thus, the two potential monopolar spindles that are trapped in a common cytoplasm form a bipolar spindle, and are, therefore, functional half-spindles.

If a cell that forms a monopolar spindle remains in mitosis for an extended period of time, a bipolar spindle will form (Sluder, 1978; Sluder and Rieder, 1985; Sluder and Begg, 1985). The resulting bipolar cell will cleave, and at the next mitosis monopolar spindles will form in each daughter cell. This confirms that splitting and replication of the mitotic centre are distinct events. While the monopolar cell (with two potential pole-forming units at the monopole) is in an extended mitosis, splitting of the pole occurs, resulting in separation of the pole-forming units and formation of a bipolar spindle. Each pole of the latter has only one pole-forming unit, replication of which may only occur in the next cell cycle, and a monopolar spindle is formed once again.

Evidence has recently been obtained to suggest that centrioles are indicators of the reproductive capacity of, or the number of potential poles present at, a mitotic pole (Paweletz *et al.*, 1984; Sluder and Rieder, 1985). When sea-urchin spindles were isolated following a β-mercaptoethanol treatment cycle and examined in the electron microscope, paired centrioles were observed at metaphase surrounded by osmiophilic pericentriolar material (Paweletz *et al.*, 1984). As the cells progress towards quadripartition, the centriole pairs in isolated spindles split, each surrounded by pericentriolar material. More direct evidence linking centrioles with the reproductive capacity of the pole has come from examination of semi-thick serial sections of eggs recovering from β-mercaptoethanol treatment (Sluder and Rieder, 1985). Only one centriole is located at each pole of a tetrapolar cell. After quadripartition, the centrosome of each monopole contains two centrioles. If the monopolar cell remains in an extended mitosis, a bipolar spindle forms and each pole again contains only one

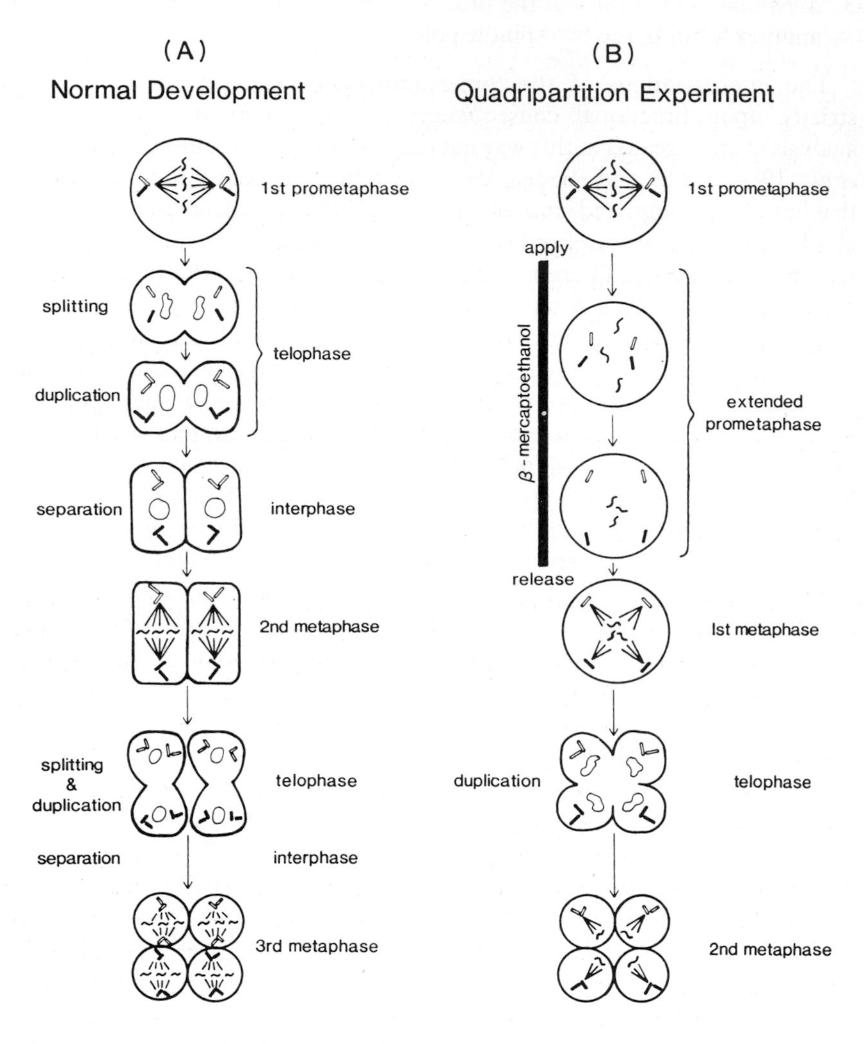

Figure 2.3. Diagram of the centrosome cycle in echinoderm eggs, redrawn from Mazia, Harris, and Bibring (1960) and Sluder and Rieder (1985). (A) Normal development through third metaphase. The splitting, duplication and separation events occur in rapid succession and are separated for diagrammatic purposes only. (B) Quadripartition is induced by β-mercaptoethanol treatment for the length of the first mitosis through early interphase of control samples. Duplication of the centrioles and potential poles is blocked by β-mercaptoethanol, while splitting and separation occur normally.

centriole. In every case the centriole number corresponds to the functional reproductive capacity of the pole.

To summarize, the centrosome reproductive cycle in echinoderm eggs has been described in both functional and structural terms. The polar organizer is spatially, mechanically and functionally associated with the centriole; however, the centriole cylinder itself is not the polar organizer (Sluder and Rieder, 1985). The presence of pericentriolar material surrounding the centriole appears to correlate with a functioning mitotic pole. When parent and daughter centrioles are paired, the pericentriolar material is associated with the parent centriole. Daughter centrioles can acquire pericentriolar material in the same cell cycle only if mitosis is prolonged (Sluder and Rieder, 1985). This allows for the splitting of the centrioles. Therefore, this association of pericentriolar material with the centriole is dependent upon centriolar splitting, but not centriole replication. The results of the functional and structural studies of centrosome replication in echinoderm eggs are presented in Fig. 2.3.

B. Cultured cells

In cultured cells the replication of the centrosome has been followed primarily by electron-microscopic structural observations (Robbins *et al.*, 1968; Kuriyama and Borisy, 1981). In these studies the replication of the centriole was used as a marker for centrosome replication. Unfortunately, a functional assay as precise as the β-mercaptoethanol experiments using sea-urchin eggs has not been described for cultured cells.

In culture, each daughter cell normally receives a pair of orthogonally orientated centrioles of approximately equal length. As the cell progresses through G_1, the centrioles become disorientated. By the beginning of S phase, a short daughter centriole appears at the proximal end of each parent centriole in an orthogonal orientation. The new daughter centriole grows in length during S and G_2 and is nearly full size at prophase when the centriole pairs separate and migrate to opposite ends of the nucleus. An orthogonally orientated centriolar pair is located at each spindle pole and each pair is segregated to one of the two daughter cells as mitosis is completed (Robbins *et al.*, 1968; Kuriyama and Borisy, 1981). The cycle is repeated with each cell generation. On the basis of these observations, several stages of centriole reproduction have been defined (Kuriyama and Borisy, 1981). Disorientation is the loss of orthogonal relationship between parent and daughter centriole during G_1. The disorientation stage is probably analogous to the splitting event described in functional terms by Mazia *et al.* (1960) using sea-urchin eggs (Mazia *et al.*, 1960). Nucleation is the initiation of procentriolar growth at late G_1 or early S phase. Elongation is the growth stage of the procentriole during S

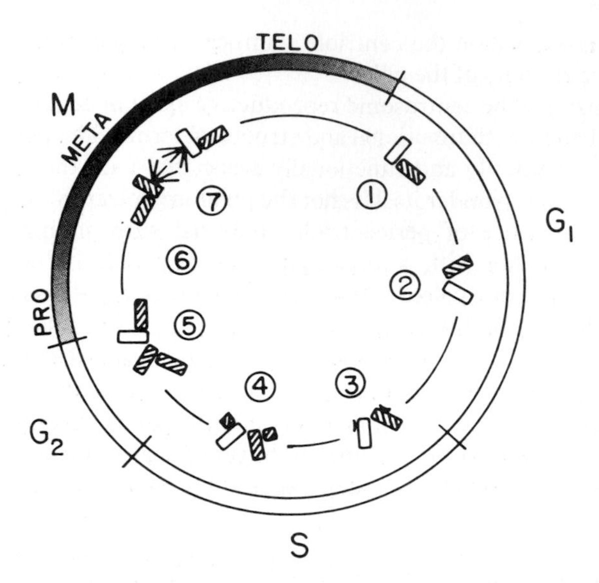

Figure 2.4. Diagram of the centriole cycle in Chinese hamster ovary cells as determined by ultrastructural observations. (1) An orthogonally orientated centriole pair is delivered to each daughter cell; (2) disorientation of the centriole pair occurs early in G_1; (3) duplication of the centrioles is initiated at the time the cell enters S phase; (4) elongation of the new daughter centrioles proceeds through S phase; (5) full-length daughter centrioles have grown when the cell enters prophase; (6) centriole pairs separated in prophase moving to opposite sides of the nucleus; (7) a spindle is organized by the replicated centrosomes. The shading intensity within the ring depicting the cell cycle position reflects the microtubule-initiating capacity of the centrosome. Nucleating capacity increases at prophase and drops again during telophase. The replication of the centrioles and the microtubule-initiating capacity of the centrosomes are distinct.

and G_2. When daughter centriole lengths were categorized, the different size categories were found with equal frequency, suggesting uniform procentriole growth (Kuriyama and Borisy, 1981). The closest functional correlate to the nucleation and elongation stages in the sea-urchin system is the duplication event. In functional terms alone, duplication refers to the doubling of potential pole units. Whether the procentriole could serve as a functional unit after nucleation, after elongation, or following further maturation events is not known. The final stage of the centriole cycle in cultured cells is separation of the centriole pairs. A direct analogy between the separation events in both systems can be postulated. The centriole cycle is shown schematically in relation to the cell cycle in Fig. 2.4.

The close association between centriole nucleation and the initiation of S phase raised the possibility that the centriole cycle may be dependent upon DNA synthesis. This has been shown not to be the case (Rattner and Phillips, 1973; Kuriyama and Borisy, 1981). The converse possibility, a dependence of DNA synthesis on centriole nucleation, also does not seeem to occur (Kuriyama *et al.*, 1986). In cells blocked with excess thymidine, nucleation proceeded normally with only a slight delay in elongation; however, additional cycles of centriole duplication did not occur owing to the cell cycle block (Kuriyama and Borisy, 1981). The centriole cycle also proceeded normally in cytoplasts from S-phase cells. Again, nucleation was not observed after removal of the nucleus prior to S phase (Kuriyama and Borisy, 1981). In contrast, treatment of cells with taxol blocked nucleation of centrioles without affecting the cell's capacity to initiate DNA synthesis (Kuriyama *et al.*, 1986). Therefore, centriole replication and DNA synthesis are events coordinated in time, but are independent processes. The question remained, whether nuclear activities were required for additional cycles of centrosomal reproduction, since in each of the above examples the cell cycle was blocked by drug treatment or manipulation of the cells. It has recently been demonstrated in enucleated sea-urchin eggs that centrosome and centriole replication occur normally, cycling with the same temporal coordination as in nucleated eggs (Sluder *et al.*, 1986). At least for specialized cells such as eggs, which store significant amounts of gene products in their cytoplasm, the mechanisms regulating centrosome reproduction can be completely under cytoplasmic control. In somatic cells, however, availability of newly synthesized centrosomal subunits may be required at each cell cycle. In this case, each complete cycle of centrosome replication requires specific nuclear events; thus, coordinated centrosomal and nuclear events control centrosome number (Sluder *et al.*, 1986).

Attempts to define the centriole cycle in functional terms by experimental perturbations, in a manner similar to the β-mercaptoethanol treatment of echinoderm eggs, have met with limited success. Mitotic inhibitory drugs, Colcemid and nocodazole, have been used to prolong mitosis (Kuriyama, 1982; Keryer *et al.*, 1984). Following drug removal, the number of organized sites for microtubule regrowth, i.e. spindle poles, generally increases (Kuriyama, 1982; Keryer *et al.*, 1984). This suggests that these drugs may induce centriole disorientation and separation. Examination of the individual poles demonstrates that in many cases centriole separation had occurred (Keryer *et al.*, 1984). However, at least one pole in a large percentage of the cells did not contain centrioles; this acentriolar pole was none the less surrounded by pericentriolar material. Thus, Colcemid induces not only centriole disorientation but fragmentation of the pericentriolar cloud (Keryer *et al.*, 1984). These experiments cannot be used to analyse directly the functional aspects of centriole reproduction, but they do show that a daughter centriole can become associated with pericentriolar material in the same cell cycle and serve as a functional spindle pole in a manner analogous to the quadripartition experiments with sea-urchin eggs. The results also indicate that the presence of a centriole is not required at a functional pole, but that centrioles may serve to focus pericentriolar material under normal conditions. Other studies have also shown that centrioles become disorientated and separate when interphase cells are exposed to Colcemid and are then allowed to recover from drug treatment (Brooks and Richmond, 1983), or when cells are cultured for extended periods in the presence of Colcemid (Kuriyama, 1982). In general, however, in those systems where centrioles are present at the centrosome, such as the mitotic pole of sea-urchin eggs and the majority of cultured cells, the centriole cycle is an accurate reflection of the centrosome cycle. In addition, the centriole number correlates with the reproductive capacity of the mitotic pole.

VI. Centrosome Function

Clearly, the centrosome functions as a site for microtubule anchorage and nucleation throughout the cell cycle, and a number of functional properties have been defined in sufficient detail to differentiate centrosomes of interphase and mitotic cells. The components involved and the mechanisms underlying the centrosomal nucleation of microtubules are, however, not understood.

Microtubule nucleation and growth from the centrosome has been demonstrated *in vivo* in cells following release and recovery from drug-induced depolymerization of microtubules (Osborn and Weber, 1976; Brinkley *et al.*, 1976), recovery from temperature-induced depolymerization (Frankel, 1976), and from the polymerization of microinjected derivatized tubulin (Salmon *et al.*, 1984; Saxton *et al.*, 1984; Soltys and Borisy, 1985; Mitchison *et al.*, 1986). Similarly, *in vitro* experiments using lysed cell systems (Snyder and McIntosh, 1975; Brinkley *et al.*, 1981), crude lysate containing centrosomes (Gould and Borisy, 1977; Borisy and Gould, 1977; Telzer and Rosenbaum, 1979; Bergen *et al.*, 1980), centrosome–nucleus complexes (Wang *et al.*, 1979; Kuriyama and Borisy, 1981; Roobol *et al.*, 1982), yeast spindle pole bodies (Hyams and Borisy, 1978), and purified centrosomes (Mitchison and Kirschner, 1984a; Bornens *et al.*, 1987), have shown nucleated microtubule growth off of centrosomes.

From *in vitro* studies using isolated centrosome-containing fractions it has been shown that the pericentriolar material is the site of microtubule nucleation (Gould and Borisy, 1977; Telzer and Rosenbaum, 1979). More microtubules converged at the end of the centriole pair with the greatest amount of associated pericentriolar material; removal of the pericentriolar material from the centrioles resulted in no microtubule nucleation; and fragments of pericentriolar material from mitotic cells nucleated microtubules in the absence of centrioles. Fragments of homogenized mitotic apparatus isolated from the eggs of the sea-urchin *Hemicentrotus pulcherimus* are also capable of nucleating an astral microtubule array (Toriyama *et al.*, 1988). Extraction of the homogenate with 0.5 M KCl destroyed the nucleating activity. Separation of the extract by phosphocellulose chromatography resulted in the identification of a fraction capable of forming a granular assembly that could support microtubule growth. A 51 kDa polypeptide is the major protein component of this granular assembly, and may represent the identification of a major protein component of the pericentriolar material.

Studies by Weisenberg and Rosenfeld (1975) have demonstrated that a change in the MTOC occurs in activated *Spisula* eggs that triggers its capacity to form an astral array of microtubules when incubated with tubulin-containing homogenates from either activated or unactivated eggs. A similar MTOC preparation from unactivated eggs is unable to form an astral microtubule array under identical incubation conditions. These initial results suggested a difference in the properties of centrosomes during the cell cycle, and have been confirmed and expanded upon in studies of tissue culture cells. In comparing interphase and mitotic centrosomes, it is apparent that there is a proliferation of microtubule nucleation sites within the pericentriolar material as the cell enters division (Kuriyama and Borisy, 1981). The initiating capacity of the centrosome increases from early prophase to late prometaphase, returning to the interphase levels at telophase. The number of microtubules initiated by a mitotic centrosome has been reported to be from 5 to 10 times greater than the number initiated from interphase centrosomes (Kuriyama and Borisy, 1981). With Chinese hamster ovary cells the number of microtubules nucleated *in vitro* increases from approximately 24 per interphase centrosome to nearly 250 per mitotic centrosome (Borisy and Gould, 1977). It is apparent, therefore, that the microtubule nucleating capacity of the centrosome is linked to the mitotic cycle and is independent of the centriole cycle (Fig. 2.4).

Centrosomes isolated from a given stage of the cell cycle will nucleate microtubules from monomeric tubulin. This microtubule initiation is proportional to the tubulin concentration, but the absolute number of microtubules initiated is saturable. This suggests that a finite number of microtubule-initiating sites are present at the centrosome (Mitchison and Kirschner, 1984a; Kuriyama, 1984). The length of centrosome-initiated microtubules is also dependent upon tubulin concentration (Mitchison and Kirschner, 1984a). The rate of microtubule growth, however, is constant, and is the same rate as that determined for plus end growth on axonemes (Bergen and Borisy, 1980). Therefore, the centrosome also controls the polarity of microtubule growth, and all microtubules nucleated off the centrosome would necessarily be of the same polarity. This is consistent with the observed polarity of anchored microtubules having the plus end distal to the centrosome (Bergen and Borisy, 1980). The nucleating activity is sensitive to treatments including low pH, salt extraction and protease digestion (Wang *et al.*, 1979; Mitchison and Kirschner, 1984a; Kuriyama, 1984). RNase apparently has no effect on nucleating activity (Kuriyama, 1984). An additional property of the centrosome is its influence upon microtubule structure. While microtubules assembled *in vitro* in the absence of centrosomes form a mixture of tubules with 13 to 15 protofilaments, those assembled off axonemal microtubules (Scheele *et al.*, 1982) or centrosomes (Evans *et al.*, 1985) *in vitro* uniformly contain 13 protofilaments. The nucleating site within the pericentriolar material, therefore, serves as a template microtubule, containing structural information constraining the microtubule into a fixed geometry of 13 protofilaments. During wing development in *Drosophila* a switch in protofilament number from 13 to 15 is observed in the microtubules comprising the trans-alar array. The loss of centrosomes from cells present in the epidermal layers coincides with this change in protofilament number (Tucker *et al.*, 1986).

The mechanism underlying centrosome stabilization of microtubules has been the subject of recent studies. In theory, stabilization could be achieved by the capping of a thermodynamically less-stable end of the microtubule or, alternatively, the centrosome may provide a localized region within the cell in which microtubule assembly is promoted. Mitchison and Kirschner (1984a,b) have proposed that, as a consequence of the dynamic properties of individual microtubules, in the long term the centrosome will be the source of most if not all of the microtubules within the cell. This is consistent with experiments by Karsenti *et al.* (1984) that demonstrated that microtubules can regrow in the cytoplasm of cytoplasts lacking centrosomes, but that when the centrosome is present a preponderence of the regrown microtubules are focused at the centrosome. In 8-cell blastomeres of the mouse embryo the organization of a focused region of microtubule growth may precede the accumulation of pericentriolar material at that site (Karsenti *et al.*, 1984). In fact, the maintenance of the microtubule network is required for the assembly of large foci of pericentriolar material. Regardless of the initial mechanism of assembly, the pericentriolar material serves as the foci for microtubule nucleation in the following cell cycles.

Are centrosomes functionally required to form a bipolar spindle? In sea-urchin eggs, the requirement for centrosomal participation in the formation of a bipolar spindle appears absolute (Sluder and Rieder, 1985). Sea-urchin eggs briefly treated with Colcemid and then fertilized exhibit male and female pronuclei that do not fuse. Following nuclear envelope breakdown, the eggs were irradiated to inactivate the Colcemid and a functional bipolar spindle developed around the male chromosomes only. In sea-urchin egg development, the sperm centrioles are components of the spindle pole (Schatten *et al.*, 1986). Centrosomes, therefore, were present near the male chromosomes following Colcemid inactivation. The female chromosomes, although present in the same single-cell cytoplasm, were not able to form a bipolar spindle in the absence of centrosomal influence (Sluder and Rieder, 1985).

Several lines of evidence suggest that the essential component of the spindle pole is the pericentriolar material. When the centriole is destroyed by laser microbeam irradiation, mitosis continues normally (Berns and Richardson, 1977). However, if the pericentriolar material is first sensitized to irradiation damage with acridine orange, leaving the centriole undamaged by irradiation, chromosome separation and anaphase movements are inhibited (Berns *et al.*, 1977). Pericentriolar material is also present in functional spindle poles that lack centrioles (Section IV, above). This has also been observed in PtK cells undergoing meiotic-like reduction divisions (Brenner *et al.*, 1977) and in multipolar CHO cells recovering from a colcemid block (Keryer *et al.*, 1984; Sellitto and Kuriyama, 1988). The latter experiments suggest that the centriole may function as an organizer of pericentriolar material. As a result, the bipolarity of the spindle is coupled to the centriole cycle. The reproduction of the centrioles ensures that a critical amount of pericentriolar material is assembled at each pole around the pairs of separated centrioles.

The absolute requirement for centrosomal participation in the formation of a bipolar spindle has recently been challenged by observations suggesting that chromosomes may be capable of forming spindles in the absence of centrosomal components. Microinjection of karyoplasts or DNA into *Xenopus* eggs results in the development of localized arrays of microtubules (Karsenti *et al.*, 1984a,b). These arrays, however, do not become sufficiently organized to resemble a functional spindle structurally unless centrosomes are also present in close proximity to the injected nuclei. Two additional reports suggest that chromosomes or their kinetochores can function as spindle-organizing centres (Debec *et al.*, 1982; Church *et al.*, 1986). Displacement of one spindle pole was achieved by flattening crane-fly spermatocytes in late stages of meiosis, yet a bipolar spindle still formed even in the absence of one pole (Debec *et al.*, 1982). Residual pericentriolar material was not detected at the acentriolar pole by electron microscopy and these aster-free poles were not stained by the anti-pericentriolar antisera 5051 (section IV) (Bastmeyer *et al.*, 1986), or by a human centrosome reactive serum that reacted with a 112 kDa protein component of the centrosome of spermatocytes (Bastmeyer and Russel, 1987).

Also, the aster-free poles were unable to nucleate microtubule assembly, a known centrosomal property (Debec *et al.*, 1982; Bastmeyer *et al.*, 1986). The authors conclude that the chromosomes themselves are capable of inducing the formation of a functional half-spindle in this insect system. The close approximation of the centrosome-free pole with the plasma membrane may also have induced secondary microtubule-organizing sites associated with the membrane. Alternatively, other centrosomal components not recognized by the 5051 serum (Bastmeyer *et al.*, 1986), or the human centrosome reactive serum (Bastmeyer and Russel, 1987) may be present at the acentriolar pole. In the second report, single chromosome bivalents were mechanically detached from the spindle of *Drosophila* primary spermatocytes and placed in the cytoplasm (Church *et al.*, 1986). An acentriolar "mini-spindle" formed in association with the displaced bivalent, and chromatid separation occurred at anaphase. The organization of microtubules may have been a property of the bivalent alone, or of pericentriolar material or membranes that became associated with the bivalent during manipulation. The exact nature of the spindle pole-organizing material or activities present in these spermatocyte systems remains to be determined, but the evidence suggests that under certain conditions centrosomes or components of centrosomes may not be required for spindle formation.

VII. Regulation of Centrosome Function

The organization of the mitotic spindle and the resulting sequential events of mitosis are under spatial and temporal control. Microtubules are specifically nucleated by the centrosomes, and the pattern of microtubule assembly and disassembly within the spindle is specifically related to progression through anaphase. The overall spatial arrangement and size of the spindle also have distinct effects on the timing of further mitotic events (Sluder, 1979; Sluder and Begg, 1983). How are these and other facets of mitosis regulated? Recent evidence suggests that the free Ca^{2+} concentration (Kiehart, 1981; Wolniak *et al.*, 1983; Izant, 1983; Ratan *et al.*, 1986; Poenie *et al.*, 1986), in part regulated by calmodulin (Welsh *et al.*, 1979; Keith, 1985; Brady *et al.*, 1986; Welsh and Sweet, this volume), may directly affect a number of mitotic processes. In addition, the phosphorylation state of some centrosomal components may regulate centrosomal activity (Vandré *et al.*, 1984; Vandré and Borisy, 1985).

As the cell enters mitosis, the microtubule initiating capacity of the centrosome increases substantially, and then decreases back to interphase levels as the cell exits mitosis. The increase in nucleating activity is correlated with an increase in the amount of pericentriolar material. The overall amount of pericentriolar material must, therefore, double with each cell cycle, and also accumulate at the centrosome prior to mitosis. Whether this accumulation is an abrupt or gradual process is not known. Possibly, the pericentriolar material becomes structurally altered at mitosis, exposing more microtubule initiating sites. Alternatively, the increase in microtubule initiating activity could result from a specific activation process, such as a post-translational modification, occurring at mitosis.

Whatever the process for activation of the pericentriolar material, a reversal of this process or inactivation must occur during the exit from mitosis. If, for example, new templates for microtubule nucleation are exposed at mitosis, they must be buried or function for a limited period of time to allow for exit from mitosis. Several alternative mechanisms can also be postulated. For example, centrosome activity could be initiated by the accumulation of factors at the centrosome creating a local environment favouring microtubule polymerization. This initiation activity could decrease owing to the diffusion of these same factors away from the centrosome. Increased microtubule nucleation could also be initiated by a specific biochemical activation of a centrosomal component, whose inhibition could result in decreased microtubule initiation. Snyder and Hamilton (1982) have concluded that a specific alteration occurs in the state of centrosomaal microtubule initiation capacity at the metaphase–anaphase transition. This was based on the results of nocodazole blockage and release experiments. If PtK_1 cells were blocked with nocodazole for 20 minutes at metaphase and then released, the mitotic state was re-established. However, if the nocodazole blockage was started just after anaphase onset, the cell continued to progress toward an interphase state. In this case, a mitotic spindle was not reformed following drug release. If the drug block was applied before anaphase onset, microtubule initiating capacity remained high after drug release. Conversely, when the drug was applied after anaphase onset, microtubule initiating capacity decreased, and was low after drug release.

The phosphorylation–dephosphorylation of certain centrosomal components can be correlated with the microtubule initiating capacity of the centrosome (Fig. 2.5). A monoclonal antibody, MPM-2, specific for a group of mitotic phosphoproteins, stains centrosomes by indirect immunofluorescence throughout mitosis (Vandré *et al.*, 1984, 1986). The MPM-2 antibody has also been shown to react with spindles and spindle poles in the syncitial nuclei of *Drosophila* (Millar *et al.*, 1987), spindle poles in *Aspergillus nidulans* (Engle *et al.*, 1988) and basal bodies in *Paramecium* (Vandré *et al.*, 1986; Keryer *et al.*, 1987), and stains spindle poles of *Caenorhabditis elegans* cells but is absent at the restrictive temperature in the temperature-sensitive embryonic arrest mutant emb-29V (Hecht *et al.*, 1987). The intensity of the MPM staining

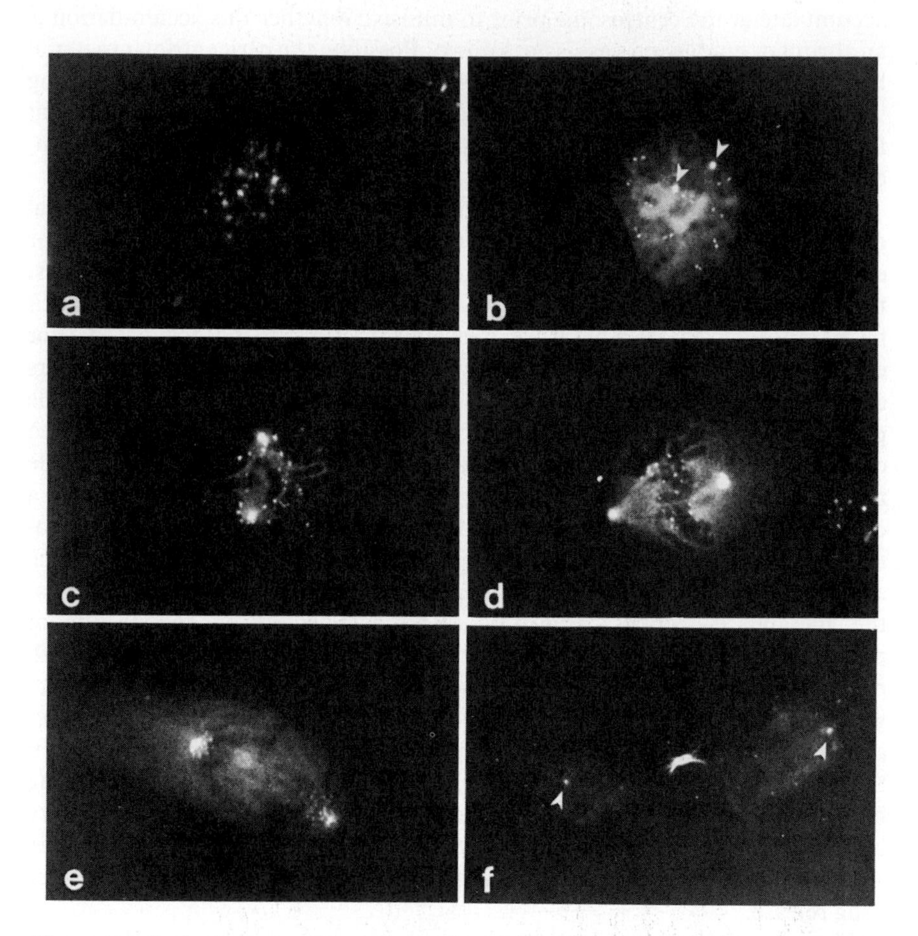

Figure 2.5. Indirect immunofluorescence staining of MPM-2 during the cell cycle of PtK_1 cells. (a) Interphase cells show weakly staining spots within the nucleus the centrosomes are not stained. (b) Late prophase prior to nuclear envelope breakdown. Centrosomes stain intensely with MPM-2 (arrowheads). (c) Prometaphase, the intensity of centrosomal staining increases. (d) Metaphase, the spindle poles react most intensely with MPM-2. (e) The intensity of centrosomal staining decreases at anaphase. (f) The centrosomes are only weakly stained by MPM-2 at telophase (arrowheads). The staining by MPM-2 reflects the appearance of immunoreactive phosphoproteins at the centrosome, and correlates with the microtubule-initiating capacity of the centrosome (compare with Fig. 2.1(a)–(f). In addition, the MPM-2 stains kinetochores and the midbody, which also serve as microtubule-organizing centres during mitosis.

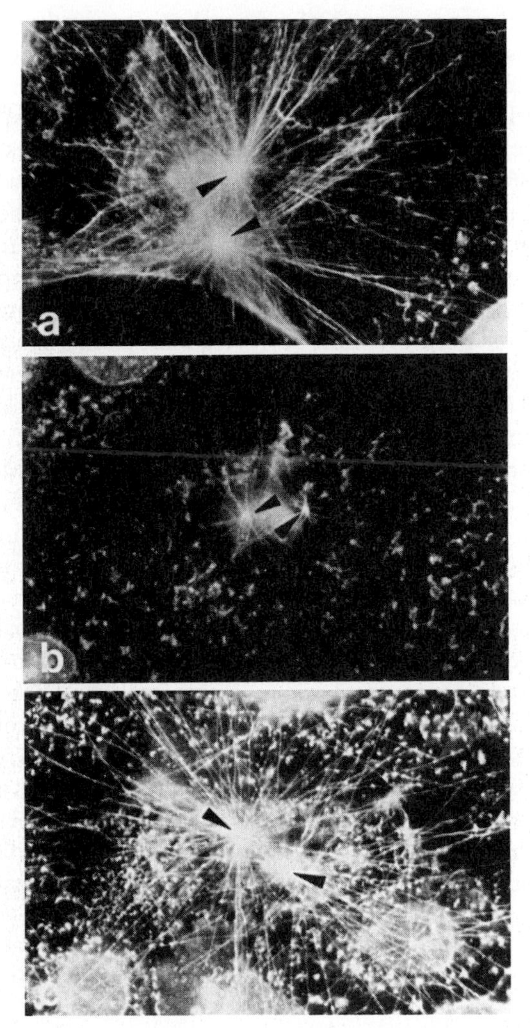

Figure 2.6. Phosphatase sensitivity of microtubule nucleation by centrosomes. A centrosome nucleation assay was developed using detergent-lysed and extracted LLC-PK pig kidney epithelial cells that had previously had their microtubule network depolymerized *in vivo* by treatment with nocodazole. (a) Microtubule regrowth at mitotic centrosomes after incubation with tubulin, detected by immunofluorescence staining with antitubulin. Control incubation in Tris-buffer prior to exogenous tubulin addition. (b) Antitubulin staining of cells pretreated with alkaline phosphatase for 30 minutes prior to tubulin incubation. The number of nucleated microtubules is decreased significantly from that of controls (a). (c) The inclusion of β-glycerolphosphate during the alkaline phosphatase pretreatment prevented the decrease in the number of centrosomally nucleated microtubules. (Arrowheads denote location of centrosomes.)

associated with the centrosomes of mammalian tissue culture cells increases through prophase, peaking at metaphase, and decreases back to interphase levels by the time cells are in late telophase.

A direct functional correlation between reactivity with the MPM-2 antibody and microtubule nucleation capacity has only recently been established (Centonze *et al.*, 1986). In addition to pretreatment of lysed cells with the MPM-2 antibody, alkaline phosphatase pretreatment also blocks the nucleation of microtubules off mitotic centrosomes (Fig. 2.6). In the first example (Centonze *et al.*, 1986), antibody is bound to phosphorylated epitopes present on centrosomal components and either directly blocks sites required for microtubule initiation or sterically inhibits nucleation by binding to adjacent sites. In the latter case (Fig. 2.6), the alkaline phosphatase pretreatment removes phosphate from centrosomal components. However, it is not known which of the phosphates removed (either the phosphate present at the MPM-2 epitope, and/or other phosphate residues not recognized by the MPM-2 antibody) is required for microtubule nucleating activity. It is possible to suggest that the phosphorylation of centrosomal components may trigger the increased microtubule nucleating capacity of the centrosome at the entry to mitosis, and further, that the dephosphorylation of these same components is required for exit from mitosis (Vandré *et al.*, 1984, 1985). In support of the latter possibility, the experiments described by Snyder and Hamilton (1982) (discussed above) were repeated to examine the level of MPM-staining at the centrosomes of metaphase and anaphase cells treated with nocodazole (Vandré and Borisy, 1987). The level of MPM-staining directly correlates with the presence of phosphorylated centrosomal components, and was a marker for the changes in centrosomal phosphorylation. The results of these experiments demonstrated that the onset of mitotic protein dephosphorylation coincided with the onset of anaphase, and did not depend on chromosomal movements.

VIII. The Centrosome Cycle: An Overview and Future Perspectives

The centrosome cycle may be viewed in the context of the reproductive cycle of the organism as well as the reproduction of individual cells. In Boveri's theory of fertilization (see Wilson, 1925), the centrosome was considered to be the fertilizing element proper. Boveri held that the centrosome was continuously passed on by division in somatic cells and the germ line but that the process came to an end in the mature egg after the second meiotic division when the egg centrosome degenerated or was rendered inactive. Further cell division was thus inhibited and parthenogenesis was avoided. In contrast, in the mature sperm , Boveri held that the centrosome was retained in preparation for fertilization.

Boveri's theory had to be modified by subsequent studies of artificial parthenogenesis in which unfertilized eggs were induced to cleave and develop. These studies demonstrated that the egg did not irreversibly lose the ability to express its centrosome, but rather was able either to form a new one or to reactivate a pre-existing one. The contribution of the sperm was also revised and Wilson (1925) concluded that the sperm either introduced a centrosome or had the power to incite the formation of one within the cytoplasm of the egg.

These early studies defined problems that remain with us today. How is the centrosome passed on in somatic cells and in the germ line up to formation of the egg? Although a number of events in the replication of centrioles and in the accumulation of pericentriolar material have been described, we still lack answers to fundamental questions. What regulates the reproduction of centrosomes? It would clearly be a disaster for the cell if centrosome duplication failed to keep pace with chromosome duplication. If centrosome duplication lagged chromosome duplication, cells would enter mitosis with a single pole and would be unable to segregate their chromosomes. If centrosome duplication led, multipolar mitosis would result, segregating the chromosomes into less than full complements and generating inviable cells.

The electron microscope has revealed a pair of centrioles within the centrosomes of most animal cells. Why are there two centrioles? Why not one? How are two and only two centrioles produced each cell cycle and how do they maintain a regular relationship with the pericentriolar material? Indeed, we know virtually nothing about the molecular composition of the pericentriolar material. What does it consist of? When is it synthesized? How does it associate with and accumulate about the centrioles? How is its activity in the initiation of microtubules regulated? What happens to this material at the conclusion of mitosis? Is it dispersed, degraded or reutilized?

The questions we ask about centrosome reproduction in the cell cycle may be reflected back on to the problems of the germ line and fertilization. How does the egg centrosome degenerate or become rendered inactive? How is the evident capacity of the unfertilized egg to generate centrosomes suppressed? In fertilization, if the sperm does not contribute a centrosome, how does it elicit the formation of one within the egg?

Our questions about the centrosome revolve about its reproduction and its function. The centrosome received its name because it was the central body of the array of fibres present at each pole of a dividing cell. Electron microscopy and antitubulin immunofluorescence have revealed these fibres to be microtubules, and the centrosome to nucleate, organize and anchor these microtubules. Are there separate molecules responsible for these three activities? How many molecules does the centrosome comprise and what are their respective roles? Are there other activities of the centrosome yet to be defined? For example, some models of mitosis hold that spindle microtubules are depolymerized at the centrosome and that the removal of microtubule subunits is coupled to translocation of kinetochore fibres. Is the centrosome a microtubule depolymerizing machine as well as a polymerizing one?

The questions enumerated here about centrosome reproduction and function indicate that we have far to go before we attain an understanding of this essential cell structure at the molecular level. A promising new approach for centrosomal analysis may lie in the genetic analysis of *Drosophila* mutants that express phenotypes characterized by defects in centrosomal function, such as the recently described *mgr* and *polo* mutations (Gonzalez *et al.*, 1988; Sunkel and Glover, 1988). It is clear that it will be necessary to identify centrosomal components, to study their synthesis, associations and regulation, to determine how they control the formation of microtubules and how centrosomal reproduction is coordinated, both in terms of the cell cycle and in development.

REFERENCES

Albrecht-Buehler, G. (1977). *Cell* **12**, 333–339.

Albrecht-Buehler, G. and Bushnell, A. (1980). *Exp. Cell Res.* **126**, 427–437.

Avers, C. J. (1976). *Cell Biology*, pp. 372–390. van Nostrand, New York.

Bajer, A. S. and Molé-Bajer, J. (1981). *Cold Spring Harbor Symp. Quant. Biol.* **46**, 263–283.

Bastmeyer, M. and Russel, D.-G. (1987). *J. Cell Sci.* **87**, 431–438.

Bastmeyer, M., Steffen, W. and Fuge, H. (1986). *Eur. J. Cell. Biol.* **42**, 305–310.

Bergen, L. and Borisy, G. G. (1980). *J. Cell Biol.* **84**, 141–150.

Bergen, L., Kuriyama, R. and Borisy, G. G. (1980). *J. Cell Biol.* **84**, 151–159.

Berns, M. W. and Richardson, S. M. (1977). *J. Cell Biol.* **75**, 977–982.

Berns, M. W., Rattner, J. B., Brenner, S. and Meredith, S. (1977). *J. Cell Biol.* **72**, 351–367.

Blackburn, G. R., Barrau, M. D. and Dewey, W. C. (1978). *Exp. Cell Res.* **113**, 183–187.

Bonifacino, J. S., Klausner, R. D. and Sandoval, I. V. (1985). *Proc. Natl. Acad. Sci. USA* **82**, 1146–1150.

Borisy, G. G. and Gould, R. R. (1977). In *Mitosis Facts and Questions* (ed. M. Little, N. Paweletz, C. Petzalt, H. Ponstingl, D.. Schroeter and H.-P. Zimmerman), pp. 78–87. Springer-Verlag, New York.

Bornens, M. and Karsenti, E. (1984). In *Membrane Structure and Function.* Vol. 6 (ed. E. E. Bittar), pp. 99–171. Wiley, New York.

Bornens, M., Paintrand, M., Berges, J., Marty, M.-C. and Karsenti, E. (1987). *Cell Motil. Cytoskel.* **8**, 238–249.

Brady, R. C., Cabral, F. and Dedman, J. R. (1986). *J. Cell Biol.* **103**, 1855–1861.

Bré, M.-H., Kreis, T. E. and Karsenti, E. (1987). *J. Cell Biol.* **105**, 1283–1296.

Brenner, S. L. and Brinkley, B. R. (1982). *Cold Spring Harbor Symp. Quant. Biol.* **46**, 241–254.

Brenner, S., Branch, A., Meredith, S. and Berns, M. W. (1977). *J. Cell Biol.* **72**, 368–379.

Brinkley, B. R. (1985). *A. Rev. Cell Biol.* **1**, 145–172.
Brinkley, B. R. and Stubblefield, E. (1970). In *Adv. Cell Biol.* **1**, 119–185.
Brinkley, B. R., Cox, S. M., Pepper, D. A., Wible, L., Brenner, S. L. and Pardue, R. L. (1981). *J. Cell Biol.* **90**, 554–562.
Brinkley, B. R., Fuller, A. M. and Highfield, D. P. (1976). In *Cell Motility* (ed. R. Goldman, T. Pollard and J. Rosenbaum), pp. 867–871. Cold Spring Harbor Laboratory, New York.
Brooks, R. F. and Richmond, F. N. (1983). *J. Cell Sci.* **61**, 231–245.
Calarco-Gillman, P. D., Siebert, M. C., Hubble, R., Mitchison, T. and Kirschner, M. (1983). *Cell* **35**, 621–629.
Centonze, V. E., Vandre, D. D. and Borisy, G. G. (1986). *J. Cell Biol.* **103**, 412a.
Chu, L. K. and Sisken, J. E. (1977). *Exp. Cell Res.* **107**, 71–77.
Church, K., Nicklas, R. B. and Lin, H.-P. P. (1986). *J. Cell Biol.* **103**, 2765–2773.
Clayton, L., Black, C. M. and Lloyd, C. W. (1985). *J. Cell Biol.* **101**, 319–324.
Cleveland, L. R. (1963). In *The Cell in Mitosis* (ed. Levine), pp. 3–53. Academic Press, New York.
Connolly, J. A. and Kalnins, V. I. (1978). *J. Cell Biol.* **79**, 526–532.
Coss, R. A. (1974). *J. Cell Biol.* **63**, 84–98.
Debec, A., Szöllösi, A. and Szöllösi, D. (1982). *Biol. Cell* **44**, 133–138.
DeBrabander, G., Geuens, G., Nuydens, R., Willlebrords, R., Aerts, F. and DeMey, J. (1986). *Int. Rev. Cytol.* **101**, 215–274.
de Harven, E. and Bernhard, W. (1956). *Z. Zellforsch.* **45**, 378–398.
DeMey, J., Aerts, F., DeRaeymaeker, M., Daneels, G., Moeremans, M., DeWever, B., Vandré, D. D., Vallee, R. B., Borisy, G. G. and DeBrabander, M. (1987). In *Nature and Function of Cytoskeletal Proteins in Motility and Transport: Progress in Zoology,* Vol. 34 (ed. K. E. Wolhfarth-Botterman), pp. 187–205. Gustav Fischer Verlag, New York.
Engle, D. B., Doonan, J. H. and Morris, N. R. (1988). *Cell Motil. Cytoskel.* **10**, 432–437.
Evans, L., Mitchison, T. and Kirschner, M. (1985). *J. Cell Biol.* **100**, 1185–1191.
Forer, A. and Zimmerman, A. M. (1974). *J. Cell Sci.* **16**, 481–497.
Frankel, F. R. (1976). *Proc. Natl. Acad. Sci. USA* **73**, 2798–2802.
Fuge, H. (1977). *Int. Rev. Cytol. Suppl.* **6**, 1–58.
Fulton, C. (1971). In *Origin and Continuity of Cell Organelles* (ed. J. Reinhart and H. Ursprung), pp. 170–221. Springer-Verlag, New York.
Fung, B. P. and Kasamatsu, H. (1985). *Cell Tissue Res.* **239**, 43–50.
Gonzalez, C., Casal, J. and Ripoll, P. (1988). *J. Cell Sci.* **89**, 39–47.
Gould, R. R. and Borisy, G. G. (1977). *J. Cell Biol.* **73**, 601–615.
Gosti-Testu, F., Marty, M.-C., Berges, J., Maunoury, R. and Bornens, M. (1986). *EMBO J.* **5**, 2545–2550.
Gosti-Testu, F., Marty, M.-C., Courvalin, J. C., Maunoury, R. and Bornens, M. (1987). *Proc. Natl. Acad. Sci. USA* **84**, 1000–1004.
Hartman, H., Puma, J. P. and Gurney, T. (1974). *J. Cell Sci.* **16**, 241–260.
Hecht, R. M., Berg-Zabelshansky, M., Rao, P. N. and Davis, F. M. (1987). *J. Cell Sci.* **87**, 305–314.
Heidemann, S. R. and Kirschner, M. W. (1975). *J. Cell Biol.* **67**, 105–117.
Heidemann, S. R., Sander, G. and Kirschner, M. (1977). *Cell* **10**, 337–350.
Houliston, E., Pickering, S. J. and Maro, B. (1987). *J. Cell Biol.* **104**, 1299–1308.
Hyams, J. S. and Borisy, G. G. (1978). *J. Cell Biol.* **78**, 401–414.

Izant, J. G. (1983). *Chromosoma* **88**, 1–10.
Johnson, U. G. and Porter, K. R. (1968). *J. Cell Biol.* **38**, 403–425.
Karsenti, E., Kobayashi, S., Mitchison, T. and Kirschner, M. (1984a). *J. Cell Biol.* **98**, 1763–1776.
Karsenti, E., Newport, J. and Kirschner, M. (1984b). *J. Cell Biol.* **99**, 47s–54s.
Karsenti, E., Newport, J., Hubble, R. and Kirschner, M. (1984c). *J. Cell Biol.* **98**, 1730–1745.
Keith, C. H. (1985). *J. Cell Biol.* **101**, 147a.
Keryer, G., Ris, H. and Borisy, G. G. (1984). *J. Cell Biol.* **98**, 2222–2229.
Keryer, G., Davis, F. M., Rao, P. N. and Beisson, J. (1987). *Cell Motil. Cytoskel.* **8**, 44–54.
Kiehart, D. P. (1981). *J. Cell Biol.* **88**, 604–617.
Kirschner, M. and Mitchison, T. (1986). *Cell* **45**, 329–342.
Kuriyama, R. (1982a). *Cell Struct. Funct.* **7**, 307–315.
Kuriyama, R. (1982b). *J. Cell Sci.* **53**, 155–171.
Kuriyama, R. (1984). *J. Cell Sci.* **66**, 277–295.
Kuriyama, R. and Borisy, G. G. (1981a). *J. Cell Biol.* **91**, 814–821.
Kuriyama, R. and Borisy, G. G. (1981b). *J. Cell Biol.* **91**, 822–826.
Kuriyama, R., Keryer, G. and Borisy, G. G. (1984). *J. Cell Sci.* **66**, 265–274.
Kuriyama, R., Dasgupta, S. and Borisy, G. G. (1986). *Cell Motil. Cytoskel.* **6**, 355–362.
Lin, W., Fung, B., Shyamala, M. and Kasamatsu, H. (1981). *Proc. Natl. Acad. Sci. USA* **78**, 2373–2377.
Luca, F. C., Bloom, G. S. and Vallee, R. B. (1986). *Proc. Natl. Acad. Sci. USA* **83**, 1006–1010.
Luykx, P. (1970). *Int. Rev. Cytol. Suppl.* **2**, 1–171.
Maro, B., Howlett, S. K. and Webb, M. (1985). *J. Cell Biol.* **101**, 1665–1672.
Mazia, D. (1984). *Exp. Cell Res.* **153**, 1–15.
Mazia, D. (1987). *Int. Rev. Cytol.* **100**, 49–92.
Mazia, D., Harris, P. J. and Bibring, T. (1960). *Biophys. Biochem. Cytol.* **7**, 1–20.
Mazia, D., Paweletz, N., Sluder, G. and Finze, E.-M. (1981). *Proc. Natl. Acad. Sci. USA* **78**, 377–381.
McDonald, B. R. and Weijer, J. (1966). *Can. J. Genet. Cytol.* **8**, 42–50.
McIntosh, J. R. (1983). *Modern Cell Biol.* **2**, 115–142.
McNiven, M. A. and Porter, K. R. (1988). *J. Cell Biol.* **106**, 1593–1605.
Millar, S. E., Freeman, M. and Glover, D. M. (1987). *J. Cell Sci.* **87**, 95–104.
Mitchison, T. and Kirschner, M. (1984a). *Nature* **312**, 232–237.
Mitchison, T. and Kirschner, M. (1984b). *Nature* **312**, 237–242.
Mitchison, T., Evans, L., Schulze, E. and Kirschner, M. (1986). *Cell* **45**, 515–527.
Nadezhdina, E. S., Fais, D. and Chentsov, Y. S. (1978). *Cell Biol. Int. Rep.* **2**, 601–606.
Neighbors, B. W., Williams, R. C., Jr. and McIntosh, J. R. (1988). *J. Cell Biol.* **106**, 1193–1204.
Nigg, E. A., Schäfer, G., Hilz, H. and Eppenberger, H. M. (1985). *Cell* **41**, 1039–1051.
Oliver, J. M., Osborne, W. R. A., Pfeiffer, J. R., Child, F. M. and Berlin, R. D. (1981). *J. Cell Biol.* **91**, 837–847.
Osborn, M. and Weber, K. (1976). *Proc. Natl. Acad. Sci. USA* **73**, 867–871.
Paweletz, N., Mazia, D. and Finze, E.-M. (1984). *Exp. Cell Res.* **152**, 47–65.

Pennell, R. I., Vendy, K. P., Bell, P. R. and Hyans, J. S. (1988). *Eur. J. Cell Biol.* **46**, 51–60.
Pickett-Heaps, J. D. (1969). *Cytobios* **1**, 257–280.
Poenie, M., Alderton, J., Steinhardt, R. and Tsien, R. (1986). *Science* **233**, 886–889.
Porter, K. R. (1966). In *Principles of Biomolecular Organization*, pp. 308–345. Ciba Foundation Symposium. Little Brown, Boston.
Randall, J. T. and Disbrey, C. (1965). *Proc. Roy. Soc. B.* **162**, 473–491.
Ratan, R. R., Shelanski, M. L. and Maxfield, F. R. (1986). *Proc. Natl. Acad. Sci. USA* **83**, 5136–5140.
Rattner, J. B. and Phillips, S. G. (1973). *J. Cell Biol.* **57**, 359–372.
Rieder, C. L. (1979). *J. Cell Biol.* **80**, 1–9.
Rieder, C. L. (1982). *Int. Rev. Cytol.* **79**, 1–58.
Rieder, C. L. and Borisy, G. (1982). *Biol. Cell* **44**, 117–132.
Robbins, E., Jentzsch, G. and Micali, A. (1968). *J. Cell Biol.* **36**, 329–339.
Roobol, A., Havercroft, J. C. and Gull, K. (1982). *J. Cell Sci.* **55**, 365–381.
Sager, P. R., Rothfield, N. L., Oliver, J. M. and Berlin, R. D. (1986). *J. Cell Biol.* **103**, 1863–1872.
Sakai, H. and Kuriyama, R. (1974). *Devl. Growth Differ.* **16**, 123–134.
Salmon, E. D., Leslie, R. J., Saxton, W. M., Karow, M. L. and McIntosh, J. R. (1984). *J. Cell Biol.* **99**, 2165–2174.
Sammak, P. J., Gorbsky, G. J. and Borisy, G. G. (1987). *J. Cell Biol.* **104**, 395–406.
Sato, C., Nishizawa, K., Nakamuru, H., Komagoe, Y., Shimada, K., Ueda, R. and Suzuki, S. (1983). *Cell Struct. Funct.* **8**, 245–254.
Sato, C., Nishizawa, K., Nakamuru, H. and Ueda, R. (1984). *Exp. Cell Res.* **155**, 33–42.
Sato, C., Tanabe, K., Nishizawa, K., Nakayma, T., Kobayashi, T. and Nakamuru, H. (1985). *Exp. Cell Res.* **160**, 206–220.
Saxton, W. M., Stemple, D. L., Leslie, R. J., Salmon, E. D., Zavortink, M. and McIntosh, J. R. (1984). *J. Cell Biol.* **99**, 2175–2186.
Schatten, G., Simerly, C. and Schatten, H. (1985). *Proc. Natl. Acad. Sci. USA* **82**, 4152–4156.
Schatten, H., Schatten, G., Mazia, D., Balczon, R. and Simerly, C. (1986). *Proc. Natl. Acad. Sci. USA* **83**, 105–109.
Scheele, R. B., Bergen, L. G. and Borisy, G. G. (1982). *J. Mol. Biol.* **154**, 485–500.
Schulze, E. and Kirschner, M. (1986). *J. Cell Biol.* **102**, 1020–1031.
Segarini, P. R., Shyamala, M., Atcheson, C. L. and Kasamatsu, H. (1983). *J. Cell. Physiol.* **116**, 311–321.
Sellitto, C. and Kuriyama, R. (1988). *J. Cell Sci.* **89**, 57–65.
Sherline, P. and Mascaro, R. N. (1982). *J. Cell Biol.* **93**, 507–511.
Sluder, G. (1978). In *Cell Reproduction* (ed. E. R. Dirksen, D. M. Prescott and C. F. Fox), pp. 563–569. Academic Press, New York.
Sluder, G. (1979). *J. Cell Biol.* **80**, 674–691.
Sluder, G. and Begg, D. A. (1983). *J. Cell Biol.* **97**, 877–886.
Sluder, G. and Begg, D. A. (1985). *J. Cell Sci.* **76**, 35–51.
Sluder, G. and Rieder, C. L. (1985a). *J. Cell Biol.* **100**, 887–896.
Sluder, G. and Rieder, C. L. (1985b). *J. Cell Biol.* **100**, 897–903.
Sluder, G., Miller, F. J. and Rieder, C. L. (1986). *J. Cell Biol.* **103**, 1873–1881.
Snell, W. J., Dentler, W. L., Haimo, L. T., Binder, L. I. and Rosenbaum, J. L. (1974). *Science* **185**, 357–360.

Snyder, J. A. and Hamilton, B. T. (1982). *Eur. J. Cell Biol.* **27**, 191–199.
Snyder, J. A. and McIntosh, J. R. (1975). *J. Cell Biol.* **67**, 744–760.
Soltys, B. J. and Borisy, G. G. (1985). *J. Cell Biol.* **100**, 1682–1689.
Sorokin, S. P. (1968). *J. Cell Sci.* **3**, 207–230.
Steffen, W., Fuge, H., Dietz, R., Bastmeyer, M. and Müller, G. (1986). *J. Cell Biol.* **102**, 1679–1687.
Stubblefield, E. and Brinkley, B. R. (1967). In *The Origin and Fate of Cell Organelles* (ed. K. B. Warren), pp. 175–218. Academic Press, New York.
Sunkel, C. E. and Glover, D. M. (1988). *J. Cell Sci.* **89**, 25–38.
Szöllösi, D. (1964). *J. Cell Biol.* **21**, 465–479.
Tassin, A.-M., Maso, B. and Bornens, M. (1985). *J. Cell Biol.* **100**, 35–46.
Telzer, B. R. and Rosenbaum, J. L. (1979). *J. Cell Biol.* **81**, 484–497.
Toriyama, M., Ohta, K., Endo, S. and Sakai, H. (1988). *Cell Motil. Cytoskel.* **9**, 117–128.
Tucker, J. B., Milner, M. J., Currie, D. A., Muir, J. W., Forrest, D. A. and Spencer, J.-J. (1986). *J. Cell Biol.* **41**, 279–289.
Tuffanelli, D. L., McKeon, F., Kleinsmith, D. K., Burnham, T. K. and Kirschner, M. (1983). *Arch. Dermatol.* **119**, 560–566.
Turksen, K., Aubin, J. E. and Kalnins, V. I. (1982). *Nature* **298**, 763–765.
Vallee, R. B., Bloom, G. S. and Luca, F. C. (1984). In *Molecular Biology of the Cytoskeleton* (ed. G. G. Borisy, D. W. Cleveland and D. B. Murphy), pp. 111–130. Cold Spring Harbor Laboratory, New York.
Vallee, R. B. and Davis, S. E. (1983). *Proc. Natl. Acad. Sci. USA* **80**, 1342–1346.
van Beneden, E. (1876). Bull. Acad. Roy. Belgique **42**, 35–97.
Vandré, D. D. and Borisy, G. G. (1985). In *Cell Motility: Mechanism and Regulation* (ed. H. Ishikawa, S. Hatano and H. Sato), pp. 389–401. University of Tokyo Press, Tokyo.
Vandré, D. D. and Borisy, G. G. (1987). *J. Cell Biol.* **105**, 174a.
Vandré, D. D., Davis, F. M., Rao, P. N. and Borisy, G. G. (1984a). *Proc. Natl. Acad. Sci. USA* **81**, 4439–4443.
Vandré, D. D., Kronebusch, P. and Borisy, G. G. (1984b). In *Molecular Biology of the Cytoskeleton* (ed. G. G. Borisy, D. W. Cleveland and D. B. Murphy), pp. 1–16. Cold Spring Harbor Laboratory, New York.
Vandré, D. D., Davis, F. M., Rao, P. N. and Borisy, G. G. (1986). *Eur. J. Cell Biol.* **41**, 72–81.
Vorobjev, I. A. and Chentsov, Y. S. (1982). *J. Cell Biol.* **98**, 938–949.
Vorobjev, I. A. and Nodezhdina, E. S. (1987). *Int. Rev. Cytol.* **106**, 227–293.
Wang, E., Connolly, J. A., Kalnins, V. I. and Choppin, P. W. (1979). *Proc. Natl. Acad. Sci. USA* **76**, 5719–5723.
Weisenberg, R. C. and Rosenfeld, A. C. (1975). *J. Cell Biol.* **64**, 146–158.
Welsh, M. J., Dedman, J. R., Brinkley, B. R. and Means, A. R. (1979). *J. Cell Biol.* **81**, 624–634.
Wheatley, D. N. (1982). *The Centriole: A Central Enigma of Cell Biology*. Elsevier, New York.
Wick, S. W., Seagull, R. W., Osborn, M., Weber, K. and Gunning, B. E. S. (1981). *J. Cell Biol.* **89**, 685–690.
Willingham, M. C. and Pastan, I. (1985). *J. Histochem. Cytochem.* **33**, 59–64.
Wilson, E. B. (1911). *The Cell in Development and Heredity*, 2nd edn. Macmillan, New York.

Wilson, E. B. (1925). *The Cell in Development and Heredity*, 3rd edn. Macmillan, New York.
Wolfe, J. (1972). In *Advances in Cell and Molecular Biology*, Vol. 2 (ed. E. J. Dupraw), pp. 151–192. Academic Press, New York.
Wolniak, S. M., Hepler, P. K. and Jackson, W. T. (1983). *J. Cell Biol.* **96**, 598–605.
Zieve, G. and Solomon, F. (1982). *Cell* **28**, 233–242.
Zeligs, J. D. and Wollman, S. H. (1979). *J. Ultrastruct. Res.* **66**, 97–108.

CHAPTER 3

The Kinetochore: Structure and Molecular Organization

B. R. BRINKLEY, M. M. VALDIVIA, A. TOUSSON
and R. D. BALCZON

Department of Cell Biology and Anatomy, University of Alabama at
Birmingham, Schools of Medicine and Dentistry, Birmingham, Alabama, USA

I. Introduction

The kinetochore is a specialized locus on eukaryotic chromosomes that functions to attach chromosomes to the mitotic spindle and is required for normal movement and distribution of chromosomes in mitosis and meiosis. In his early monograph on mitosis, Schrader (1953) identified the kinetochore as "an element of fundamental importance in the movement of chromosomes", but concluded that little was known about it owing to its "homogeneous and structureless" appearance and weak staining properties. During the three decades that have followed since Schrader's statement, much has been learned about genes, chromosomes and the biology of mitotic cells, but the kinetochore itself still remains an enigma. Until recently, knowledge of the kinetochore was confined largely to its ultrastructure and cytochemistry and its behaviour during chromosome movement as revealed by cine and video analysis (for reviews see Rieder, 1982 and Godward, 1985). Recently, however the application of the techniques of immunocytochemistry, biochemistry and molecular biology has led to a renewed interest in the kinetochore and its relationship to chromosome structure and gene segregation. The availability of techniques for analysing kinetochore–microtubule interactions in lysed cell models and in living cells has made it possible to compare and contrast the kinetochore with other microtubule-organizing centres in mitotic cells such as the centrosomes. In this chapter, we will review recent work on the structure and molecular composition of the kinetochore and summarize its role in mitosis and chromosome distribution.

MITOSIS: Molecules and Mechanisms
ISBN 0-12-363420-2

II. Centromere and Kinetochore: Are They Synonymous Structures?

The terms *centromere* and *kinetochore* have been used synonymously in the cell biology and genetics literature for many years (see reviews of Schrader, 1953; Ris and Witt, 1981; Rieder, 1982). In view of our current knowledge of DNA and chromosome structure and the interaction of mitotic and meiotic chromosomes with spindle microtubules, is it still correct to view *centromeres* and *kinetochores* as one and the same? The term kinetochore implies a specific structure on the chromosomes where spindle fibres attach and through which mitotic forces act to pull chromosomes to the pole. Indeed, many electron-microscopic studies have identified unique structural components at the primary constriction of metaphase chromosomes to which spindle microtubules are attached. In many eukaryotic cells the kinetochore appears as a trilamellar plaque or disc with microtubules attached to the outermost lamella (Figs. 3.1 and 3.2). In "ball and cup" kinetochores, common to the chromosomes of higher plants, microtubules attach to a diffuse fibrous mass localized at the primary constriction (Fig. 3.3). In some organisms, such as the budding yeast *Saccharomyces cerevisiae*, a discrete structure between microtubules and centromeric chromatin is not apparent (Peterson and Ris, 1976). In general, however, the kinetochore can be seen as a structurally distinct interface between spindle fibres (kinetochore microtubules) and the chromosome.

The term centromere, on the other hand, generally refers to the entire region of the chromosome where spindle microtubules attach, including the attachment site, primary constriction and associated heterochromatin. The centromeric DNA in many mammalian chromosomes is composed of highly repetitive base sequences (satellite DNA). Centromeric heterochromatin and satellite DNA, however, are not found in all chromosomes and may be absent from the primary constriction of many plant and animal chromosomes, including those of some mammals (see Rieder, 1982). The relationship between centromeric heterochromatin and kinetochore structure and function remains unclear. Centromeric chromatin has unique properties that distinguish it from the rest of the chromatin. It is highly condensed, genetically "silent", and resistant to hypotonic swelling (Brinkley *et al.*, 1980; Ris and Witt, 1981) and nuclease digestion (Rattner *et al.*, 1978; Valdivia and Brinkley, 1985).

The morphological distinction between centromeres and kinetochores as described above can be misleading. The yeast chromosome has no distinct kinetochore and microtubules appear to attach directly to the chromatin fibres (Peterson and Ris, 1976). Experiments involving cloned fragments of yeast centromeric DNA show convincingly that the nucleotide sequence of the DNA fragments convey mitotic stability and meiotic behaviour to replicating

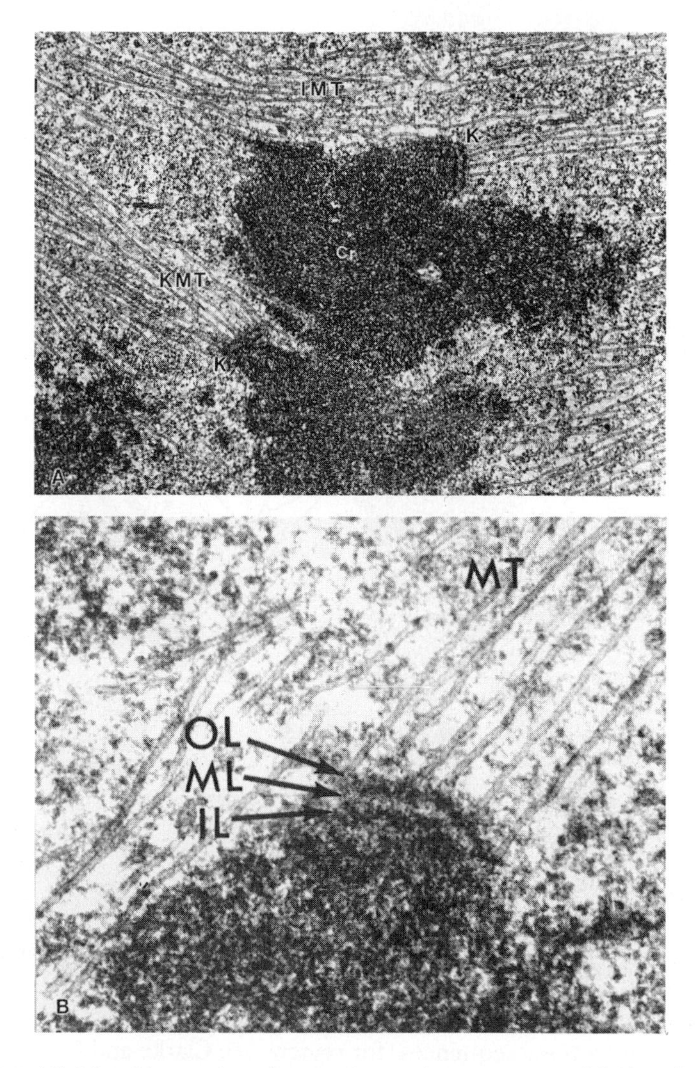

Figure 3.1. (A) Ultrathin section of metaphase chromosomes of PtK_1 cell showing trilaminar plates of kinetochores (K). Note bundles of kinetochore microtubules (kMT) and interpolar microtubules (IMT) of the spindle. (From Brinkley and Cartwright, 1971.) (B) Higher magnification of kinetochore showing details of outer layer (OL), middle layer (ML) and inner layer (IL) with attached microtubules (MT). (From Pepper and Brinkley, 1979.)

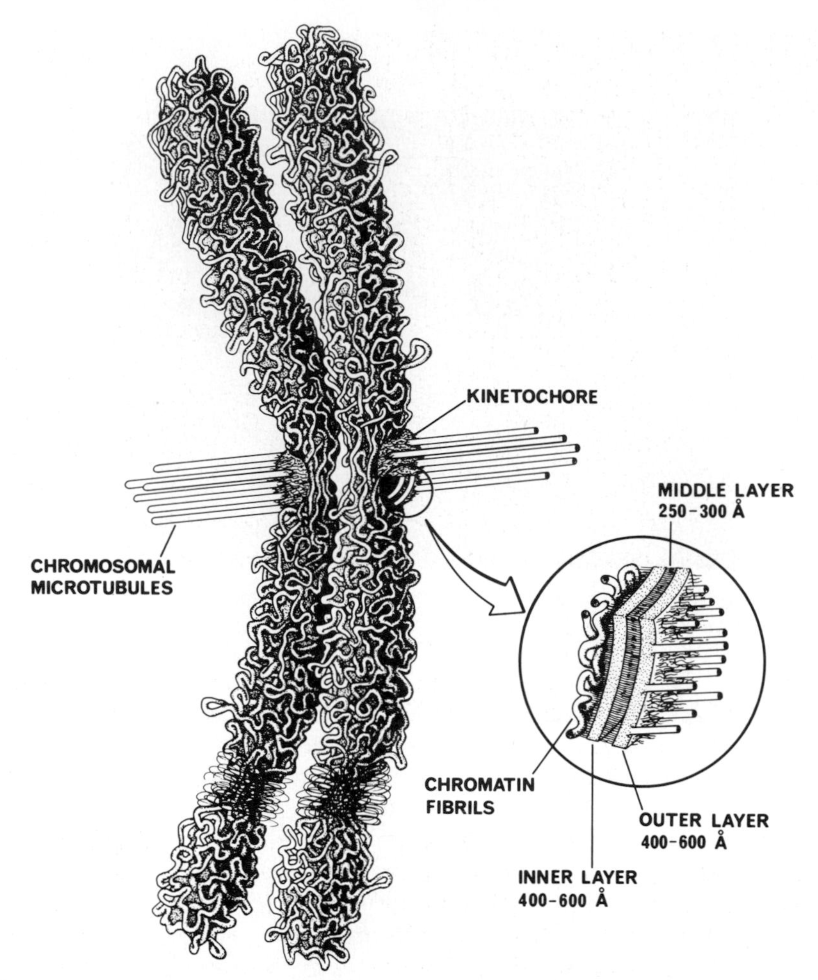

Figure 3.2. Diagram of metaphase chromosome and kinetochore showing dimensions of trilaminar plate. (From Brinkley *et al.*, 1985. Drawing modified from S. L. Wolfe (1972) *Biology of the Cell*, Wadsworth, Belmont, Ca., fig. 11.13.) California.

plasmids carrying these sequences (for review, see Clarke and Carbon, 1985; Bloom *et al.*, 1984). There is growing evidence that the centromeric chromatin is organized into distinct nucleosomes and that the unique biochemistry of this chromatin facilitates both attachment of spindle microtubules and segregation of chromosomes to spindle poles. In yeast, therefore, each chromosome appears to contain a nuclease-protected centromere core containing DNA of approximately 220–250 base pairs. This length of DNA coincides with the attachment of one microtubule to each chromosome (see Bloom *et al.*, 1984).

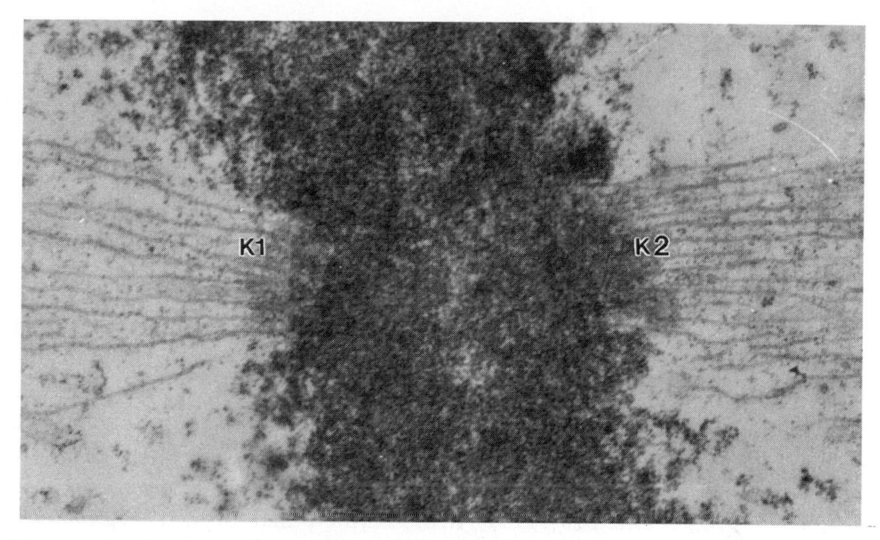

Figure 3.3. "Ball and cup" kinetochores (K1 and K2) of *Haemanthus* chromosome. (Micrograph courtesy Dr. Andrew Bajer.)

Presumably, the DNA sequence contained in the centromere provides the necessary information for the binding of specific centromeric proteins, which, in turn, bind to microtubule protein and anchor microtubules to the kinetochore. Thus, all components appear to be structurally and functionally interrelated.

In a recent review, Murray and Szostak (1985) offered a useful definition in which the term kinetochore is used to describe the structural components (proteins, etc.) that are assembled on the centromeric DNA, reserving the term centromere for the DNA itself. This definition, inspired largely by the work on yeast centromeres, seems adequate for all chromosomes until more is learned about centromeres/kinetochores.

III. Structure and Molecular Organization

All eukaryotic chromosomes are attached to spindle microtubules during mitosis, and in most the site of attachment can be identified by its unique morphology and staining characteristics as revealed by light and electron microscopy. Interesting exceptions include some insects, fungi, yeasts and Protista (see review by Rieder, 1982) as well as microchromosomes of birds (Brinkley *et al.*, 1974), where only one or a few microtubules insert into each chromosome. If a specialized structure (kinetochore) exists in these chromosomes, it might not be detected because so few microtubules are involved in the attachment site. In the chromosomes of most organisms, however, the kinetochores can be identified morphologically and are displayed in one of two principle designs: *diffuse* or *localized*.

A. Diffuse kinetochores

In some insects, such as the arthropods, and in many higher plant groups (monocotyledons) the chromosomes lack a primary constriction and their kinetochore microtubules attach all along the length of the chromosome. Consequently, as the chromosomes move to the poles in anaphase, they remain parallel to each other as opposed to the bent or V-shape of more conventional chromosomes being pulled by microtubules attached to a single localized site. Schrader (1935) termed these "diffuse" or "non localized" kinetochores and Hughes-Schrader and Ris (1941) demonstrated that broken fragments of chromosomes generated by ionizing radiation remained attached to the spindle and continued to undergo movement at anaphase. It was therefore assumed that the kinetochore was distributed throughout the chromosome from one end to the other. Indeed, electron microscopy showed that these kinetochores could be recognized as distinct lightly stained "plates" that extend along the length of the chromosome with microtubules attached all along their lengths (for review, see Rieder, 1982; Godward, 1985).

Chromosomes with the above type of kinetochore are said to be holocentric (Buck, 1967; Comings and Okada, 1972; Bokhari and Godward, 1982). In some organisms with diffuse kinetochores, microtubules attach at multiple sites all along the length of the chromosomes and are referred to as polycentric chromosomes (Ris and Kubai, 1970; Maeki, 1981; Friedlander and Wahrman, 1970; Mughal and Godward, 1973; also see review by Godward, 1985). Very little is known about the molecular organization of kinetochores in general, and especially about diffuse kinetochores. Rieder (personal communications) reported that polycentric chromosomes of the leafhopper exist as small discrete chromosomes in prophase but fuse into one composite chromosome with many kinetochores (Fig. 3.4). Presumably, the determinants of the kinetochore exist on small chromosomes but appear polycentric when end-to-end fusion occurs in metaphase.

Figure 3.4. Polycentric or "diffuse" kinetochore of the leafhopper chromosomes. Note multiple sites along chromosome where microtubules attach (arrow heads). (Micrograph courtesy of Dr. Conley Rieder.)

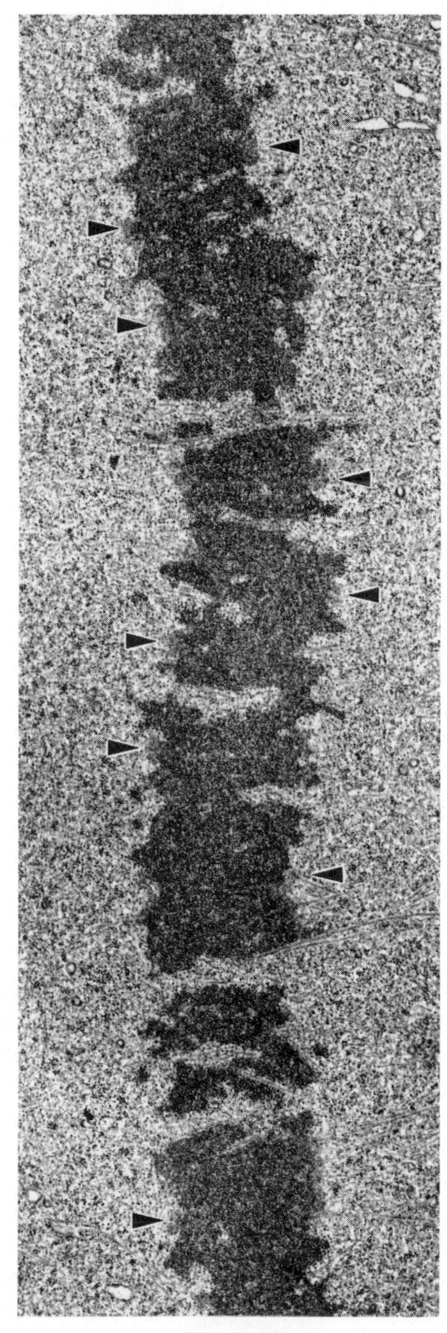

Figure 3.4.

B. Localized kinetochores

Localized kinetochores, as the term implies, are those that are confined to a specific site along the chromosome. The position of the kinetochore is easily identified by a prominent primary constriction at metaphase. Localized kinetochores may be near the middle of the chromosome with arms of equal length flanking each side, as in *metacentric chromosomes*, or near one end positioned between a short arm and a long arm, as in *submetacentric* or *acrocentric chromosomes*. The kinetochore may also be at the very end of the chromosome, as in *telocentric chromosomes*.

When localized kinetochores are examined by electron microscopy (EM) in ultrathin sections, they appear either as crescent-shaped, trilaminar plates or as diffuse balls of less densely staining material positioned in the clefts of the primary constriction, depending on the species examined. The trilaminar type kinetochores (Figs. 3.1 and 3.2) are highly conserved and appear in a wide variety of organisms from algae to mammals. The more diffuse "ball and cup" kinetochores (Fig. 3.3) are characteristic of metaphase chromosomes of higher plants and some insects (see Rieder, 1982). However, even mammalian chromosomes with typical trilaminar plate-type kinetochores may display a spherical or ball-shaped mass during early prophase (Roos, 1973). Since most of the knowledge of kinetochore ultrastructure comes from studies of the trilaminar kinetochore at metaphase, the remainder of this section will be concerned with the structure of the mammalian kinetochore as originally described at the EM level in our laboratory (Brinkley and Stubblefield, 1966) and which has been further characterized by ourselves and numerous other investigators during the past two decades (see Rieder, 1982, for review).

C. Ultrastructure

When the primary constriction of metaphase chromosomes is observed in ultrathin sections by electron microscopy, the kinetochore can easily be recognized as a plate located within the clefts of a constriction to which many microtubules are attached (Figs. 3.1 and 3.2). When examined at higher magnification, the kinetochore plate displays a trilaminar organization consisting of an inner layer 40–60 nm thick, a lightly staining middle layer approximately 25–30 nm thick, and an outer layer also 40–60 nm thick. The inner layer of the kinetochore plate is firmly attached to centromeric chromatin. Thin fibrils extend across the middle layer and appear to cross-link the inner and outer layers. The outer layer contains a fuzzy surface or corona (Brinkley and Stubblefield, 1966; Jokelainen, 1967) that is particularly apparent when cells are arrested in mitosis with colcemid or other mitotic inhibitors. The composition of the corona is not known, but the fine fibrils have been shown to bind tubulin as indicated by immunogold staining with tubulin antibodies (Mitchison and Kirschner, 1985a). Microtubules attach to the outer kinetochore plate, where they are embedded and perhaps entwined within the corona fibrils.

The overall size and three-dimensional organization of the kinetochore varies with the size of the individual chromosome and the stage of mitosis (for a review of kinetochore structure during mitosis, see Rieder, 1982). The structure of the kinetochore may be influenced by drugs that disrupt microtubules, such as Colcemid and related mitotic poisons (see Brinkley *et al.*, 1985). After Colcemid treatment the kinetochore plate becomes attenuated (Fig. 3.5). Figure 3.6 shows a drawing of the outer kinetochore plate of the large X-chromosome of the Indian muntjac (*Muntiacus muntjak vaginalis*) derived from serial-section reconstruction of a Colcemid-arrested metaphase cell. Note that the plate is rectangular with an overall length of 2.0 μm, a width of 0.7 μm, and a thickness of 0.04 μm. The face, or coronal surface, of the plate is punctuated with several small clefts and holes that give the impression of sliced cheese. Although it was generally believed that Colcemid reduced the kinetochore to a single-layered plate, Brenner *et al.* (1981) demonstrated by immunoperoxidase staining with antikinetochore antiserum that the three layers are maintained after Colcemid treatment, but that the inner layer is actually slightly embedded into the centromeric chromatin, giving the appearance of only a bilayer. Structural alterations induced by microtubule inhibitors may have major significance concerning the molecular composition of kinetochores as well as important functional implications, since many of these agents are potent inducers of aneuploidy (Brinkley *et al.*, 1985).

How does the trilaminar kinetochore organization, which resembles a pair of stacked discs, relate structurally to the general organization of DNA fibres of the centromere and chromosome arms? Several investigators have proposed that the trilaminar plates are indeed composed of DNA fibres (Brinkley and Stubblefield, 1966, 1970; Ris and Witt, 1981; Rattner, 1986). Hamkalo and co-workers (Rattner *et al.*, 1978; Rattner *et al.*, 1975; Lica and Hamkalo, 1983) found that the centromere is considerably resistant to nuclease digestion when compared to the chromosome arms. Moreover, an electron-dense plate can be seen when centromere preparations are examined in the electron microscope (in both side and face-on views). Improved resolution of the kinetochore plate was obtained when Rattner (1986) digested metaphase chromosomes with the restriction enzymes *Eco*R1 followed by *Sau*96-1. As shown in Fig. 3.7, these preparations show dense patches at each centromere that appear to be composed of pleated folds of 20–25 nm fibres. Owing to their location, unique structure, and staining properties, he concludes that these are residual

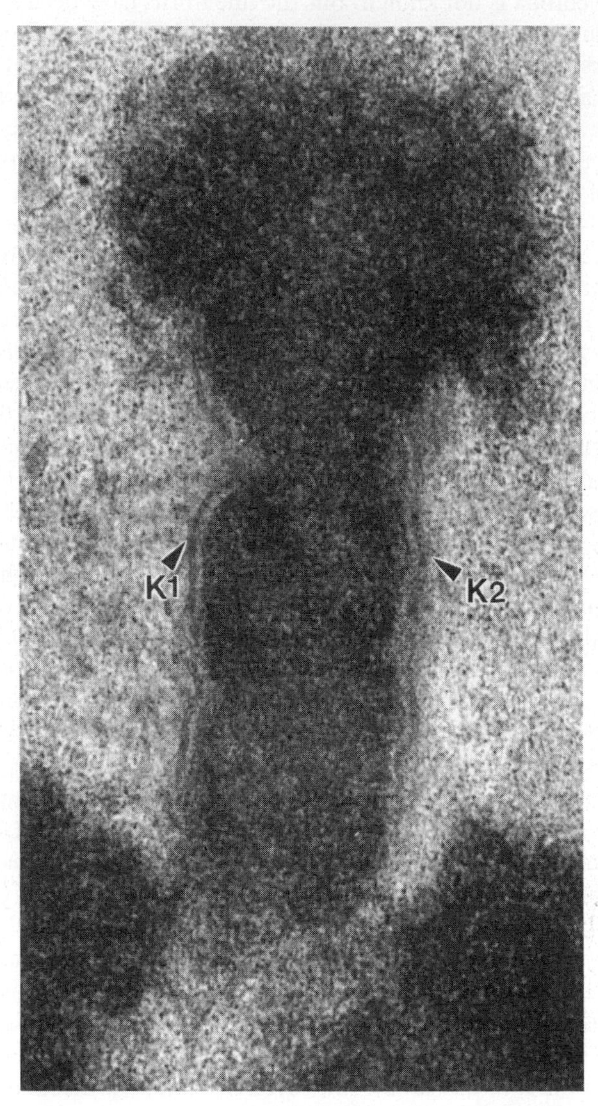

Figure 3.5. Kinetochore of X-chromosome of the Indian muntjac after exposure to colcemid (0.6 μ/ml) for 4 hours. Note extensive kinetochore plate (K1 and K2).

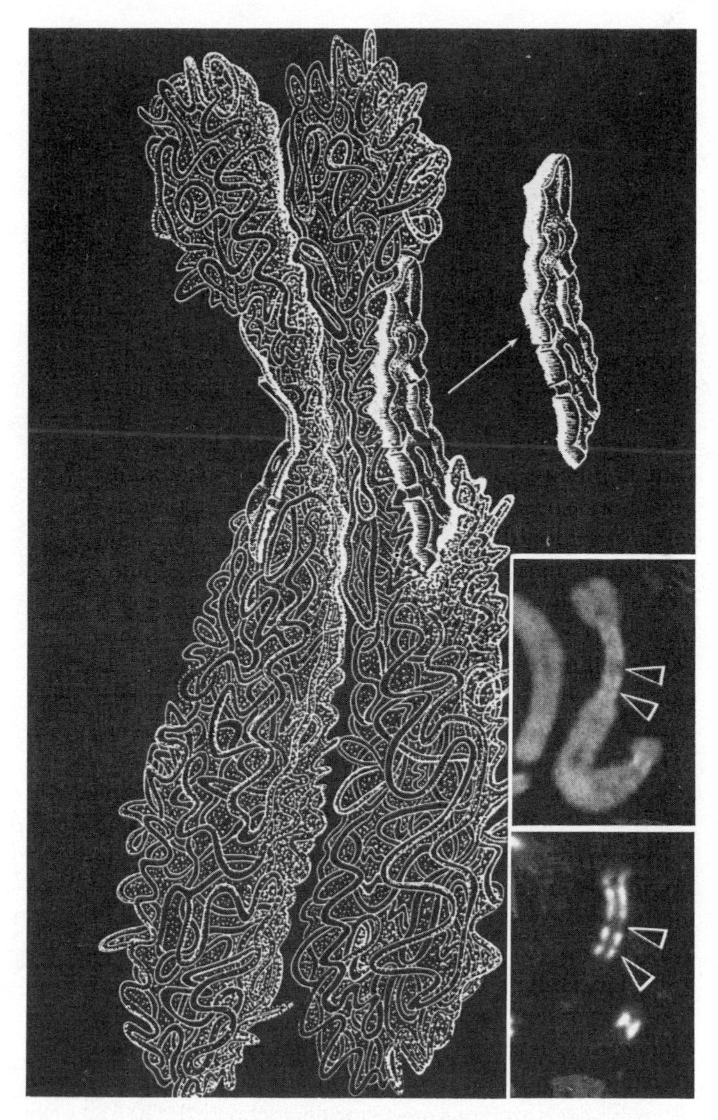

Figure 3.6. Diagram of X-chromosome of the Indian muntjac showing details of the outer plate. This image was derived from serial-section reconstruction. Inset shows X-chromosome double-stained with Hoechst (upper) and with kinetochore antibodies (lower). Note segmented nature of kinetochore (arrow heads). (From Brinkley *et al.*, 1984.)

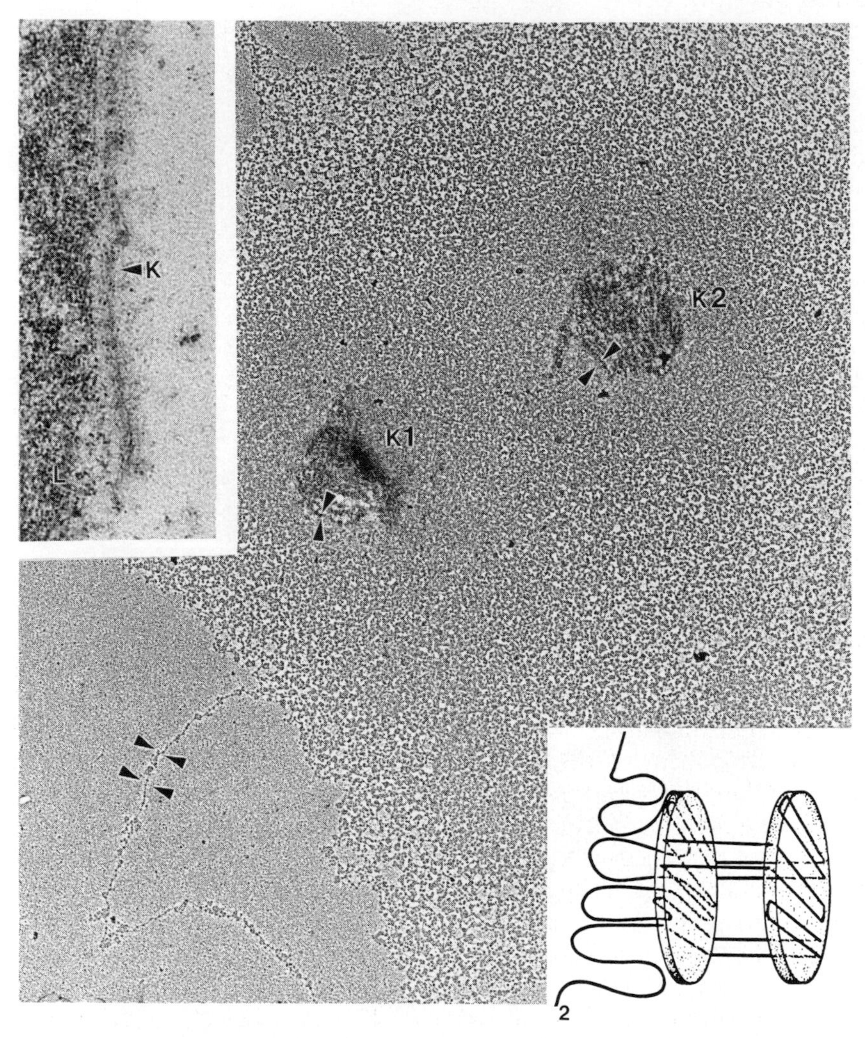

Figure 3.7. Electron micrograph of whole mount, shadowed preparation of centromeres. Metaphase chromosome was digested by restriction enzymes *Eco*RI and *Sau*96-1. Electron-dense regions are sister kinetochores (K1 and K2). Note pleated folds of 20–25 nm fibres in the kinetochores. Compare with beaded chromatin fibre of centromere with those of the kinetochore (arrow heads). Upper inset shows ultrathin section through kinetochore fibres revealing a plate-like structure (K). Lower inset is a diagrammatic interpretation of these images. (From Rattner, 1986.)

kinetochore plates. If so, the kinetochore plate is formed by a series of folded chromatin fibres and could be continuous with the DNA fibre of the chromosome, as shown in Fig. 3.7. Rattner's model is attractive because it is consistent with the interpretation of yeast centromeric chromatin, which also exists as a nuclease-resistant core that is continuous but structurally distinct from the flanking nucleosomal chromatin (see Bloom *et al.*, 1984). A similar but more simplified model was proposed much earlier by Brinkley and Stubblefield (1970).

An alternative interpretation of a folded-fibre model was proposed by Ris and Witt (1981), who concluded from high-voltage EM studies that the outer plate of the kinetochore is actually formed by the association of a series of 3 nm chromatin fibre loops arranged perpendicular to the plane of the plate. Although additional studies are necessary before kinetochore organization can be understood, these results suggest that the kinetochore plates are not separate structures, but may indeed be composed of, or integrated with, specialized chromatin fibres that are continuous with the chromosomal DNA fibres.

D. Cytochemistry and immunofluorescence

Relatively little is known about the chemical composition of the kinetochore, but several cytochemical studies suggest that the kinetochore is composed of DNA, RNA and proteins. Pepper and Brinkley (1980) reported that DNase I treatment of mammalian chromosomes specifically condenses the kinetochore plate, while RNase has little or no effect. Brinkley and co-workers (1980) and Ris and Witt (1981) found that the stability of the kinetochore is influenced by ionic strength. In the latter study, Ris and Witt reported that the outer layer is dissolved in low-molarity KCl, with microtubules terminating into a mass of chromatin fibres that stain positively for DNA. Most of the evidence for RNA in the kinetochore has come from the use of Bernhard's (1969) uranyl acetate–EDTA staining procedure, which preferentially stains cellular components that contain ribonucleoprotein (RNP). Rieder (1979a,b) used Bernhards' procedure to demonstrate the presence of RNP in the kinetochore of newt cells and PtK_1 cells. In the newt cells, the RNP staining was specifically localized to the inner plate of the trilaminar kinetochore and was eliminated by RNase treatment. Rieder has also shown that the corona of PtK_1 kinetochore contains an RNP staining component. To date, no apparent function for the kinetochore RNP has been identified.

In contrast to the above findings, Pepper and Brinkley (1980) were unable to identify an RNA component in the kinetochore. These investigators studied the effects of RNase T1 treatment both on the ability of the kinetochore to nucleate microtubules and on the ultrastructure of the kinetochore. However, as discussed by Pepper and Brinkley (1980), this study was limited to investigation of the outer kinetochore plate and did not rule out the possibility of RNA as a component of the inner plate.

Through the use of immunocytochemical staining, Pepper and Brinkley (1977) discovered that a polyclonal antibody against 6 S tubulin stained both dense layers of the kinetochore. Although the localization of tubulin in kinetochores might actually be due to soluble tubulin adhering to the plates artifactually, Pepper and Brinkley (1979) found that antitubulin blocked site-specific nucleation of microtubules and concluded that tubulin is indeed a structural and functional component of the kinetochore.

Recently, Mitchison and Kirschner (1985a) noted that the kinetochores of isolated mammalian chromosomes were brightly stained with tubulin antibody if the chromosomes were isolated from colcemid-arrested cells. In contrast, chromosomes isolated from cells arrested with vinblastine did not stain by antitubulin immunofluorescence. However, when these chromosomes were incubated with tubulin at concentrations necessary for spontaneous microtubule assembly, tubulin was bound specifically by the kinetochore. Using immunogold labelling, these investigators demonstrated tubulin localization on the fibrous corona at the outer plate of the kinetochore. Moreover, it was found that the bound tubulin remained even when kinetochore-associated microtubules were depolymerized, suggesting that the tubulin was not part of the microtubule structure. When the kinetochores were incubated with tubulin at concentrations above the critical concentration, microtubule nucleation occurred with a decreased lag and an increased slope, indicating that the kinetochore-bound tubulin facilitated microtubule nucleation at the kinetochore.

Thus a non-microtubule form of tubulin may indeed constitute a structural and functional component of the kinetochore, but additional experiments are necessary before strong conclusions can be made. Further comments on the interaction of microtubules with the kinetochore will be presented in a later section.

E. Biochemistry and molecular biology

Although kinetochores have been well characterized at the light- and electron-microscopic level, their molecular composition remains uncertain. This has been due in part to the absence of suitable techniques for obtaining enriched fractions of centromeres from metaphase chromosomes and the lack of specific probes for identifying kinetochore-specific molecules. From a limited knowledge based largely on cytochemistry, it is assumed that the kinetochore is composed of a complex assortment of macromolecules that play both a structural and functional role. Only recently has limited progress been made in characterizing the biochemistry and molecular biology of the centromere/kinetochore region and the initial studies have been confined largely to yeast chromosomes.

1. *Centromeric DNA*

Although cytochemical experiments have hinted at the possible presence of DNA in kinetochores (for review, see Rieder, 1982), a milestone in the field was accomplished when centromeric DNA (CEN-DNA) from the yeast *Saccharomyces cerevisiae* was cloned and sequenced in the laboratory of Carbon and co-workers (for references, see Clark and Carbon, 1985; Bloom *et al.*, 1984). When fragments of CEN-DNA from chromosomes I and III (CEN 1 and CEN 3) were introduced into yeast on autonomously replicating plasmids, the plasmids were maintained at low copy number, indicating mitotic stability. The CEN-DNAs isolated from five chromosomes of yeast have now been studied and all have the common ability to stabilize autonomously replicating plasmids in mitosis and to direct segregation in a Mendelian fashion during meiosis. The nucleotide sequences of yeast DNA fragments carrying CEN 3, CEN 4, CEN 6 and CEN 11 have now all been determined. All contain four elements that indicate sequence homologies between the different chromosomes. The four elements appear to be conserved in their nucleotide sequence and all display a "core" 220–250 base pair (bp) segment that is resistant to nuclease digestion and is completely homologous in three regions referred to as elements I (14 bp), III (11 bp) and IV (10 bp). The structure of the nuclease-resistant core has been partially defined by chromatin mapping studies (Bloom *et al.*, 1983) and the 220–250 bp segment was found to be in a unique chromatin configuration. The nucleosome core particle of conventional chromatin consists of 160 bp of DNA wrapped around histone proteins to give rise to a cylindrical particle with a diameter of 11 nm and a height of approximately 5 nm. Bloom and co-workers estimated the yeast centromeric core to be folded into a chromatin particle 15–20 nm in diameter. Given that microtubules are about 20–24 nm in diameter, and in yeast make direct contact with the chromatin fibres (Peterson and Ris, 1976), Bloom *et al.* (1983) proposed that the nuclease-resistant core of yeast might serve as a microtubule attachment site and thereby as a "primitive" kinetochore.

Centromeric DNA sequences have also been obtained for two of the three chromosomes (chromosomes I and II) of the fission yeast *Schizosaccharomyces pombe* (Nakaseko *et al.*, 1986). Information on the centromere of chromosome III has also been obtained by creating "mini-chromosomes" consisting essentially of just the centromere region (Niwa *et al.*, 1986). The chromosomes of *S. pombe* are, on average, 6-fold larger than those of *S. cerevisiae* and condense at mitosis. Perhaps not surprisingly, therefore, the centromeres are also much larger and more complex. Interestingly, they contain at least two classes of repeated sequence, one present within the centromere of a particular chromosome, the other located in the centromere of all three chromosomes (Nakaseko *et al.*, 1986).

2. *Kinetochore-associated proteins: analysis by human autoantibodies*

In 1981, a discovery in the rheumatology clinic had a serendipitous and significant impact on mitosis research and specifically on our knowledge of the centromere/kinetochore of eukaryotic chromosomes. Moroi *et al.* (1980) reported that sera from patients with a variant of the autoimmune disease progressive systemic sclerosis (scleroderma) (CREST variant) contained antibodies that bound selectively to the centromere of metaphase chromosomes and to foci within interphase nuclei (see also, Fritzler *et al.*, 1980; McCarty *et al.*, 1983; Ayer and Fritzler, 1984; Spowart *et al.*, 1985). Initially, CREST serum was identified as having "anticentromere antibodies" (ACA) because of the localization of stain on metaphase chromosomes as seen by indirect immunofluorescence (Fig. 3.8A). However, Brenner *et al.* (1981) demonstrated by immunoelectron microscopy that CREST antisera bind specifically to the inner and outer layers of the trilamellar plates of the kinetochore (Fig. 3.8B). Thus, CREST serum should be more appropriately identified as containing "antikinetochore antibodies" (AKA). In the same study, Brenner *et al.* also identified specific foci in interphase nuclei and defined them as "prekinetochores" owing to the numerical relationship to kinetochores on metaphase chromosomes (Fig. 3.9).

The binding of CREST antibodies to the kinetochore plates was significant because it clearly differentiated the centromere from the kinetochore; a demarcation indicated earlier on the basis of morphology and function but one often ignored in the literature (see Section II, above). The availability of antibodies that reacted specifically with the kinetochore plates opened a number of new approaches. For the first time, it was possible to identify the centromere/kinetochore region of chromosomes in interphase and to determine the approximate time of replication of this region during the cell cycle (Brenner *et al.*, 1981), to identify the centromere/kinetochore region in highly differentiated cells such as sperm (Brinkley *et al.*, 1986), and to define cyclic changes in the prekinetochores of some mammalian cells during the cell cycle (Brinkley *et al.*, 1984).

The CREST antiserum also provided a probe to begin to investigate the biochemistry of the kinetochore, and a family of centromere-associated proteins was discovered by this approach. Using immunoblotting and, in some cases, the affinity purification technique of Olmsted (1981), a number of polypeptides ranging in molecular weight from 14 to 140 kDa have been

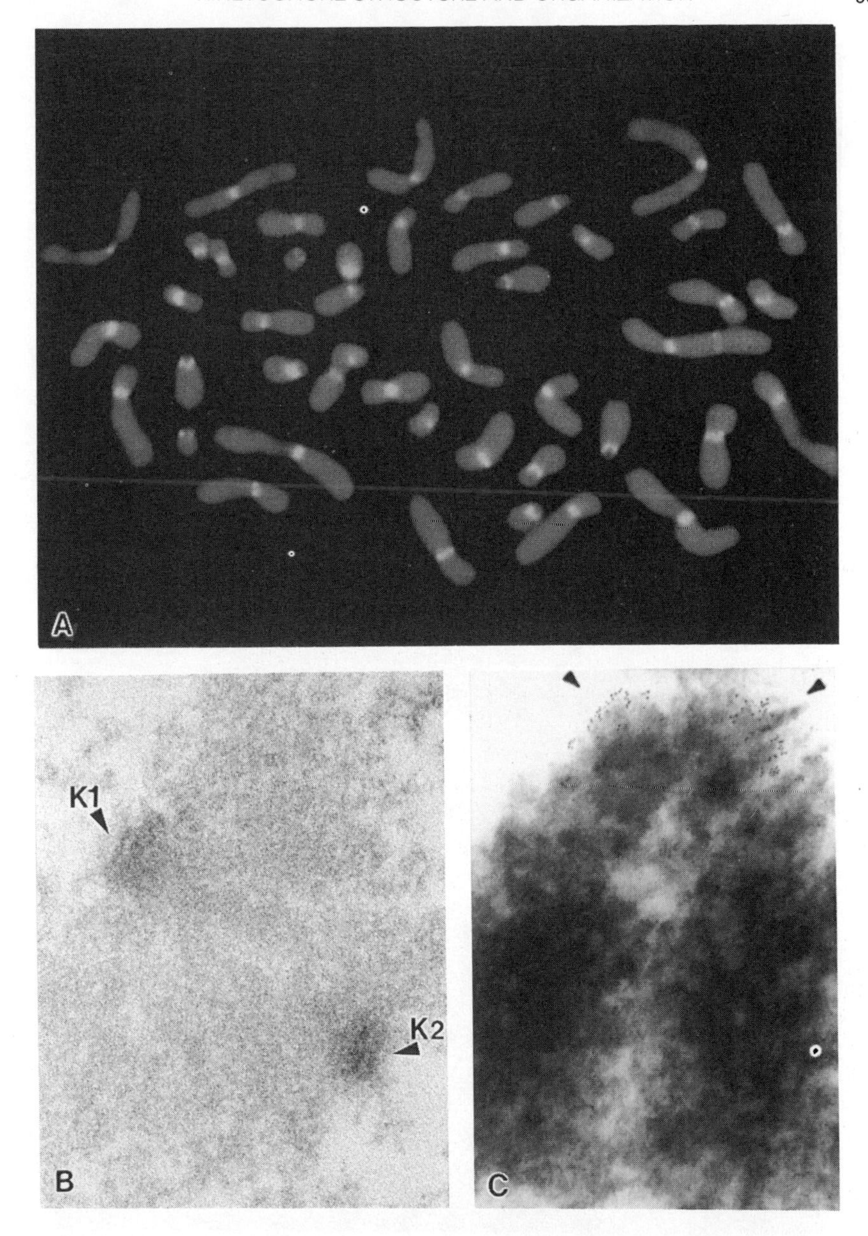

Figure 3.8. Chromosomes stained with antikinetochore serum from scleroderma CREST patients. (A) Human metaphase chromosomes with a 45 XX -9, -11 tdic (9p:11p) composition. (From Merry *et al.*, 1985.) (B) CREST immunoperoxidase staining confined to trilamellar kinetochore plates (K1 and K2). (From Brenner *et al.*, 1981.) (C) Semi-thick sections of CHO metaphase chromosome CREST immunogold staining of kinetochore plate (arrow heads). (From Valdivia *et al.*, 1986.)

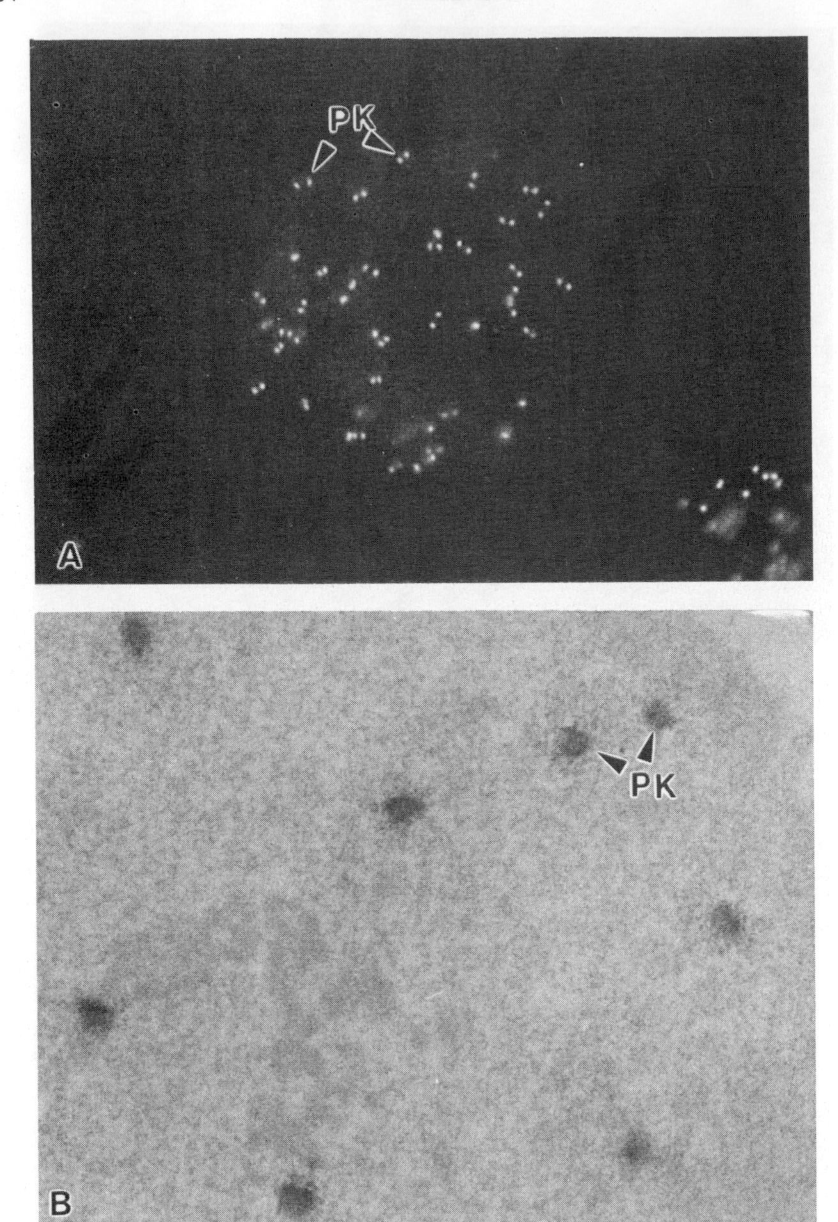

Figure 3.9. Interphase "prekinetochores" (PK) stained with antikinetochore antiserum. (A Double spots show replicated prekinetochore from Chinese muntjac nucleus. (B) Immunoperoxidase antikinetochore stained prekinetochores as seen by transmission electron microscopy of PtK_1 cell nucleus. (From Brenner *et al.*, 1981.)

Table 3.1.

Polypeptides (kDa)	Tissue	References
14, 20, 23, 34	HeLa cells	Cox *et al.* (1983)
17, 80, 140	HeLa cells	Earnshaw *et al.* (1985)
50	Rat	Earnshaw *et al.* (1985)
15, 33	Testis (trout) Thymus (rabbit)	Ayer and Fritzler (1984)
19.5	HeLa cells	Guldner *et al.* (1984)
18, 80	HeLa cells	Valdivia *et al.* (1986)
70	Thymus (rabbit, calf)	Nishikai *et al.* (1984)

described (Table 3.1). Although the localization of all of these antigens in the kinetochore has not been determined, three polypeptides, 17–18 kDa (CENP-A), 80 kDa (CENP-B), and 140 kDa (CENP-C) share antigenic determinants and are probably true centromeric proteins (Earnshaw and Rothfield, 1985). The CENP-A and CENP-B antigens are consistently detected in isolated chromosomes and CENP-B is found in a nuclease-resistant fraction containing isolated and partially purified kinetochores (Valdivia and Brinkley, 1985). Palmer and Margolis (1985) and Palmer *et al.* (1987) determined the 18 kDa CREST antigen to be tightly bound to interphase chromatin of mononucleosome size and characterized CENP-A as a histone-like component. Although the function of the CREST antigens in kinetochores is not known, one report indicates that CREST serum interferes with microtubule nucleation by kinetochores in a lysed cell system (Cox *et al.*, 1983) and Balczon and Brinkley (1987) identified the 80 kDa antigen as a nearest neighbour to kinetochore-bound tubulin.

Cloned cDNAs for the CENP-B antigen have recently been isolated by screening a human endothelial cell λgt11 expression library with scleroderma CREST antiserum (Earnshaw *et al.*, 1987). Sequence analysis and genomic blotting of the cDNA indicated that CENP-B contained two highly acidic subdomains and was encoded by a single human chromosomal locus. Although the location of CENP-B in the kinetochore has not yet been established, the acidic nature of the molecule would argue against it being a major DNA-binding protein. Its two large acidic domains may facilitate binding to histones or a histone-like protein such as CENP-A (Earnshaw *et al.*, 1984, 1985; Valdivia and Brinkley, 1985; and Palmer *et al.*, 1987). Obviously, further cDNA analysis of the CREST antigens could be extremely valuable in determining the structure and function of this interesting family of centromeric proteins.

The identification of kinetochore-associated antigens by CREST antiserum is a significant achievement that should ultimately lead to characterization of the biochemistry and molecular biology of the centromere/kinetochore region of the mammalian chromosome. Efforts are underway in several laboratories to select cDNAs for other CREST antigens. Hopefully, such an approach will lead to a detailed characterization of other proteins in the kinetochore and the identification of their corresponding genes.

3. *Fractionation of the kinetochore/centromere*

The centromere region of mammalian metaphase chromosomes, like that of *S. cerevisiae*, is composed of chromatin that is highly resistant to digestion with DNase (Rattner *et al.*, 1978). In addition, the mammalian centromere is resistant to disruption by hypotonic solutions (Brinkley *et al.*, 1980; Ris and Witt, 1981), EDTA, NaOH and trypsin (Comings *et al.*, 1971). Only recently have investigators succeeded in selectively fractionating this region of the metaphase chromosome. The procedure of Valdivia and Brinkley (1985), as shown in Fig. 3.10, begins with metaphase chromosomes isolated from synchronized mitotic cells in hexylene glycol buffer according to the procedure

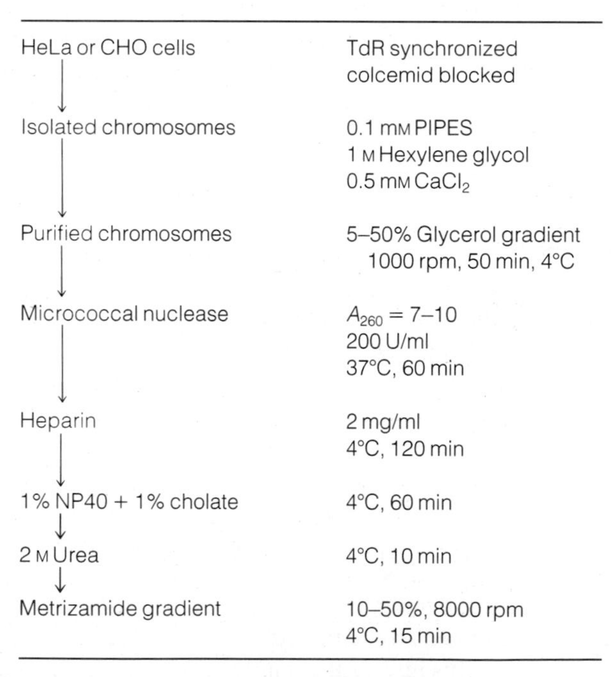

Figure 3.10. Protocol for the fractionation of kinetochores from isolated metaphase chromosomes. (From Valdivia and Brinkley, 1985.)

described by Wray and Stubblefield (1970). Low-speed glycerol gradient centrifugation was used to free chromosomes from cytoplasmic components and more rapidly sedimenting material such as nuclei, aggregated chromosomes and cytoplasmic debris. After purification, chromosomes were extensively digested with micrococcal nuclease to render mononucleosome-size DNA. Later, heparin extraction was carried out essentially as described by Bornens (1977) to release all histones, leaving chromosomal scaffolds or cores. The remaining pellet was further treated with a mixture of Nonidet P-40 and cholate to release aggregates of dehistonized chromosomes and to remove non-histone proteins resistant to heparin treatment. Finally, a brief treatment with 2M urea was used to dissociate the remaining chromosomal structures that contained kinetochores as revealed by immunofluorescent staining with CREST antiserum. The latter material was layered onto a metrizamide density gradient to further enrich for immunoreactive components and to separate out non-reactive chromatin.

During various steps of the fractionation procedure, DNA, histones and non-histone proteins were released (Fig. 3.11D), but the remaining material strongly reacted with CREST antiserum as shown in Fig. 3.11B. When examined by EM immunoperoxidase, each kinetochore is seen as a discrete darkly stained mass associated with less-dense fibres of the chromosome (Fig. 11C). After urea treatment, the kinetochores could still be detected but their size and shapes varied considerably, indicating probable rearrangement of morphology. Nevertheless, these observations attest to the amazing structural stability of the kinetochore, which even after extensive treatment with up to 4M urea for 1 hour and separation in metrizamide, still reacted with CREST antiserum and appeared relatively similar in morphology to structures seen originally on intact chromosomes.

Through quantitative analysis, Valdivia and Brinkley (1985) determined the amount of protein and DNA present in their most-enriched kinetochore fraction. Table 3.2 shows a typical preparation with more than 50% of total chromosomal protein solubilized after micrococcal nuclease and heparin treatment. This is not surprising, since it is well known that both treatments together release most of the DNA and histones from the chromatin (Bornens, 1977; Hay and Candido, 1983). The final pellet after urea treatment contained only 10% of the total starting protein and after sedimentation on the metrizamide gradient the fractions exhibited free kinetochores that represented less than 4% of total chromosomal protein and less than 1% total DNA with a small amount of RNA also present in the final pellet. Analysis by one-dimensional SDS gel electrophoresis showed that the kinetochore fraction contained very little, if any, histones (Fig. 3.11D). Obviously, the absence of histones in such preparations is by no means an indication that histones are not part of the intact functional kinetochore. In fact, an 18 kDa protein, one of two major

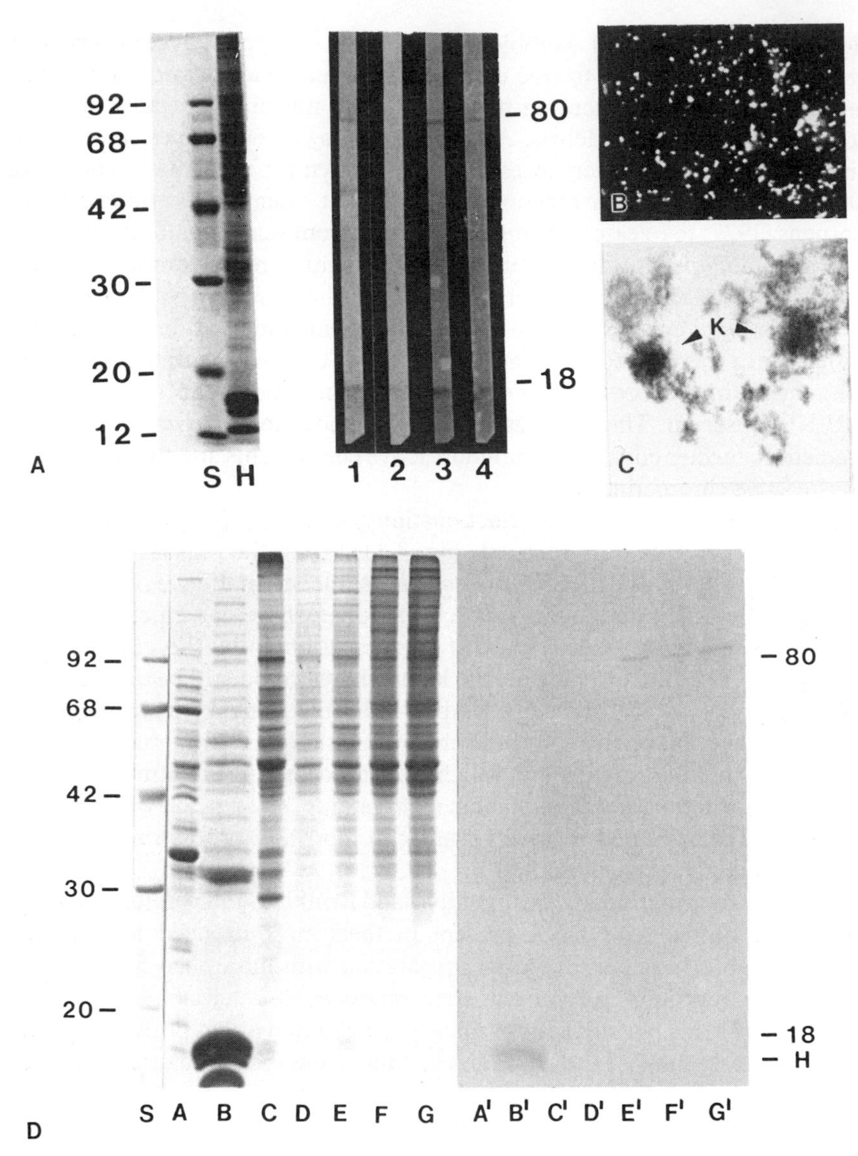

Figure 3.11. Western blot analysis of HeLa chromosomes. (A) Lane S, molecular weight markers. Lane H proteins from whole HeLa chromosomes. Lanes 1–4 indicate proteins that react with four different human CREST sera respectively. Two proteins of 80 and 18 kDa are recognized. (From Valdivia *et al.*, 1986.) (B)–(C) Immunofluorescence and immunoperoxidase-stained isolated kinetochores. (D) SDS electrophoretic analysis of kinetochore purification scheme (lanes A–G). Lanes A′–G′ show Western blots using antikinetochore antiserum. Note histones and 18 kDa proteins are solubilized after heparin extraction step, but 80 kDa remains with final pellet. (From Valdivia and Brinkley, 1985.)

Table 3.2. Quantification of a typical chromosomal preparation (from Valdivia and Brinkley, 1985)

Fraction	Total protein (mg)	Percent of total chromosomal proteins
Whole chromosomes	92.00	100.0
MNase supernatant	5.25	5.7
Heparin supernatant	45.75	49.7
Detergent supernatant	22.20	24.1
Urea supernatant	9.00	9.7
Urea pellet	10.00	10.8
Metrizamide fractions	3.20	3.4

kinetochore peptides recognized by autoantibodies from scleroderma CREST serum, is released into the supernatant after treatment with nuclease and heparin and appears to be chromatin-like in its solubility properties (Valdivia and Brinkley, 1985). The other major CREST antigen, an 80 kDa peptide, remains associated with the residual kinetochore fraction and resists solubilization with nuclease, heparin, detergent or urea. Therefore, the 18 kDa peptide appears to be histone-like while the 80 kDa protein is part of the metaphase chromosome "scaffold" (see also Earnshaw *et al.*, 1984).

Similar polypeptide compositions were observed in fractionated kinetochores from both Chinese hamster ovary (CHO) and HeLa cells. The major protein in the kinetochore fraction was actin (42 kDa), which is often found in isolated chromosome preparations (Sanger, 1975; Gooderham and Jeppesen, 1983) and in highly purified chromosomes and chromosome scaffold preparations (Lewis and Laemmli, 1982). Obviously, the presence of actin in kinetochores is uncertain at this time and it is difficult to rule out the possibility that kinetochore-associated actin is a cytoplasmic contaminant. Actin-like microfilaments have been observed on the outer plate of kinetochores and on kinetochore microtubules of chromosomes of filamentous green algae (Schibler and Pickett-Heaps, 1980).

IV. Kinetochore–Microtubule Interactions in the Mitotic Spindle

The conclusion by Schrader (1953) that the kinetochore is an "element of fundamental importance in movement of chromosomes" is widely supported today, but its role in movement is still not very well understood. It is obvious from studies of both living as well as fixed cells that the kinetochore is the principal site on chromosomes for the attachment of spindle fibres and that such attachment is essential for chromosome alignment at metaphase and segregation at anaphase (see Rieder, 1982; Brinkley *et al.*, 1985). Aside from this obvious conclusion, however, many questions remain unanswered concerning the role of the kinetochore in force production and movement. The kinetochore is only one component of a complex mitotic apparatus consisting of spindle poles, interpolar fibres, kinetochore fibres, and a cascade of regulatory and force-producing molecules and ions that are only beginning to be characterized. Proceeding with the well-founded notion that microtubules are an essential feature of kinetochore function, we will evaluate the kinetochore as a microtubule organizing centre (MTOC), as a site favourable for tubulin binding and microtubule assembly, and as a region of chromosomes which may attract and "capture" the growing ends of microtubules that have their origins at the spindle poles or elsewhere.

The dissolution of the nuclear envelope at the end of prophase signals the initial association of microtubules with the kinetochore and the onset of prometaphase or congressional movement. Such movements are often erratic and oscillatory and are apparently characterized by unstable kinetochore–microtubule interactions. The entire period of congressional movement is characterized by a "trial-and-error" phase in which chromosomes undergo orientation and reorientation until they finally achieve the proper alignment on the metaphase plate. Apparently, many orientations are inappropriate and the attachment of microtubules to the kinetochore is frequently unstable throughout prometaphase (Dietz, 1958; Nicklas, 1985). When alignment is finally achieved at metaphase, the coupling between spindle microtubules and the kinetochore becomes much more persistent and stable. According to Nicklas (1985), "The most frequent outcome of the initial encounter of chromosomes with the spindle is an imperfect arrangement of kinetochore microtubules . . . . Equal chromosome distribution results from a combination of (a) dynamics that favour stumbling on the appropriate state more or less by chance with (b) conditions that promote it more directly, such as polarized kinetochore position and, in some instances, by chromosome movement." Thus, any model designed to characterize kinetochore–microtubule interactions must take into account the labile nature of this attachment during congressional chromosome movement.

A. Origin of kinetochore microtubules

Examination of the dividing cell by electron microscopy shows that the mitotic spindle contains at least three classes of microtubules based upon their point of origin (Witt *et al.*, 1980):

1. those that have one end attached to the kinetochore and the other free or attached to a spindle pole;
2. microtubules that have one end attached to the centrosome or polar structures of the spindle and the other end free (asters) or directed to the opposite pole (interpolar fibres);
3. occasional microtubules with both ends free.

Recent assembly/disassembly kinetics of free-ended microtubules suggests that they may be relatively rare and short lived (Mitchison and Kirschner, 1984b).

In spite of a considerable depth of knowledge of spindle ultrastructure, controversy still exists concerning the origin of kinetochore microtubules (McDonald, this volume). Whether microtubules arise at the kinetochore or elsewhere is a central problem in understanding kinetochore function. Several investigators have concluded that kinetochores attach to microtubules that have their origins at other locations, usually the centrosome (Inoué and Ritter, 1978; LaFountain and Davidson, 1979; Nicklas *et al.*, 1979; Pickett-Heaps and Tippit, 1978; Paweletz, 1974). According to this view, the kinetochore is a receptor with unique properties to attract and bind the growing ends of microtubules that originate elsewhere. Others argue that kinetochores maintain the capacity to initiate tubulin polymerization and are therefore capable of generating, in part, their own microtubules (Inoué, 1964; Roos, 1976; Brinkley *et al.*, 1967; Brinkley and Stubblefield, 1970; Witt *et al.*, 1980). A third view holds that kinetochores attach to short preformed microtubules which then serve as nucleating seeds for the assembly of longer kinetochore microtubules (Bajer and Mole-Bajer, 1972; DeBrabander *et al.*, 1980). Actually, all three mechanisms have now been demonstrated experimentally.

Several experimental approaches have been used to investigate the issue of the origin of kinetochore microtubules. Efforts to disrupt the centrosome by physical displacement (Dietz, 1966) or by laser irradiation (Berns *et al.*, 1977) have generally supported the notion that kinetochores nucleate their own microtubules. However, such experiments are inconclusive because they fail to rule out the possibility that residual polar components remain that could continue to nucleate microtubules. Other experimental approaches have included disruption of spindle fibres with physical agents such as cold temperatures, hydrostatic pressure or colchicine-like drugs and then following the origin of microtubules during recovery (Pease, 1946; Roth, 1967; Brinkley *et al.*, 1967; Brinkley and Stubblefield, 1970; Goode, 1973; Alov and Lyubski, 1974; Witt *et al.*, 1980; Rieder and Borisy, 1981). These investigations have generally supported the notion that kinetochores can serve as microtubule assembly sites.

A study by Witt *et al.* (1980), was especially convincing. These investigators used high-voltage electron microscopy (HVEM) to analyse thick sections of Chinese hamster ovary cells that had been arrested with Colcemid and then allowed to recover for various periods of time prior to fixation. The experiments were designed so that kinetochores had never been associated with microtubules, thereby eliminating the argument that they were "experienced" or conditioned prior to nucleation. When examined briefly after recovery, short microtubules were oriented at various angles with respect to the kinetochore plates. Later, long parallel bundles formed from the kinetochores and these were clearly independent of the centrosome. Although there can be little doubt that these experiments clearly showed that kinetochores are capable of initiating microtubule polymerization, they can be challenged on the basis that high concentrations of drug were used and that conditions of the experiment departed significantly from what might be expected in normal cells.

Utilizing more gentle conditions, Rieder and Borisy (1981) subjected PtK_1 cells to low temperatures (6°C) that resulted in the disruption of the nuclear envelope and blockage of spindle microtubule formation. By re-warming the cells to 18°C, the formation of spindle fibres could be observed by HVEM in cells fixed at various times after recovery. This study showed that the centrosome generated microtubule arrays well before microtubules became associated with the kinetochore. Moreover, the order in which kinetochores attached to microtubules was influenced by the proximity to a centrosome, with kinetochores nearest to a centrosome being the first to attach. Similar observations have also been made in untreated PtK_1 cells during mitosis (Roos, 1973, 1976). In most cases, microtubules extending from kinetochores were oriented along an axis between the kinetochore and a centrosome. Kinetochore-to-kinetochore orientations seen briefly in cells recovering from drug treatments (Witt *et al.*, 1980; Brinkley *et al.*, 1985) were not reported in cells recovering from cold arrest. In general, these experiments supported the involvement of the centrosome in kinetochore microtubule associations, thereby favouring the capture hypothesis.

The notion that kinetochores can nucleate the assembly of microtubules was given additional experimental support in the mid 1970s when several laboratories demonstrated that isolated chromosomes and chromosomes in lysed cells could initiate the growth of microtubules from their kinetochores when incubated with exogenous brain tubulin in an appropriate *in vitro* assembly buffer (McGill and Brinkley, 1975; Telzer *et al.*, 1975; Snyder and McIntosh, 1975; Gould and Borisy, 1978; Pepper and Brinkley, 1979). Although these experiments were technically impressive, many argued that the conditions for isolating chromosomes or extracting cells left residual tubulin or microtubule fragments on kinetochores, which favoured nucleation. Nevertheless, these pioneering experiments opened the way for the *in vitro* analysis of microtubule assembly sites in cells.

B. Polarity of kinetochore microtubules

Microtubules are polar organelles and their association with MTOCs are likewise polar. Using techniques for morphological detection of microtubule polarity developed by Heidemann and McIntosh (1980), Euteneuer and McIntosh (1981), Borisy (1978) and Haimo *et al.*(1979), it was clearly shown that microtubules in the half-spindle of the mitotic apparatus, including both kinetochore and interpolar tubules, were all of the same polarity. The minus, or slow-growing ends, of microtubules were embedded in the centrosome while the plus, fast-growing ends extended toward the equatorial plate. Consequently, those microtubules that were bound to the chromosomes all had their plus ends embedded in the kinetochore. Obviously, this uniform polar arrangement of spindle microtubules favours the capture hypothesis since the most reasonable explanation for the binding of plus ends to the kinetochores is that they had their origin at the centrosome. However, this observation does not rule out the possibility that microtubules can be assembled at the kinetochore either by adding tubulin subunits to the ends of centrosome-derived microtubules or by the spontaneous nucleation of new microtubules that could elongate with their negative ends distal to the kinetochore. As described below, both modes of assembly have been demonstrated experimentally in *in vitro* models.

Although the question of the origin of kinetochore microtubules remains unresolved, recent innovations in the study of kinetochores *in vitro* provide considerable new insight into the dynamics of kinetochore-associated microtubules. Utilizing a modification of the isolation procedure of Lewis and Laemmli (1982), Mitchison and Kirschner (1985a,b) extracted chromosomes from CHO cells with "functional" kinetochores capable of organizing microtubules *in vitro*. This superb model not only enabled these investigators to analyse the interaction of phosphocellulose purified tubulin with kinetochores, it also led to an experimental analysis of both microtubule *nucleation* and microtubule *capture*.

C. Microtubule nucleation by kinetochores

When isolated chromosomes were incubated with tubulin at concentrations above the critical concentration necessary for spontaneous microtubule assembly, microtubules grew from the kinetochores (Mitchison and Kirschner, 1985a). These investigators noted that kinetochores, unlike centrosomes, fail to nucleate assembly below the steady-state concentration for free microtubules in their system. The polarity of nucleation was determined by briefly pulsing the preparations with biotinylated tubulin followed by fixation and processing for indirect immunofluorescence using mouse antibiotin antibodies to detect the new subunits. Unlike centrosomes, which displayed 95% of their plus ends distal to the organelle, kinetochore-nucleated microtubules were of mixed polarity with both plus and minus ends distal to the kinetochore plates. Another important feature of these experiments was the convincing demonstration of irreversible tubulin binding by the corona. Tubulin binding by the kinetochore may have important implications concerning reactions with microtubules as suggested by the earlier experiments of Pepper and Brinkley (1977, 1979). Mitchison and Kirschner (1985a) concluded, however, that their experiments argued against kinetochore nucleation as an efficient means of generating microtubules at prometaphase for the following reasons:

1. the mixed polarity of kinetochore microtubules nucleated *in vitro* is different from that reported in intact spindles (see Section IV.B, above); and
2. the inefficiency of kinetochore nucleation as compared to centrosomes, which nucleate well below the steady-state concentration (Mitchison and Kirschner, 1984a) and with uniform polarity.

D. Microtubule capture by kinetochores

In a companion paper, Mitchison and Kirschner (1985b) investigated the interaction of preformed microtubules with kinetochores *in vitro*. Capitalizing on the fact that various organelles, such as centrosomes, fragmented axonemes and preformed microtubule "seeds", can nucleate microtubule assembly at tubulin concentrations well below the critical concentration required for kinetochore nucleation, these investigators demonstrated the ability of kinetochore to capture and bind the ends of preformed microtubules. The capture of microtubules by kinetochores resulted in their becoming stabilized to dilution, suggesting a capping of the microtubule ends bound to the kinetochore. Preferential stability of kinetochore microtubules to cold temperatures has also been demonstrated in intact metaphase cells (Brinkley and Cartwright, 1975).

Microtubules assembled under the conditions of capture displayed a mobilized dynamic process at their captured ends when ATP was added to the system. Using biotin-labelled seeds to follow assembly dynamics at the kinetochore, these investigators demonstrated the translocation of labelled segments away from the kinetochores in the presence of unlabelled tubulin and ATP. Translocation was shown to be independent of microtubule assembly, leading the investigators to propose that kinetochore-associated ATPase activity could be involved in force production and movement (see Mitchison and Kirschner, 1985b, for detailed model).

E. Poleward movement at anaphase: dynamics at the kinetochore–microtubule interface

Although it is well established that the forces that move chromosomes to the pole act through the kinetochore, the actual source and location of the mitotic "motor" remains unknown. Kinetochore-to-pole microtubules get shorter as chromosomes move from the equatorial plate to the poles at anaphase. Thus, chromosomes could be passively "pulled" by forces located at the poles, "pushed" by forces acting along the spindle microtubules, or moved by forces generated at the kinetochore itself. Indeed, recent experiments have suggested that the kinetochore may play a much more direct role than previously thought in promoting chromosome movement (for review, see Nicklas, 1987). If kinetochores function as motors to power chromosome movement, the depolymerization of spindle microtubules along which movement occurs would probably take place at the kinetochore. The question of which end of the spindle microtubule subunits is added to or removed from was addressed by Mitchison *et al.* (1986), who microinjected biotinylated tubulin into cultured cells during mitosis. After appropriate intervals of time, the cells were permeabilized and exposed to 5 nm gold-labelled biotin and examined by transmission electron microscopy.

When cells were injected at metaphase and permeabilized 10 seconds later, microtubules displayed short labelled segments at their kinetochore ends. If the cells were permeabilized 100–300 seconds after injection, the labelled segments were even longer. If the cells were injected in metaphase and then allowed to progress to anaphase before analysis, a significant portion (60%) of the shortened microtubules were unlabelled at their kinetochore ends. These elegant experiments suggest that tubulin subunits can be added into microtubules at the kinetochore ends when chromosomes are aligned on the equatorial plate. When poleward motion began at anaphase, tubulin subunits depolymerized at the kinetochore. Therefore, microtubule shortening, which must accompany chromosome movement to the poles, appears to be achieved by the loss of tubulin subunit at the kinetochore, as proposed earlier by Forer (1976) and Pickett-Heaps *et al.* (1982). In order to explain their observations, Mitchison and co-workers postulate two separate forces: one acting directly on the kinetochore, moving the chromosomes toward the metaphase plate, and the other a poleward force produced by molecules like the mechanochemical ATPase kinesin (Vale *et al.*, 1985a,b; Scholey *et al.*, 1985), acting at many sites along the microtubules. This interpretation is consistent with earlier data that suggest that two separate forces and two motors are necessary to move chromosomes (Pickett-Heaps and Bajer, 1978).

Further evidence that kinetochore microtubules depolymerize at the kinetochore end as chromosomes move poleward at anaphase was obtained by Gorbsky *et al.* (1987). Fluoresceinated tubulin was microinjected into mitotic cells *in vitro* during prophase and metaphase and the cells were allowed to progress into anaphase. As cells entered anaphase, they were irradiated by a bar-shaped laser beam in order to photobleach a small strip across the half-spindle between the kinetochore and pole. This treatment did not break microtubules in the half-spindle, as shown by antitubulin immunofluorescence, but merely bleached the fluorescein molecules in a small region. After the cells were lysed and fixed, the photobleached band was identified by staining with anti-fluorescein antibodies that would not bind to photobleached fluorescein. From these experiments Gorbsky *et al.* (1987) were able to show that chromosomes "approached and invaded" the marked domain, which remained stationary with respect to the poles. Thus, chromosomes moved along depolymerizing microtubules while the latter remained stationary distal to the kinetochore.

Kinetochores can function as autonomous "organelles" when experimentally detached from the centromeres of mammalian chromosomes. "Naked" kinetochores consisting essentially of the trilaminar plates and short fragments of centromeric chromatin interact with microtubules and undergo the complete repertoire of mitotic movements when uncoupled from the chromosome by clastogenic drugs (Brinkley *et al.*, 1988).

Collectively, these observations have suggested a "PacMan" analogy (Cassimeris *et al.*, 1987; Salmon, this volume) in which kinetochores attach to and power their way along microtubules that are being depolymerized at their plus ends as chromosomes are driven to the poles (see also Salmon, 1989, this volume).

Whether kinetochores contain a mechanochemical ATPase analogous to kinesin but capable of moving chromosomes in a (plus) to (minus) (retrograde) direction along spindle microtubules, remains to be determined. Recently, Paschal and Vallee (1987) have isolated and characterized MAP-1C, a dynein-like molecule that translocates vesicles along microtubules in a retrograde direction (see also Lye *et al.*, 1987) but as yet no such translocator has been identified at the kinetochore. Although it is attractive to think of the kinetochore as containing the mitotic motor, it could also function indirectly as a force transducer. Recent *in vitro* experiments by Koshland *et al.* (1988) support the early hypothesis of Inoué (1964) that the energy for anaphase chromosome movement is derived solely from microtubule depolymerization. Hill (1985) proposed a model whereby depolymerization at the plus end of microtubules at the kinetochore results in a biased diffusion of chromosomes to the poles. Koshland *et al.* (1988) suggested that a conformational change in the microtubule at the kinetochore-associated end, i.e. curling up of protofilaments, might exert a poleward force on the kinetochore. Obviously, a good deal more information is needed concerning the biochemical nature of the kinetochore–microtubule interface before the mechanism of chromosome movement can be determined.

The kinetochore, like the microtubules associated with it, is a polarized component that on one face binds tubulin and microtubules and on the other binds specifically to centromeric DNA. This bipartite interaction must require a unique arrangement of molecules that can achieve such polarized interactions and, at the same time, facilitate force production (or transmission) and chromosome movement. Based on current knowledge, the simplest molecular arrangement going from the outside face of the kinetochore inward might include (A) microtubules → (B) tubulin → (C) kinetochore protein → (D) DNA-binding protein → (E) centromere DNA. At present we know very little about C, D and E, but the experiments described in this section provide exciting new insight into A and B. Kinetochore microtubules bind to the outer face of the kinetochore by their plus ends. Moreover, microtubules that are anchored to the kinetochore can be either stable or labile and undergo dynamic assembly/disassembly at their bound ends. Therefore, models for kinetochore structure–function must accommodate simultaneous anchorage and microtubule assembly and disassembly (Hill, 1985; Gorbsky *et al.*, 1987; Koshland *et al.*, 1988). Interjected upon this complex arrangement is the possibility that force-producing molecules (kinetochore ATPase?) reside on or in the kinetochore to interact with the microtubule lattice or other cytoskeletal component to drive the chromosome to the poles at anaphase.

In order to identify proteins of the kinetochore that are nearest neighbour to bound tubulin, Balczon and Brinkley (1987) used the reversible cross-linking reagent dithiobis(succinimidyl propionate) (Lomant's reagent) to couple exogenous tubulin to the kinetochore of isolated metaphase chromosomes. After cross-linking, the chromosomes were digested with DNase I, solubilized in sodium dodecyl sulphate (SDS) and subjected to antitubulin affinity chromatography. A tubulin–kinetochore complex was specifically eluted and analysed by polyacrylamide gel electrophoresis (PAGE) and immunoblotting with scleroderma CREST antiserum. In addition to the tubulin probe, bands of 80, 110, 24 and 54 kDa were resolved in the complex. Moreover, the 80 kDa polypeptide reacted with CREST serum and probably represents the major centromeric antigen CENP-B (Earnshaw and Rothfield, 1985). The absence of the 18 kDa (CENP-A) centromeric antigen from the complex along with the fact that it can be extracted in high salt concentrations suggests that this peptide is a histone-like component that is not associated with microtubule binding to the kinetochore (Balczon and Brinkley, 1987). This initial study provided the first direct biochemical analysis of the kinetochore–microtubule interface and could lead to the identification of other peptides required for kinetochore function.

Much remains to be learned, but the availability of new techniques and probes for analysing microtubule dynamics at the kinetochore *in vitro* as well as *in vivo* should provide a much better understanding of the molecular organization and function of the kinetochore.

V. Compound Kinetochores: Evolution by Linear Fusion

As described in an earlier section, the kinetochore of the yeast *S. cervisiae* is characterized by the binding of a single microtubule to one site on metaphase chromosomes that is composed of a 3.5 kb of centromeric DNA of which 220–250 base pairs are thought to be sufficient to define a functional kinetochore (see references in Clarke and Carbon, 1985; Bloom *et al.*, 1984). Since only one microtubule attaches to the metaphase chromosome, the yeast may be thought of as having a "unit" kinetochore. That is, one in which a finite unit of DNA specifies the binding of a single microtubule. In contrast, most chromosomes of higher plant and animal cells bind dozens of microtubules and by simple analogy may be thought of as compound structures composed of multiple "units".

Recent studies of the chromosomes of a family of Asiatic deer may help explain how large complex kinetochores might have evolved from simpler units by a system of genomic reorganization and linear fusions (Brinkley *et al.*, 1984). The chromosomes of the Indian muntjac (*Muntiacus muntjak vaginalis*) are unique among mammals owing to their low diploid number ($2N = 6$♀, 7♂; Wurster and Benirschke, 1970) and large size. The kinetochore of the X chromosome is especially large owing to an X-autosome translocation, with the short arm being the actual X and the long arm a translocated autosome. The male has a sex chromosome complement of XY_1Y_2 with the Y_1 being the abnormal homologue of the autosome translocated to the X. The female contains an XX complement with two identical X chromosomes.

Another species, the Chinese muntjac (*M. m. reevesi*), has a chromosome number of $2N = 46$, with all numbers of the complement being small telocentric elements, strikingly different from the Indian muntjac. Phenotypically, however, the two deer look almost identical and are capable of producing healthy F_1 hybrids when bred in captivity (Shi *et al.*, 1980).

Based on a variety of taxonomic, cytogenetic and biochemical criteria, it has been argued that the Chinese muntjac is the ancestral species and through multiple centric and tandem fusions has given rise to the karyotype of the Indian muntjac (Wurster and Atkin, 1972; Matthey, 1973; Shi *et al.*, 1980; Schmidtke *et al.*, 1981; Merry *et al.*, 1983). Since most of the fusions have involved centromeric DNA, the structural genes are apparently maintained and therefore the morphology of the deer has remained largely unchanged. There have, however, been dramatic changes in the karyotype.

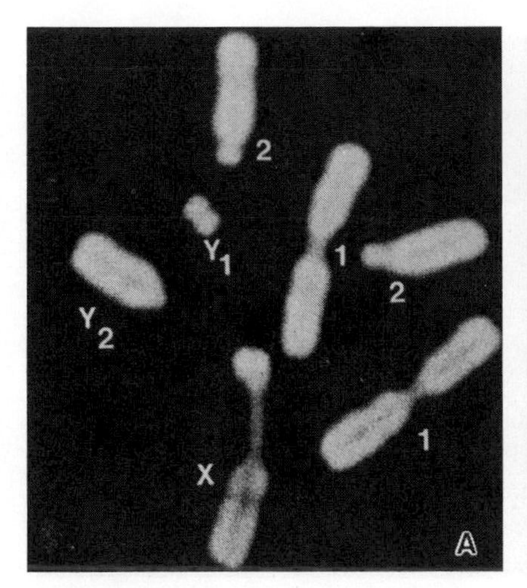

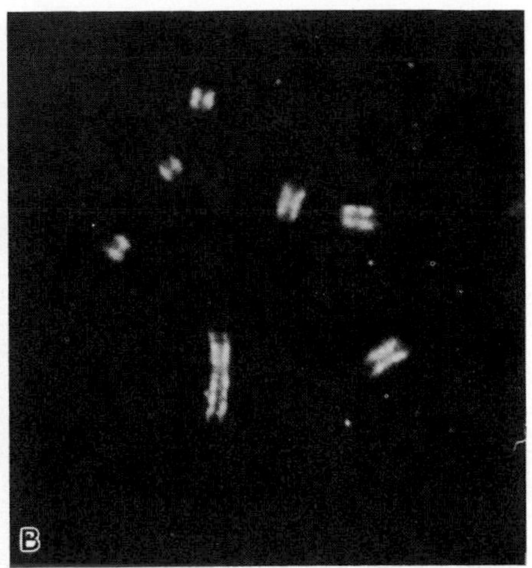

Figure 3.12. (A) Chromosomes of the male Indian muntjac deer $2N = 7$; and (B) the same chromosomes stained with CREST autoantibodies for kinetochores. (From Brinkley *et al.*, 1984.)

Using kinetochore-specific antiserum derived from scleroderma CREST patients, Brinkley and co-workers (1984) demonstrated the existence of what they termed "compound kinetochores" on chromosomes of the Indian muntjac. In addition, these investigators identified a "kinetochore cycle" in dividing muntjac cells that corresponded to stages in the cell cycle.

As shown in Figs. 3.12 and 3.13, the fluorescently stained metaphase chromosomes of the Indian muntjac are strikingly different in size and number from those of the Chinese muntjac. Examination of the kinetochores of the large X chromosomes suggests a segmented organization (Fig. 3.6). The most convincing argument for the subunit organization of kinetochores came from studies of prekinetochores in the nuclei of interphase cells. Shortly after telophase, when the Indian muntjac cells had entered the G_1 phase of the cell cycle, six (♀) or seven (♂) discrete fluorescent spots could be seen in each nucleus (Fig. 3.14A). Later, as the cells progressed into S-phase, each of the fluorescent prekinetochores appeared to "puff" and unravel, revealing multiple small subunits arranged like a string of beads (Fig. 3.14B,C). By the end of S-phase, many more fluorescent spots were seen than would be expected on the basis of the chromosome number (Fig. 3.14D). The spots were clustered and frequently aligned end to end. As cells entered G_2-phase in preparation for mitosis, the prekinetochores re-condensed into six or seven double fluorescent spots (Fig. 3.14E). The simplest interpretation of these images is that the decondensation of centromeric chromatin during interphase revealed multiple subunits associated with each of the prekinetochores.

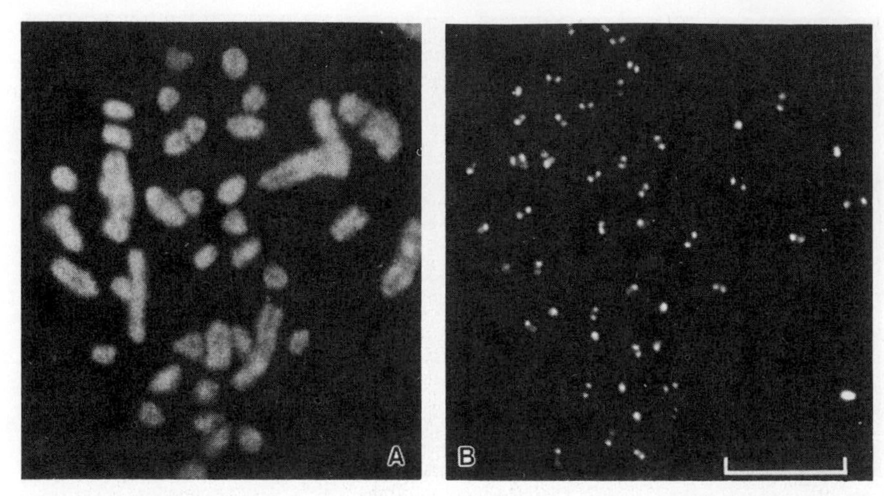

Figure 3.13. (A) Chromosomes of the female Chinese muntjac deer $2N = 46$; and (B) the same chromosomes stained with CREST autoantibodies for kinetochores. (From Brinkley *et al.*, 1984.)

When the interphase nuclei of the Chinese muntjac were stained, the cells in G_1-phase displayed approximately 46 randomly placed spots (Fig. 3.14F). During a later stage of the cell cycle, the prekinetochores aggregated into fewer, but larger, clusters, some of which looked very similar to the bead-like aggregates seen in the nuclei of Indian muntjac (compare Fig. 3.14D with 3.14I). Prior to mitosis, the clustered arrays disaggregated into randomly displaced double spots (Fig. 3.14J). Of particular interest was the one stage in interphase when the arrangements of the fluorescent spots were essentially identical in each species.

Examination of the metaphase kinetochores of the two species by electron microscopy failed to show any ultrastructural features that might suggest a fusion process (Brinkley *et al.*, 1984; Comings and Okada, 1970). When the total volume occupied by the kinetochores was measured, however, there was no significant difference between the two species, indicating little or no loss of kinetochore material during evolution.

Figure 3.14. Prekinetochore cycle in nuclei of Indian muntjac (A–E) and Chinese muntjac (F–J). Note 7 prominent spots in (A), each of which becomes enlarged in (B) and bead-like in (C) and (D). In (E) prekinetochores are double following replication. A similar cycle is noted in nuclei of the Chinese muntjac. Prekinetochores begin to cluster in (G) and (H) and become bead-like in (I) (note similarity to arrangement in (D)). Prekinetochores become double and randomly dispersed in (J). Linear fusion of "unit" kinetochores clustered as in (I) could have produced compound kinetochore as shown in Fig. 3.15. (From Brinkley *et al.*, 1984.)

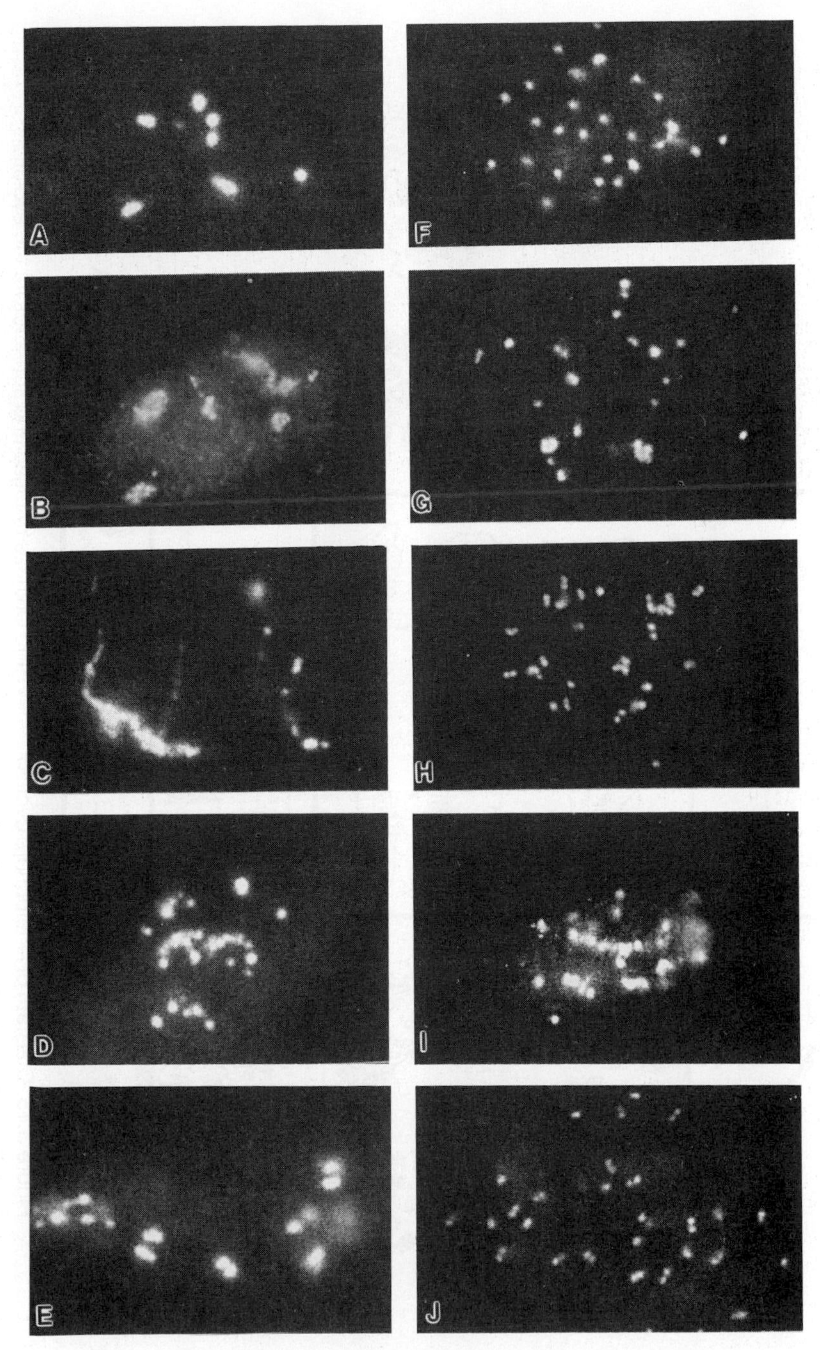

Figure 3.14.

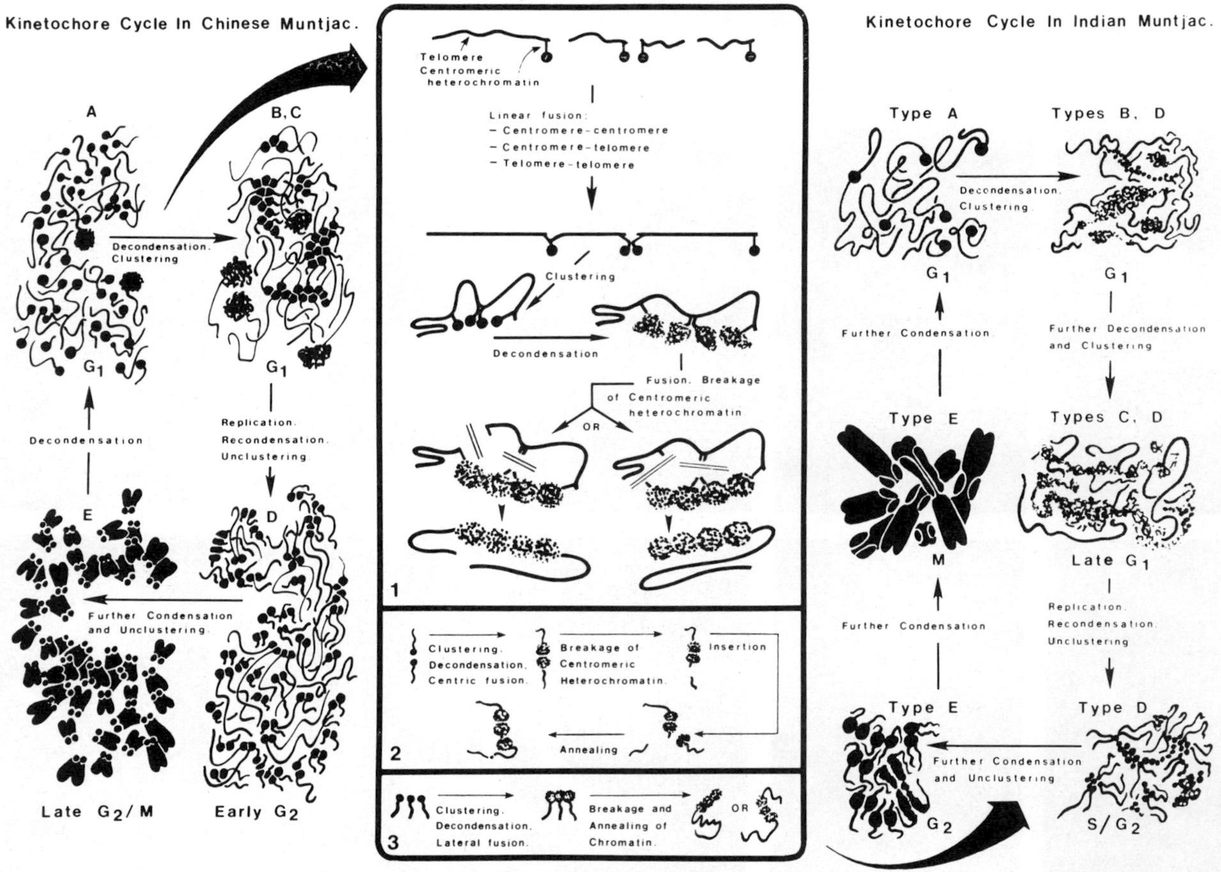

Figure 3.15. Diagram of kinetochore cycle of Indian and Chinese muntjac and schemes for the evolution of compound kinetochore. (From Brinkley *et al.*, 1984.)

As shown in Figure 3.15, several possible mechanisms can be proposed for the evolution of compound kinetochores without seriously disturbing genomic composition. The non-random movement and clustering of prekinetochores in interphase nuclei as seen in muntjac cells may have facilitated selective breakage and fusion of centromeres in a Robertsonian manner. If the breaks occurred preferentially within the inert centromeric heterochromatin, gene linkage would be altered, but the composition of the genome would be left intact. The resulting karyotype would contain the same genes carried on fewer, but larger chromosomes. Although the muntjac kinetochores may be unique, we propose that they may help to explain how "unit" kinetochores of small yeast chromosomes with only one microtubule may have evolved into "compound" kinetochores of larger chromosomes with numerous microtubules like those seen in mammalian cells (see model in Fig. 3.16). Recently, Bajer (1987)

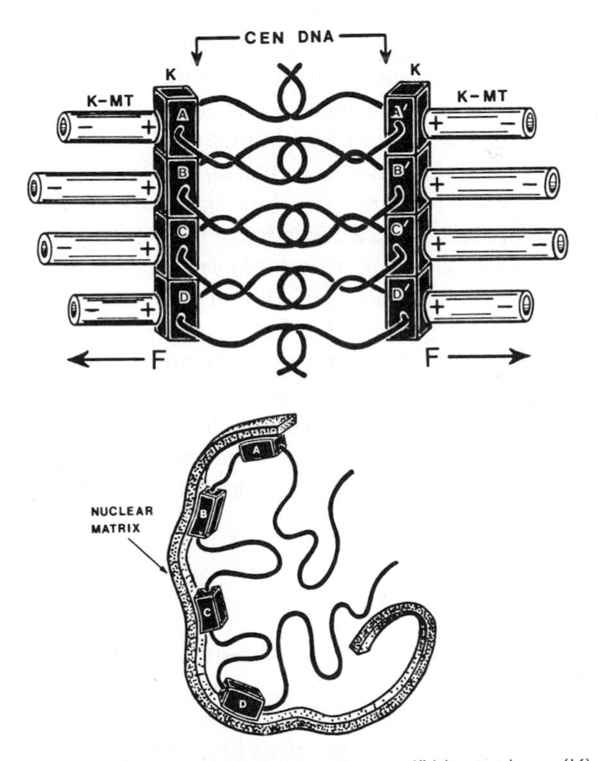

Figure 3.16. A schematic diagram of a "compound" kinetochore (K) of metaphase chromosomes. Microtubules (K-MT) are bound to "unit" kinetochore proteins A–D and A′–D′ by their plus ends. Proteins are bound to specific nucleotide sequence of centromeric (CEN-) DNA. Sister kinetochores are attached to each other by catenation of CEN-DNA loops (for rationale see Murray and Szostak, 1985). Forces (F) are applied to each sister kinetochore at metaphase. Lower diagram illustrates possible binding of prekinetochore proteins to the nuclear matrix during interphase.

reported that kinetochores of *Haemanthus* may also be composed of multiple subunits. Thus, in higher animal and plant cells, compound kinetochores may be the rule rather than the exception.

VI. Summary and Perspective

In many respects, the centromere/kinetochore region of eukaryotic chromosomes remains a black box whose contents must be identified and characterized before the mechanisms of force productions chromosome movement and segregation can be understood. A thorough knowledge of the structure–function relationship of this unique chromosomal locus would go far in defining the mechanism of mitosis. Progress has been slow, but the recent success in cloning CEN-DNA in yeast and the convincing demonstration that these sequences are essential for normal segregation in mitosis and meiosis has been a major milestone. The capacity to construct minichromosomes with functional centromere and telomere sequences may have considerable impact in the practical realm of gene therapy. This technology will very likely prove to be a powerful research tool in the hands of future mitosis researchers.

Obviously, much work remains to be done before we can link the emerging biology of yeast centromeres to a knowledge of the structure and behaviour of the complex centromere/kinetochore in higher eukaryotic cells. The development of new techniques for analysis of kinetochore microtubule assembly *in vitro* and *in vivo* may lead to a much better understanding of kinetochore function. The discovery of human autoantibodies with specificity for the kinetochores of mammalian chromosomes may help to characterize kinetochore-associated proteins and to screen libraries for cDNAs specific for these proteins. This should lead to the production of newer probes with even greater specificity with which to dissect the kinetochore and to define its function in mitosis and meiosis.

Acknowledgement

The authors wish to thank Cynthia Webster for typing the manuscript and Ann Harrell for editorial assistance and proofreading. We also acknowledge many friends and colleagues for helpful comments and stimulating discussions on kinetochore structure, function and evolution, especially Daniel Pepper, Conley Rieder, Andrew Bajer, T. C. Hsu and Raymond Zinkowski. This work was supported by a grant from HHS, NIH CA41424.

REFERENCES

Alov, I. A. and Lyubski, S. L. (1974). *Bull. Exp. Biol. Med.* **7**, 91–94.
Ayer, L. and Fritzler, M. (1984). *Mol. Immunol.* **21**, 761–770.
Bajer, A. S. (1987). *J. Cell Biol.* **43**, 23–34.
Bajer, A. S. and Mole-Bajer, J. (1972). *Int. Rev. Cytol. Suppl.* **3**, 1–271.
Balczon, R. D. and Brinkley, B. R. (1987). *J. Cell Biol.* **105**, 855–862.
Bernhard, W. (1969). *J. Ultrastruct. Res.* **27**, 250–265.
Berns, M. W., Rattner, J. B., Brenner, S. and Meredith, S. (1977). *J. Cell Biol.* **72**, 351–367.
Bloom, K. S., Fitzgerald-Hayes, M. and Carbon, J. (1983). *Cold Spring Harbor Symp. Quant. Biol.* **47**, 1175–1185.
Bloom, K. S., Amaya, E., Carbon, S., Clarke, J., Hill, A. and Yea, E. (1984). *J. Cell Biol.* **99**, 1559–1568.
Bokhari, F. A. and Godward, M. B. E. (1980). *Chromosoma* **79**, 125–136.
Borisy, G. G. (1978). *J. Mol. Biol.* **124**, 565–570.
Bornens, M. (1977). *Meth. Cell Biol.* **15**, 163–175.
Brenner, S., Pepper, D., Berns, M. W., Tan, E. and Brinkley, B. R. (1981). *J. Cell Biol.* **91**, 95–102.
Brinkley, B. R. and Cartwright, J., Jr. (1971). *J. Cell Biol.* **50**, 416–431.
Brinkley, B. R. and Cartwright, J., Jr. (1975). *Ann. NY Acad. Sci.* **253**, 428–439.
Brinkley, B. R. and Stubblefield, E. (1966). *Chromosoma* **19**, 28–43.
Brinkley, B. R. and Stubblefield, E. (1970). *Adv. Cell Biol.* **1**, 119–185.
Brinkley, B. R., Stubblefield, E. and Hsu, T. C. (1967). *J. Ultrastruct. Res.* **19**, 1–18.
Brinkley, B. R., Mace, M. L. and McGill, M. (1974). In *Electron Microscopy 1974* (ed. J. V. Sanders and D. J. Goodchild), pp. 248–249. Australian Academy of Science, Canberra.
Brinkley, B. R., Cox, S. M. and Pepper, D. A. (1980). *Cytogenet. Cell Genet.* **26**, 165–174.
Brinkley, B. R., Valdivia, M. M., Tousson, A. and Brenner, S. L. (1984). *Chromosoma* **91**, 1–11.
Brinkley, B. R., Tousson, A. and Valdivia, M. M. (1985). In *Aneuploidy* (ed. V. L. Dellarco, P. E. Voytek and A. Hollaender), pp. 243–267. Plenum, New York.
Brinkley, B. R., Brenner, S. L., Hall, J. M., Tousson, A., Balczon, R. D. and Valdivia, M. M. (1986). *Chromosoma* **94**, 309–317.
Brinkley, B. R., Zinkowski, R. P., Mollon, W. L., Davis, F. M., Pisegna, M. A., Pershouse, M. and Rao, P. N. (1988). *Nature* **336**, 251–254.
Buck, R. (1967). *J. Ultrastruct. Res.* **18**, 489–501.
Cassimeris, L. U., Walker, R. A., Pryer, N. K. and Salmon, E. D. (1987). *Bioessays* **7**, 149–154.
Clarke, L. and Carbon, J. (1985). *A. Rev. Genet.* **19**, 29–56.
Comings, D. E. and Okada, T. A. (1970). *Cytogenetics* **9**, 436–449.
Comings, D. E. and Okada, T. A. (1972). *Chromosoma* **37**, 177–192.
Comings, D. E., Avelino, E., Okada, T. A. and Wyandt, H. E. (1971). *Exp. Cell Res.* **67**, 97–110.
Cox, J. V., Schenk, E. A. and Olmsted, J. B. (1983). *Cell* **35**, 331–339.
DeBrabander, M., Geuens, G., Nuydens, R., Willebrords, R. and DeMey, J. (1980). In *Microtubule and Microtubule Inhibitors* (ed. M. DeBrabander and J. DeMey), pp. 255–268. North-Holland, New York.

Dietz, R. (1958). *Chromosoma* **9**, 359–440.
Dietz, R. (1966). *Chromosomes Today* **1**, 161–166.
Earnshaw, W. C. and Rothfield, N. (1985). *Chromosoma* **91**, 313–321.
Earnshaw, W. C., Halligan, N., Cooke, C. and Rothfield, N. (1984). *J. Cell Biol.* **98**, 352–357.
Earnshaw, W. C., Bordwell, C., Marino, C. and Rothfield, N. (1985). *J. Clin. Invest.* **77**, 426–429.
Earnshaw, W. C., Sullivan, K. F., Machlin, P. S., Cooke, Carol A., Kaiser, D. A., Pollard, T. D., Rothfield, N. R. and Cleveland, D. W. (1987). *J. Cell Biol.* **104**, 817–829.
Euteneuer, U. and McIntosh, J. R. (1981). *J. Cell Biol.* **89**, 338–345.
Forer, A. (1976). *J. Cell Biol.* **25**, 95–117.
Friedlander, M. and Wahrman, J. (1970). *J. Cell Sci.* **7**, 65–89.
Fritzler, M. J., Kinsella, T. D. and Garbutt, E. (1980). *Am. J. Med.* **69**, 520–526.
Godward, M. B. E. (1985). *Int. Rev. Cytol.* **94**, 77–105.
Goode, D. (1973). *J. Mol. Biol.* **80**, 531–538.
Gooderham, K. and Jeppesen (1983). *Exp. Cell Res.* **144**, 1–14
Gorbsky, G. J., Sammak, P. J. and Borisy, G. G. (1987). *J. Cell Biol.* **104**, 9–18.
Gould, R. R. and Borisy, G. G. (1978). *Exp. Cell Res.* **113**, 369–374.
Guldner, H. H., Lakomek, H. J. and Bautz, F. A. (1984). *Clin. Exp. Immunol.* **58**, 13–20.
Haimo, L. T., Telzer, B. R. and Rosenbaum, J. L. (1979). *Proc. Natl. Acad. Sci. USA* **76**, 5759–5763.
Hay, C. W. and Candido, P. M. (1983). *J. Biol. Chem.* **258**, 3726–3734.
Heidemann, S. and McIntosh, J. R. (1980). *Nature* **286**, 517–519.
Hill, T. L. (1985). *Proc. Natl. Acad. Sci. USA* **82**, 4404–4408.
Hughes-Schrader, S. and Ris, H. (1941). *J. Exp. Zool.* **87**, 429–456.
Inoué, S. (1964). In *Primitive Motile Systems in Cell Biology* (ed. R. D. Allen and N. Kamiya), pp. 549–594. Academic Press, New York.
Inoué, S. and Ritter, H., Jr. (1978). *J. Cell Biol.* **77**, 655–684.
Jokelainen, P. T. (1967). *J. Ultrastruct. Res.* **19**, 19–44.
Koshland, D. E., Mitchison, T. M. and Kirschner, M. W. (1988). *Nature* **331**, 499–504.
LaFountain, J. R., Jr. and Davidson, L. A. (1979). *Chromosoma* **75**, 293–308.
Lewis, C. D. and Laemmli, U. K. (1982). *Cell* **29**, 171–181.
Lica, L. and Hamkalo, B. (1983). *Chromosoma* **88**, 42–49.
Lye, R. J., Porter, M. E., Scholey, J. M. and McIntosh, J. R. (1987). *Cell Sci.* **51**, 309–318.
Maeki, K. (1981). *Proc. Japan Acad. Ser. B* **57**, 71–76.
Matthey, R. (1973). In *Cytotaxonomy and Vertebrate Evolution* (ed. A. B. Chiarelli and E. Capanna), pp. 531–553. Academic Press, London.
McCarty, G. A., Rice, J. R., Bembe, M. L. and Barada, F. A., Jr. (1983). *Arthritis Rheum.* **26**, 1–7.
McGill, M. and Brinkley, B. R. (1975). *J. Cell Biol.* **67**, 189–199.
Merry, D. E., Pathak, S, Hsu, T. C. and Brinkley, B. R. (1985). *Am. J. Hum. Genet.* **37**, 425–430.
Mitchison, T. and Kirschner, M. W. (1984a). *Nature* **312**, 232–237.
Mitchison, T. and Kirschner, M. W. (1984b). *Nature* **312**, 237–242.
Mitchison, T. and Kirschner, M. W. (1985a). *J. Cell Biol.* **101**, 755–765.

Mitchison, T. and Kirschner, M. W. (1985b). *J. Cell Biol.* **101**, 766–777.
Mitchison, T., Evans, L., Schultz, E. and Kirschner, M. (1986). *Cell* **45**, 515–527.
Moroi, Y., Peeples, C., Fritzler, M. J., Steigerwald, J. and Tan, E. M. (1980). *Proc. Natl. Acad. Sci. USA* **77**, 1627–1631.
Mughal, S. and Godward, M. B. E. (1973). *Chromosoma* **44**, 213–229.
Murray, A. W. and Szostak, J. W. (1985). *A. Rev. Cell Biol.* **1**, 289–315.
Nakaseko, Y., Adachi, Y., Shin-ichi, F., Niwa, O. and Yanasida, M. (1986). *EMBO J.* **5**, 1011–1021.
Nicklas, R. B. (1985). In *Aneuploidy* (ed. L. Dellarco, P. E. Voytek and A. Hollaender), pp. 183–195. Plenum, New York.
Nicklas, R. B. (1987). In *Chromosome Structure and Function: The Impact of New Concepts* (ed. J. P. Gustaffson, R. Appels and R. J. Kauffman). Plenum, New York.
Nicklas, R. B., Brinkley, B. R., Pepper, D. A., Kubai, D. A. and Rickards, G. K. (1979). *J. Cell Sci.* **35**, 87–104.
Nishikai, M., Okano, U., Yamashita, H. and Watanabe, M. (1984). *Ann. Rheum. Dis.* **43**, 819–824.
Niwa, O., Matsumoto, T. and Yanasida, M. (1986). *Mol. Gen. Genet.*
Olmsted, J. B. (1981). *J. Biol. Chem.* **856**, 11955–11957.
Palmer, D. K. and Margolis, R. L. (1985). *Mol. Cell Biol.* **5**, 173–186.
Palmer, D. K., O'Day, K., Wener, M. H., André, B. S. and Margolis, R. L. (1987). *J. Cell Biol.* **104**, 805–815.
Paschal, B. M. and Vallee, R. B. (1987). *Nature* **330**, 181–183.
Paweletz, N. (1974). *Cytobiol.* **9**, 368–390.
Pease, D. C. (1946). *Biol. Bull.* **91**, 145–169.
Pepper, D. A. and Brinkley, B. R. (1977). *Chromosoma* **60**, 223–235.
Pepper, D. A. and Brinkley, B. R. (1979). *J. Cell Biol.* **82**, 585–591.
Pepper, D. A. and Brinkley, B. R. (1980). *Cell Motil.* **1**, 1–15.
Peterson, J. B. and Ris, H. (1976). *J. Cell Sci.* **22**, 219–242.
Pickett-Heaps, J. D. and Bajer, A. S. (1978). *Cytobiol.* **19**, 171–180.
Pickett-Heaps, J. D. and Tippit, D. H. (1978). *Cell* **14**, 455–467.
Pickett-Heaps, J. D., Tippit, D. H. and Porter, K. R. (1982). *Cell* **29**, 729–744.
Rattner, J. B. (1986). *Chromosoma* **93**, 515–520.
Rattner, J. B., Branch, A. and Hamkalo, B. A. (1975). *Chromosoma* **52**, 329–338.
Rattner, J. B., Krystal, G. and Hamkalo, B. A. (1978). *Chromosoma* **66**, 259–268.
Rieder, C. L. (1979a). *J. Cell Biol.* **80**, 1–9.
Rieder, C. L. (1979b). *J. Ultrastruct. Res.* **66**, 109–119.
Rieder, C. L. (1982). *Int. Rev. Cytol.* **79**, 1–58.
Rieder, C. L. and Borisy, G. G. (1981). *Chromosoma* **82**, 693–716.
Rieder, C. L., Rupp, G., Peterson, A., Marki, M., Nuss, D. and Bowser, S. (Personal communication).
Ris, H. and Kubai, D. F. (1970). *A. Rev. Genet.* **4**, 263–294.
Ris, H. and Witt, P. L. (1981). *Chromosoma* **82**, 153–170.
Roos, U.-P. (1973). *Chromosoma* **44**, 195–220.
Roos, U.-P. (1976). *Chromosoma* **54**, 363–385.
Roth, L. E. (1967). *J. Cell Biol.* **34**, 47–59.
Salmon, E. D. (1989). In *Mitosis: Molecules and Mechanisms* (ed. J. S. Hyams and B. R. Brinkley). Academic Press, London.
Sanger, J. (1975). *Proc. Natl. Acad. Sci. USA* **72**, 2451–2455.

Schibler, M. J. and Pickett-Heaps, J. D. (1980). *Eur. J. Cell Biol.* **22**, 487–498.
Schmidtke, J., Brennecke, H., Schmid, M., Nietzel, H. and Sperling, K. (1981). *Chromosoma* **84**, 187–193.
Scholey, J. M., Porter, M. E., Grissom, P. M. and McIntosh, J. R. (1985). *Nature* **318**, 483–486.
Schrader, F. (1935). *Cytologia* **6**, 422–431.
Schrader, F. (1953). *Mitosis. The Movement of Chromosomes in Cell Division* (ed. L. C. Dunn), pp. 1–170. Columbia Univ. Press, New York.
Shi, L., Ye, Y. and Duan, X. (1980). *Cytogenet. Cell Genet.* **26**, 22–27.
Snyder, J. A. and McIntosh, J. R. (1975). *J. Cell Biol.* **67**, 744–760.
Spowart, G., Forster, P., Dunn, N. and Cohen, B. B. (1985). *Disease Markers* **3**, 103–112.
Telzer, B. R., Moses, M. J. and Rosenbaum, J. L. (1975). *Proc. Natl. Acad. Sci. USA* **72**, 4023–4027.
Valdivia, M. M. and Brinkley, B. R. (1985). *J. Cell Biol.* **101**, 1124–1134.
Valvidia, M. M. *et al.* (1986). *Meth. Achiev. Exp. Pathol.* **2**, 200–223.
Vale, R. D., Reese, T. S. and Sheetz, M. P. (1985a). *Cell* **42**, 39–50.
Vale, R. D., Schnapp, B. J., Mitchison, T., Stever, E., Reese, T. S. and Sheetz, M. P. (1985b). *Cell* **43**, 623–632.
Witt, P. L., Ris, J. and Borisy, G. G. (1980). *Chromosoma* **81**, 483–505.
Wray, W. and Stubblefield, E. (1970). *Exp. Cell Res.* **59**, 469–478.
Wurster, D. H. and Atkin, N. B. (1972). *Experientia* **28**, 972–973.
Wurster, D. H. and Benirschke, K. (1970). *Science* **168**, 1364–1366.

CHAPTER 4

Microtubule Dynamics and Chromosome Movement

E. D. SALMON[1]

Department of Biology, University of North Carolina, Chapel Hill, North Carolina, USA

I. Introduction

Traditionally, five stages of mitosis have been defined (Figs. 4.1 and 4.2): prophase, prometaphase, metaphase, anaphase and telophase (Wilson, 1925; Schrader, 1953; Mazia, 1961; Inoué and Sato, 1967; Nicklas, 1971; Bajer and Mole-Bajer, 1971, 1972). There are four phases of chromosome movement: attachment, congression, anaphase A and anaphase B. Chromosomes condense, kinetochores mature, and duplicated spindle poles are usually established during prophase. Prometaphase begins after nuclear envelope breakdown (NEB). At some period after the breakdown of the nuclear envelope, chromosome duplexes usually form a chromosomal fibre between one of their kinetochores and the closest spindle pole; a process termed attachment. These chromosomes are said to be mono-oriented. After attachment, mono-oriented chromosomes move (kinetochore leading) towards the pole. Later, the other kinetochore complex becomes attached to the opposite pole of a bi-polar spindle by a chromosome fibre. These chromosomes are said to be bi-oriented. Bi-oriented chromosomes usually move to a position within the spindle approximately equatorial between the poles, the metaphase plate. The movement of the chromosomes to the metaphase plate is termed congression.

---

[1]For mailing address see list of contributors at the front of this volume.

MITOSIS: Molecules and Mechanisms
ISBN 0-12-363420-2

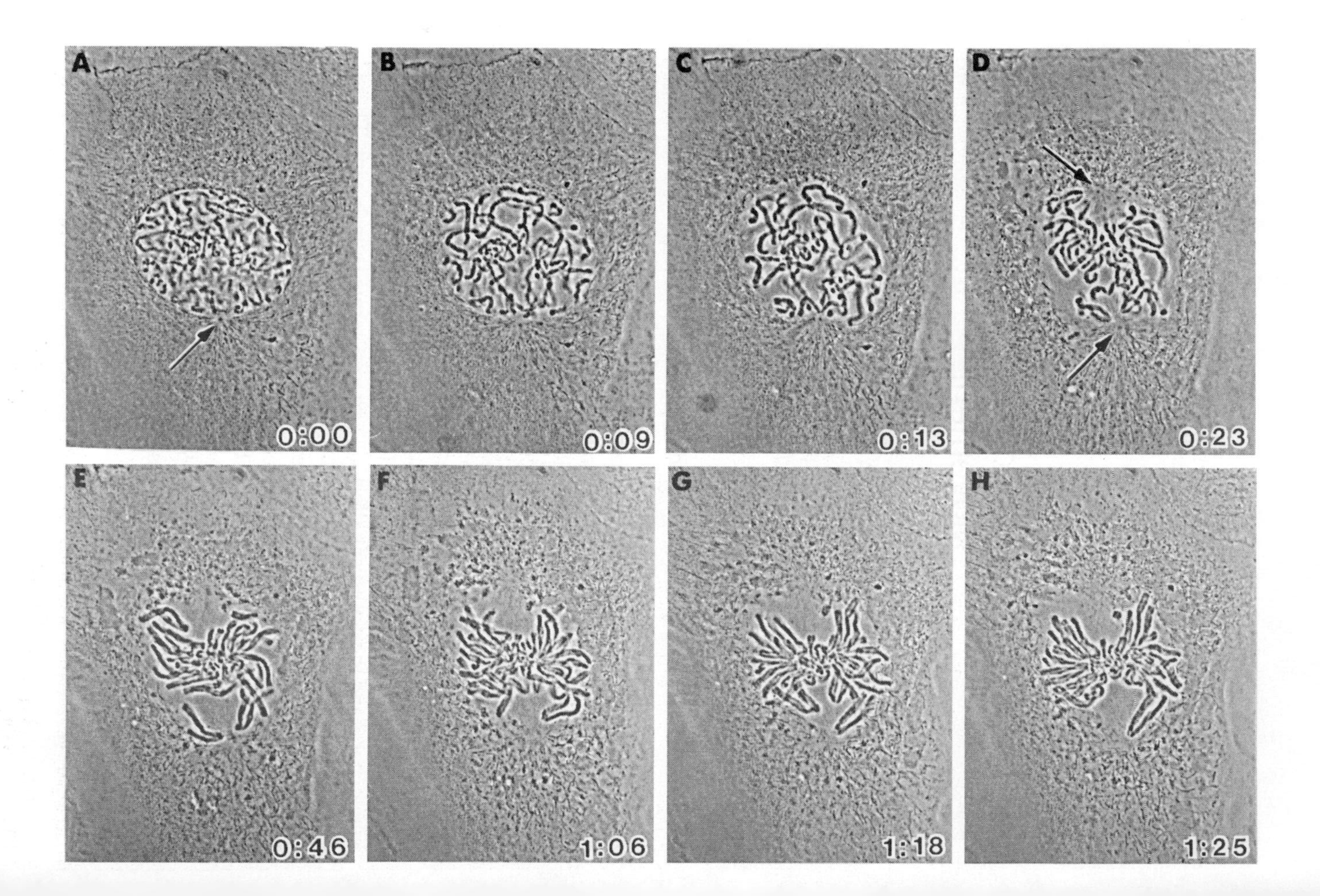
A
0:00
B
0:09
C
0:13
D
0:23
E
0:46
F
1:06
G
1:18
H
1:25

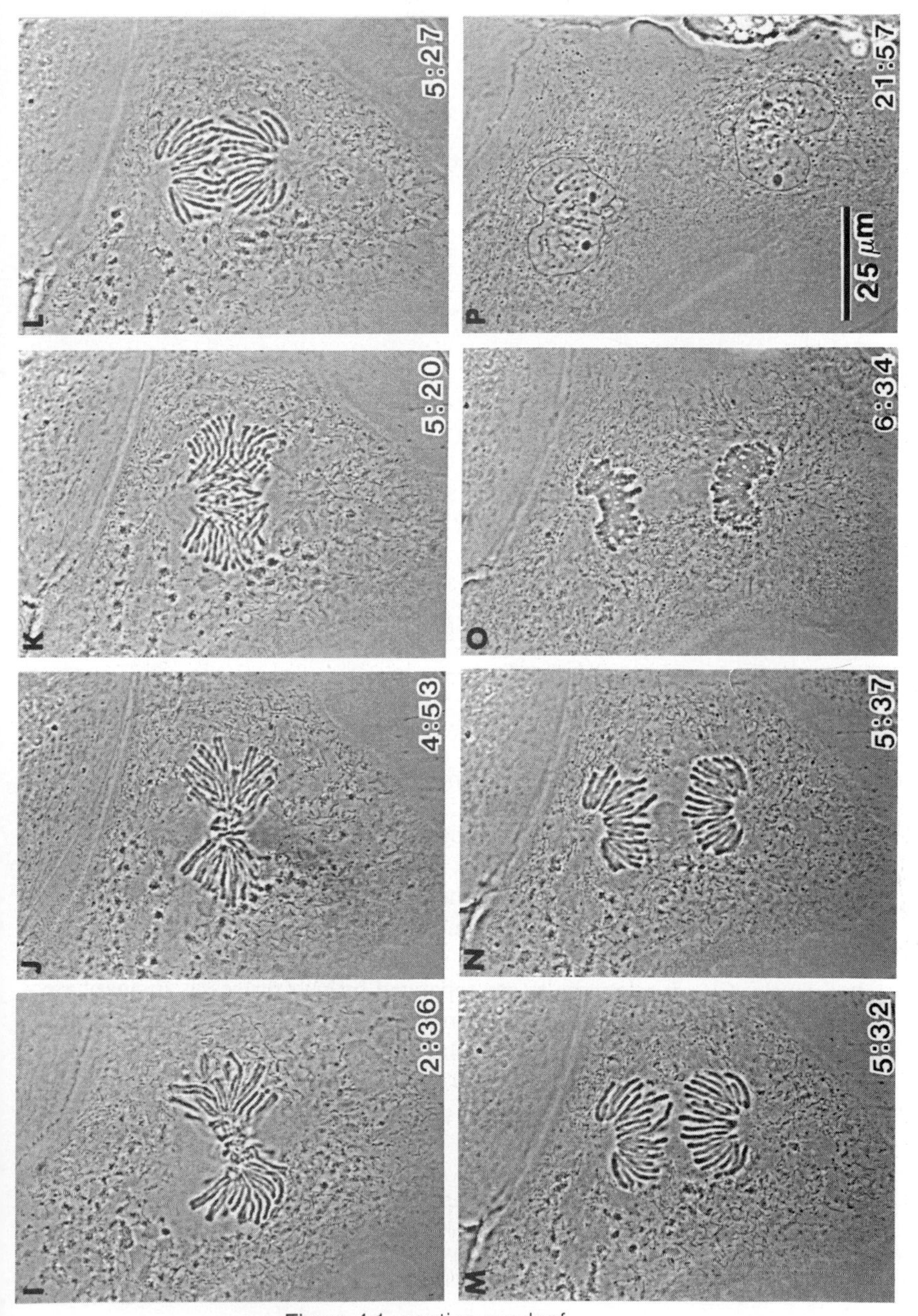

Figure 4.1 caption overleaf

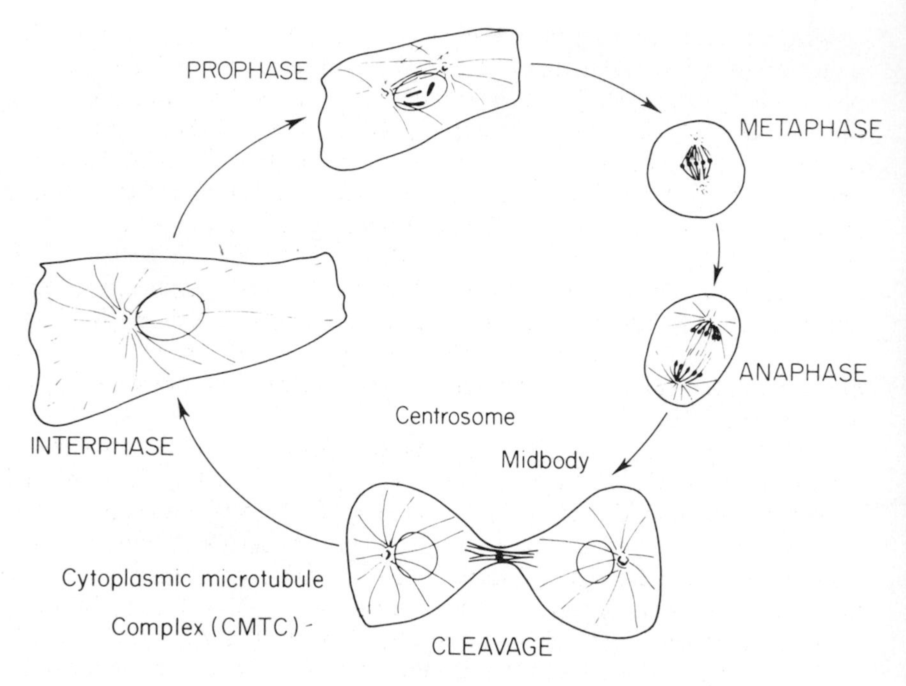

Figure 4.2. The rearrangements of microtubule arrays during the cell cycle in a typical mammalian tissue culture cell. This diagram shows the major features of the typical distribution of microtubules inferred from polarization, immunofluorescence and electron micrographs. It does not provide a comprehensive picture of all microtubules in the cell. (From Salmon and Wadsworth, 1986.)

Bi-oriented chromosomes usually do not sit statically at the spindle equator, but oscillate several micrometres back and forth between the poles. Concurrently, their chromosomal fibres lengthen and shorten. Metaphase is achieved when all chromosomes have bi-oriented and moved close to the spindle equator. Anaphase onset (AO) occurs when chromosome duplexes split into chromatids. Chromosome pairs are separated by two processes in anaphase. Anaphase A is the movement of chromosome pairs to opposite poles as their chromosomal fibres shorten and anaphase B is the elongation of the spindle interpolar length (Inoué and Ritter, 1975). In telophase, the nuclear envelope is re-formed around the chromosomes and cytokinesis occurs.

Figure 4.1. Phase-contrast micrographs of mitosis in a living newt lung epithelial cell. (A) prophase; (B–H) prometaphase; (I–J) metaphase; (K–N) anaphase; (O–P) telophase. In (A) and (D), the arrows point to the centrosomes at the spindle poles. Cytokinesis was abortive in these very flat cells. Time on each frame is given in hours:minutes. Bar = 25 μm. (Unpublished micrographs provided by Conly Rieder.)

With respect to the cytoplasmic compartment of the cell, there are two key points of transition in the cell cycle between interphase and mitosis which are not obviously defined by the traditional stages of mitosis. The first occurs at the time of NEB. Before this, the structure and activities of the cytoplasm are typical of interphase. NEB is only one aspect of the global changes that occur in the cell at that time. Extensive rearrangements occur in the organization of microtubules (Fig. 4.2) and other cytoskeletal and cytoplasmic components (Brinkley *et al.*, 1980; Vandré *et al.*, 1984; DeBrabander *et al.*, 1986). Cells generally round-up, pulling away from their neighbours in tissues, and cease much of their synthetic activity. The second major transition occurs at anaphase onset. This is marked by the splitting of paired chromosomes, but, in general, the cell returns to the structural organization and activities typical of interphase cells.

There are three essential structural components required for chromosome segregation: mitotic centres (centrosomes or spindle poles) (Vandré and Borisy, this volume), kinetochores on the chromosomes (Brinkley *et al.*, this volume), and microtubules. The mitotic centres nucleate the growth of spindle microtubules and the microtubules of the chromosome fibres connect the kinetochores on chromosomes to the spindle poles (Inoué, 1981b; Ellis and Begg, 1981; Mazia, 1984, 1987; McIntosh, 1983, 1984, 1985; Nicklas, 1985, 1987a). Thus far, the only known spindle protein essential for chromosome movement is tubulin. A variety of other motility-associated proteins have been found to be associated with the mitotic spindle fibres. These include dynein-like ATPases, kinesin, several other microtubule-associated proteins, calmodulin and actin (Sanger, 1977; Browne *et al.*, 1980; Pratt *et al.*, 1980; Zieve and Solomon, 1982; McIntosh, 1984, 1985; Scholey *et al.*, 1985; Bloom *et al.*, 1986; Leslie *et al.*, 1987; Dinsmore and Sloboda, 1988). Evidence rules out a function for actin–myosin in the generation of mitotic movements (Inoué, 1981b) and there is yet no direct evidence that any of these other spindle-associated proteins actively produce or regulate chromosome movement or spindle assembly. During the past 38 years, various techniques have been developed for the isolation of the mitotic spindle and attached chromosomes free from the cell (Fig. 4.3) (Mazia and Dan, 1952; Kane, 1962; Sakai, 1978; Salmon, 1982; Dinsmore and Slobada, 1988; Rebhun and Palazzo, 1988). Unfortunately, conditions that permit reactivation of lifelike chromosome movements in isolated spindles have yet to be discovered and the molecular mechanisms that generate the poleward forces for chromosome movement are still uncertain. More progress has been made in partially lysed cell models (Spurck and Pickett-Heaps, 1987) and in the reactivation of interpolar spindle elongation (anaphase B) in isolates (Cande, this volume; Rebhun and Palazzo, 1988).

In the past several years, much has been learned from cellular studies and *in*

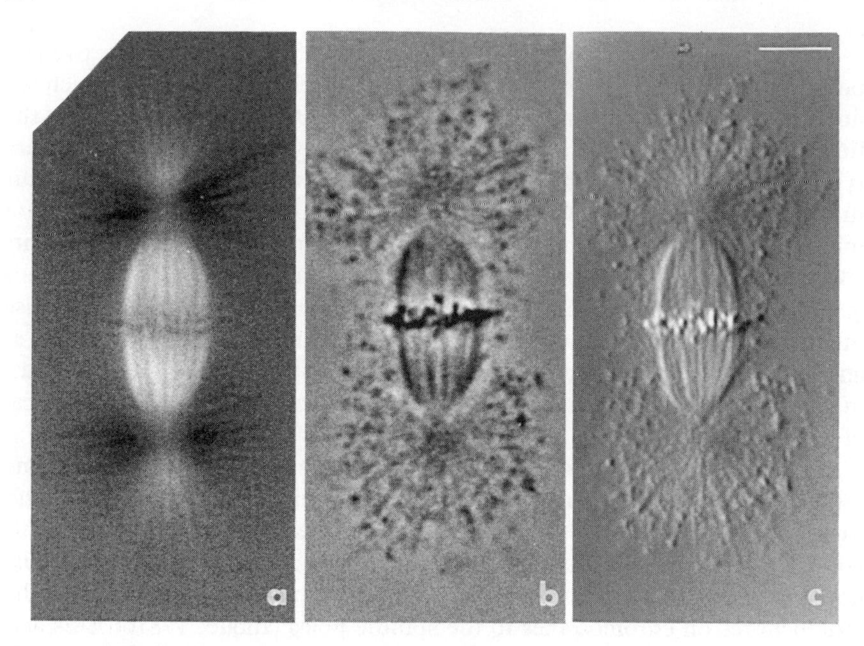

Figure 4.3. Mitotic spindle in metaphase isolated from the first-division embryos of the sea urchin *Lytechninus variegatus* photographed using (a) polarization microscopy, (b) phase contrast microscopy, and (c) differential interference microscopy. Bar = 10 μm. (From Salmon and Segall, 1980.)

vitro reconstitution experiments about the structural organization and dynamics of microtubule assembly within the spindle and the dynamics of chromosome movement. These results have reoriented the conceptual thinking about the mechanism and regulation of spindle microtubule assembly, the structural basis for the attachment of chromosomes to the spindle fibres, the generation of poleward movements of the chromosomes, the congression of the chromosomes to the spindle equator in metaphase, and the poleward movement of chromosomes in anaphase. It is clear from these studies that the assembly dynamics of microtubules is directly involved with chromosome attachment to the spindle and the production of chromosome movements.

In this chapter, I attempt to integrate recent information on the assembly dynamics of spindle microtubules and the coupling between microtubule dynamics and chromosome movements. I will neglect in-depth discussions of the structure and composition of centrosomes and kinetochores, the mechanisms of interpolar elongation during anaphase, and molecular mechanisms that may regulate changes in microtubule assembly during the cell cycle and mitosis, since they are covered in depth in other chapters in this volume. Mitosis is a complex process and varies significantly between different organisms. The discussion here is primarily focused on the mechanism of mitosis in higher eukaryotic animal cells. There are several comprehensive reviews dealing with mitosis to which the reader is also addressed (Wilson, 1925; Schrader, 1953; Östergren *et al.*, 1960; Mazia, 1961, 1984, 1987; Inoué and Sato, 1967; Luykx, 1970; Nicklas, 1971, 1987a; Bajer and Mole-Bajer, 1971, 1972; Inoué, 1981b; Pickett-Heaps *et al.*, 1982; Rieder, 1982; McIntosh, 1981, 1983, 1984, 1985; DeBrabander *et al.*, 1986; Kirschner and Mitchison, 1986a).

II. The Spindle is Formed From Polarized Arrays of Microtubules

A. Microtubule polarity

Microtubule polarity is important in consideration of the structure of the spindle and the mechanisms of chromosome movement because polarity orients the direction of force production by translocators such as kinesin or dynein (Sale and Satir, 1977; McIntosh, 1984, 1985; Vale *et al.*, 1985b; Mitchison, 1986; Cassimeris *et al.*, 1987a). In addition, the plus and minus ends of microtubules are likely to have different assembly/disassembly kinetics (Leslie and Pickett-Heaps, 1983; Purich and Kristofferson, 1984; Mitchison and Kirschner, 1984c; Horio and Hotani, 1986; Tao *et al.*, 1988; Walker *et al.*, 1988) and binding affinities for other molecular complexes, such as those associated with kinetochores (Huitorel and Kirschner, 1988) or with the centrosome complex.

Microtubules are polarized polymers of tubulin (Mandelkow and Mandelkow, 1985, 1986). Polarity can be identified by growth rate. One end, the plus end, usually grows faster than the minus end at the same tubulin concentration in reconstituted preparations (Bergen and Borisy, 1980; Purich and Kristofferson, 1984; Mitchison and Kirschner, 1984c; Horio and Hotani, 1986; Walker *et al.*, 1988). In addition to growth rate, techniques have been developed to identify microtubule polarity using structural probes (Heidemann and McIntosh, 1980; Telzer and Haimo, 1981; Mcintosh and Euteneuer, 1984; McIntosh, 1984, 1985).

B. Orientation of microtubule growth and polarity by MTOCs

Microtubule organization and orientation in both interphase and mitosis (Fig. 4.2) have been shown to be governed by nucleated assembly from microtubule organizing centres (MTOCs) like the centrosome (Borisy and Gould, 1977; Brinkley *et al.*, 1981; Brinkley, 1986; DeBrabander *et al.*, 1980, 1981a, 1986; Evans *et al.*, 1985; McIntosh, 1983; Vandré and Borisy, this volume). The cloud of amorphous material that surrounds the pair of centrioles within the centrosome nucleates the polymerization of microtubules into polar

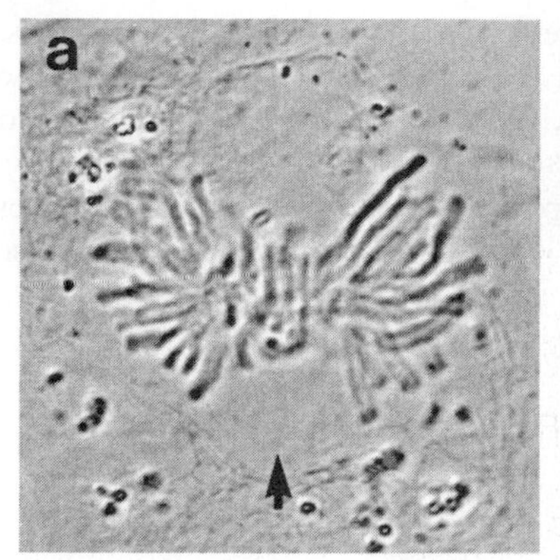

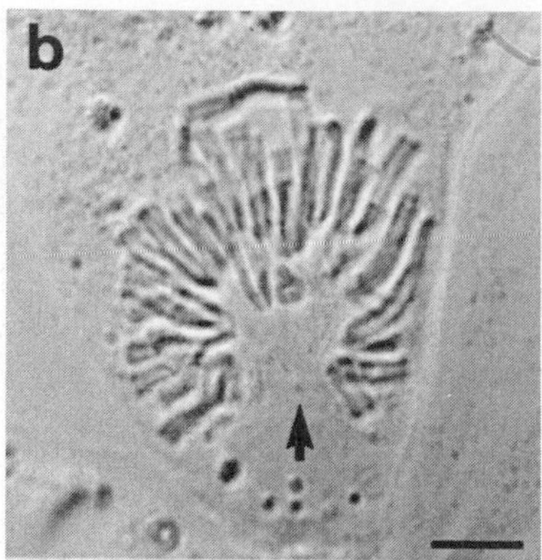

Figure 4.4. Comparison of monopolar and bipolar spindles in newt cells. Phase-contrast micrographs of living cells with (a) a bipolar spindle and (b) a monopolar spindle. Arrows point to centrosomes at spindle poles. Bar = 10 μm. (Provided by Lynne Cassimeris.)

arrays (McIntosh, 1983; Evans *et al.*, 1985). Polar microtubules have been shown to be oriented with their plus or fast-growing ends distal to the centrosome (Heidemann and McIntosh, 1980; Telzer and Haimo, 1981; McIntosh and Euteneuer, 1984). Immunofluorescence studies have shown that during interphase, most microtubules radiate outward towards the cell surface from a centrosome located near the nucleus (Brinkley *et al.*, 1980; Brinkley, 1986). For example, a mammalian culture cell, such as BSC1, contains several hundred microtubules within the cytoplasmic microtubule complex (CMTC) with mean length near 20 μm (Schulze and Kirschner, 1986, 1987).

Accurate segregation of chromosome duplexes requires a bipolar spindle (Mazia, 1961, 1984, 1987; Mazia *et al.*, 1981; Bajer, 1982; Rieder *et al.*, 1986; Sluder and Rieder, 1985; Sluder *et al.*, 1986). In preparation for mitosis, the centrosome complex is reproduced to establish the bipolarity of the spindle and to generate a bipolar spindle; the centres must also separate (Vandré and Borisy, this volume). This requires some form of microtubule assembly (Brinkley *et al.*, 1967; Brinkley, 1986; Mazia, 1984, 1987). If reproduction or separation fail to occur, then at NEB a monopolar spindle forms (Fig. 4.4).

C. Spindle microtubule organization during mitosis

At NEB, the CMTC disappears and microtubule assembly occurs mainly within the region occupied by the nucleus before NEB (Figs. 4.1 and 4.2). A bipolar spindle is formed from overlapping arrays of microtubules extending

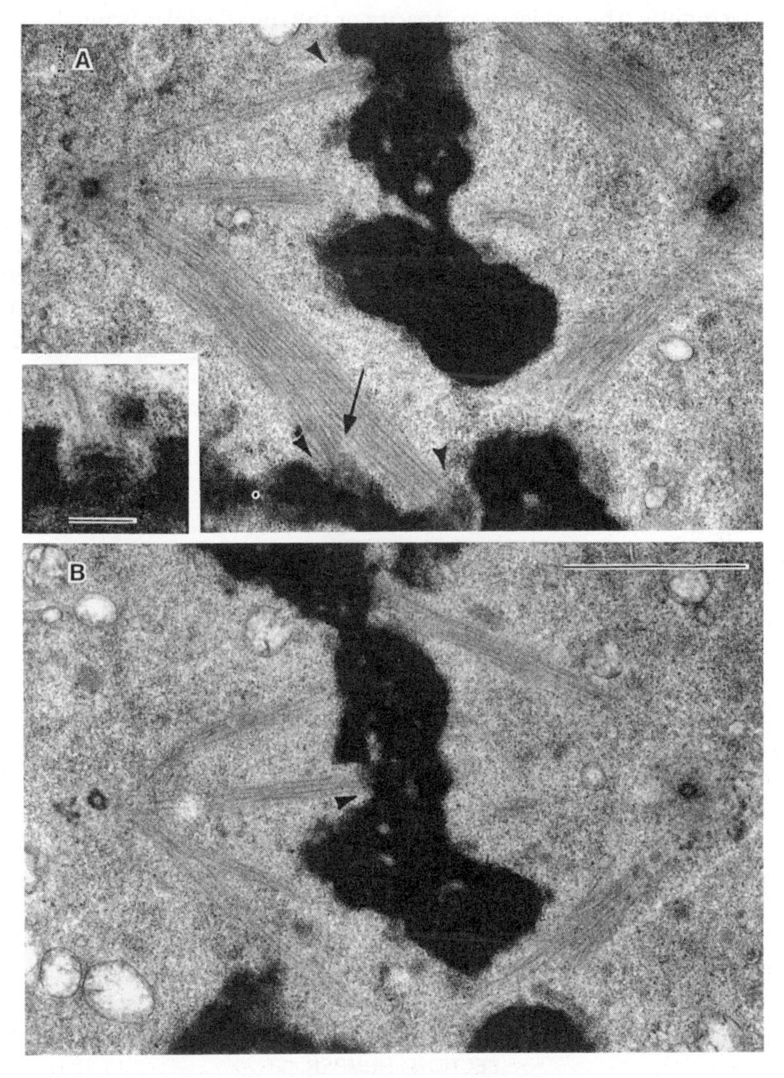

Figure 4.5. Electron micrographs of adjacent thick sections (A) and (B) showing the cold stable chromosomal fibre microtubules in PtK_1 cells. Cells were cooled to 6°C before fixation and processing for electron microscopy. Bar = 5 μm. Insert is higher magnification view of a kinetochore showing microtubule attachment. Scale = 0.25 μm. (From Rieder, 1981.)

from opposite spindle poles. All polar microtubules are oriented with their plus ends distal from their poles (McIntosh and Euteneuer, 1984; McIntosh, 1984, 1985). In mitosis, the nucleation capacity of the spindle poles is much greater than for the centrosomes during interphase (McIntosh, 1983). There are typically one thousand to several thousand polar microtubules in a metaphase spindle. These are much shorter than the microtubules in the interphase CMTC.

Microtubules attached to kinetochores are termed kinetochore microtubules (McDonald, this volume) (Fig. 4.5). These have been shown to be oriented with their plus ends proximal to the kinetochores (McIntosh, 1984, 1985). The number of kinetochore microtubules at metaphase is typically 15–30 in mammalian cells, but the number can vary considerably between non-homologous chromosomes and cell types (reviewed by Rieder, 1982). Yeast

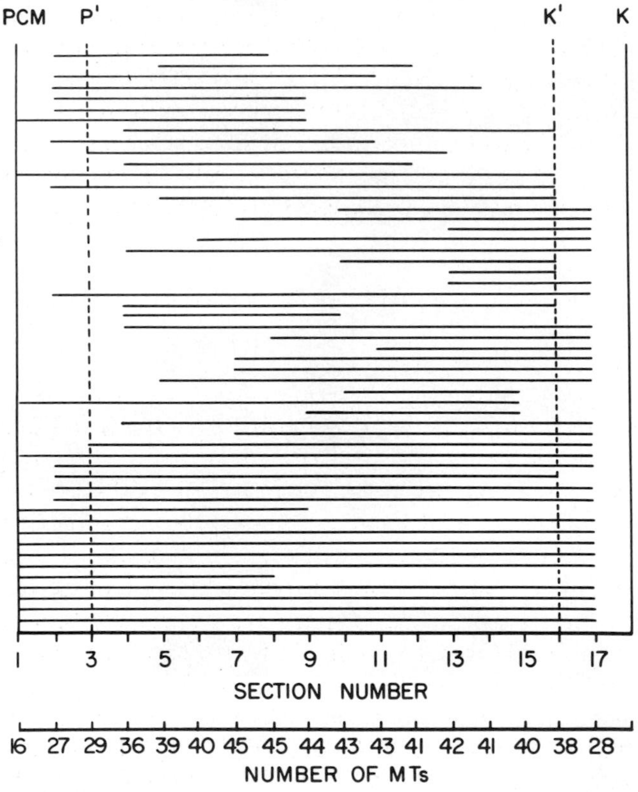

Figure 4.6. A serial-section reconstruction of the microtubule distribution in a single cold-stable chromosomal fibre from a metaphase PtK_1 cell. K′ to K represents the kinetochore region and P′ to PCM represents the pericentriolar material. (From Rieder, 1981.)

chromosomes appear to have only one kinetochore microtubule, while large plant chromosomes can have several hundred. In metaphase PtK_1 (Rieder, 1981) and CHO (Witt *et al.*, 1981) cells, ultrastructural analysis (Figs. 4.5 and 4.6) has shown that most kinetochore microtubules extend all the way between the pole and the kinetochore. However in PtK_1, about 25% of the kinetochore microtubules may have a free minus end (see Fig. 4.6). There are only a few microtubules that are attached neither to the kinetochore or the pole.

At AO, the kinetochore microtubules shorten and the mean length of the majority of polar microtubules decreases as the separated chromatids move poleward (Fig. 4.2) (Inoué, 1964; Nicklas, 1975; Salmon and Begg, 1980). Concurrently, the spindle asters begin to elongate out to the cell periphery to re-form the new CMTCs in the daughter cells. It has been shown that the extension of microtubules to the cell surface results in stimulation and orientation of cytokinesis (Rappaport, 1988; Schroeder, 1987; Inoué, 1981b; White and Borisy, 1983), but how this occurs is not well understood. Between the separating chromosomes at the spindle interzone, bundles of microtubules form that are called "stem bodies" (McIntosh *et al.*, 1975; Salmon *et al.*, 1976). These are gathered into a larger bundle, termed the midbody complex, during cytokinesis. Midbody microtubules are oriented with plus ends overlapping in the centre (McIntosh and Euteneuer, 1984). Stem bodies and midbodies may be involved with interpolar elongation (Saxton and McIntosh, 1987).

III. Polar Microtubules Exhibit Rapid Dynamic Instability

A. Spindle dynamic equilibrium assembly

In 1967, Inoué and Sato published their dynamic equilibrium model of spindle fibre assembly and chromosome movement, which was updated to deal more specifically with microtubules by Inoué (Inoué and Ritter, 1975; Inoué, 1976). The dynamic equilibrium model was based upon studies of spindle assembly in living cells, using polarization microscopy techniques to measure the assembly of the spindle fibres. These and subsequent studies showed that the majority of microtubules in the spindle can be rapidly and reversibly depolymerized in the absence of protein synthesis by a variety of physical and chemical agents, including cooling, high hydrostatic pressure, calcium, and tubulin-binding drugs like colchicine and its analogues colcemid and nocodazole (Fig. 4.7) (Inoué, 1952, 1964, 1976, 1981a; Inoué and Sato, 1967; Inoué *et al.*, 1975; Marsland, 1970; Goode, 1973; Salmon, 1975a,b,c; 1976; Salmon *et al.*, 1976, 1984b; Hamaguchi, 1975; Sluder, 1976; Salmon and Segall, 1980; Kiehart, 1981; Izant, 1983; Silver, 1986; Spurck and Pickett-Heaps, 1986a,b). Agents such as D_2O, glycols and DMSO were shown to

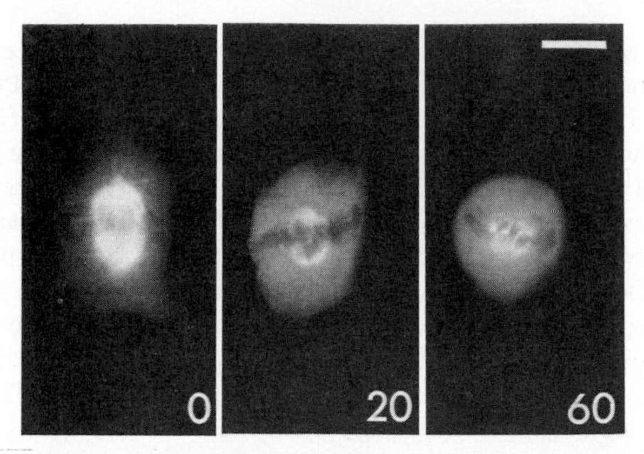

Figure 4.7. The rapid rate of microtubule depolymerization and spindle shortening induced by nocodazole in metaphase PtK_1 cells at 37°C. Cells were treated with 10 μM nocodazole, then fixed for immunofluorescence using antibodies to tubulin (Cassimeris *et al.*, 1986) at the times in seconds given on each frame. Bar = 10 μm.

promote spindle assembly reversibly at low to moderate concentrations (Inoué and Sato, 1967; Rebhun *et al.*, 1975). Observations like these initially led Inoué (1964) to propose that spindle fibre microtubules were reversibly self-assembled from a cellular pool of tubulin subunits and that microtubule assembly, initiated and organized by the mitotic centres, occurred by entropically driven reactions. These concepts have been largely confirmed by other biochemical and biophysical approaches following the discovery by Weisenberg (1972) of methods for the reversible self-assembly of microtubules *in vitro*. Inoué suggested that tubulin subunits were able to exchange rapidly at sites within microtubules in order to account for the dynamic behaviour of the spindle fibre assembly. It now appears that microtubule dynamic instability is mainly responsible for the dynamic behaviour of spindle fibres.

B. Microtubule dynamic instability *in vitro*

In recent years, dynamic instability, as originally conceived by Mitchison and Kirschner (1984a,b,c; Kirschner and Mitchison, 1986a), has been established as the fundamental mechanism of polymerization of microtubules from pure tubulin *in vitro* (reviewed by Cassimeris *et al.*, 1987a). Real-time observation of the polymerization of individual microtubules demonstrates that dynamic instability involves persistent and distinctly different phases of microtubule elongation and rapid shortening; transitions between these phases are infrequent and abrupt (Horio and Hotani, 1986; Walker *et al.*, 1986, 1988). Microtubules exhibiting dynamic instability never grow to a steady-state length. Instead, they either elongate or rapidly shorten. Following nucleation, a microtubule end usually elongates at constant velocity for a period of time until an abrupt transition (catastrophe) occurs and the end then rapidly shortens. Rapid shortening usually occurs at constant velocity until there is an abrupt transition back to elongation (rescue) or the microtubule depolymerizes all the way back to the nucleation centre. Microtubule elongation and shortening *in vitro* occur by tubulin association and dissociation reactions at microtubule ends (Kirschner, 1979; Bergen and Borisy, 1980; Purich and Kristofferson, 1984; Kristofferson *et al.*, 1986; Walker *et al.*, 1988). The rate of tubulin dissociation during rapid shortening *in vitro* is typically an order of magnitude or more faster than the rate of dissociation during elongation. The frequencies of nucleation, catastrophe and rescue are stochastic and dependent on tubulin concentration. The parameters of dynamic instability can be distinctly different for plus and minus ends (Walker *et al.*, 1988). The plus ends of microtubules have been termed the "active" ends because they grow faster and rescue less frequently than the minus ends in reassembly buffers *in vitro* (Horio and Hotani, 1986).

The basic mechanism of instability is controversial, but it is thought to involve GTP hydrolysis (Mitchison and Kirschner, 1984b; Kirschner and Mitchison, 1986a,b; Carlier *et al.*, 1984; Pantaloni and Carlier, 1986; Caplow, 1986; Walker *et al.*, 1988). The polymerizing subunit is a tubulin–GTP complex. Some time after incorporation into a microtubule end, the GTP is hydrolysed to GDP producing a core of tubulin–GDP capped at the elongating end by a region of tubulin–GTP. It has been proposed that this "GTP cap" stabilizes a labile "core" of tubulin–GDP. A catastrophe occurs when the GTP cap is lost. Rapid shortening is produced by the rapid dissociation of tubulin–GDP subunits. Rescue occurs by the infrequent re-formation of a tubulin–GTP cap.

The most direct evidence for microtubule dynamic instability has come from real-time observations of the polymerization of individual microtubules. *In vitro*, this has been accomplished for self-assembled microtubules using dark-field light microscopy (Horio and Hotani, 1986) and for nucleated microtubule assembly using differential interference contrast microscopy, video contrast enhancement and digital image-processing techniques (Allen *et al.*, 1981; Inoué, 1981a, 1986; Walker *et al.*, 1986, 1988; Salmon *et al.*, 1987). These techniques have allowed accurate determination of the rate constants and transition frequencies for microtubules assembled *in vitro* from pure tubulin and from mixtures of tubulin and brain microtubule associated proteins (MAPs) (Horio and Hotani, 1986).

C. Microtubule dynamic instability *in vivo*

Cassimeris *et al.* (1988b) have directly observed in real time the dynamic instability of the polar microtubules in the CMTC of cells explanted from newt lung (Fig. 4.8). In these large and very flat interphase cells, linear filaments

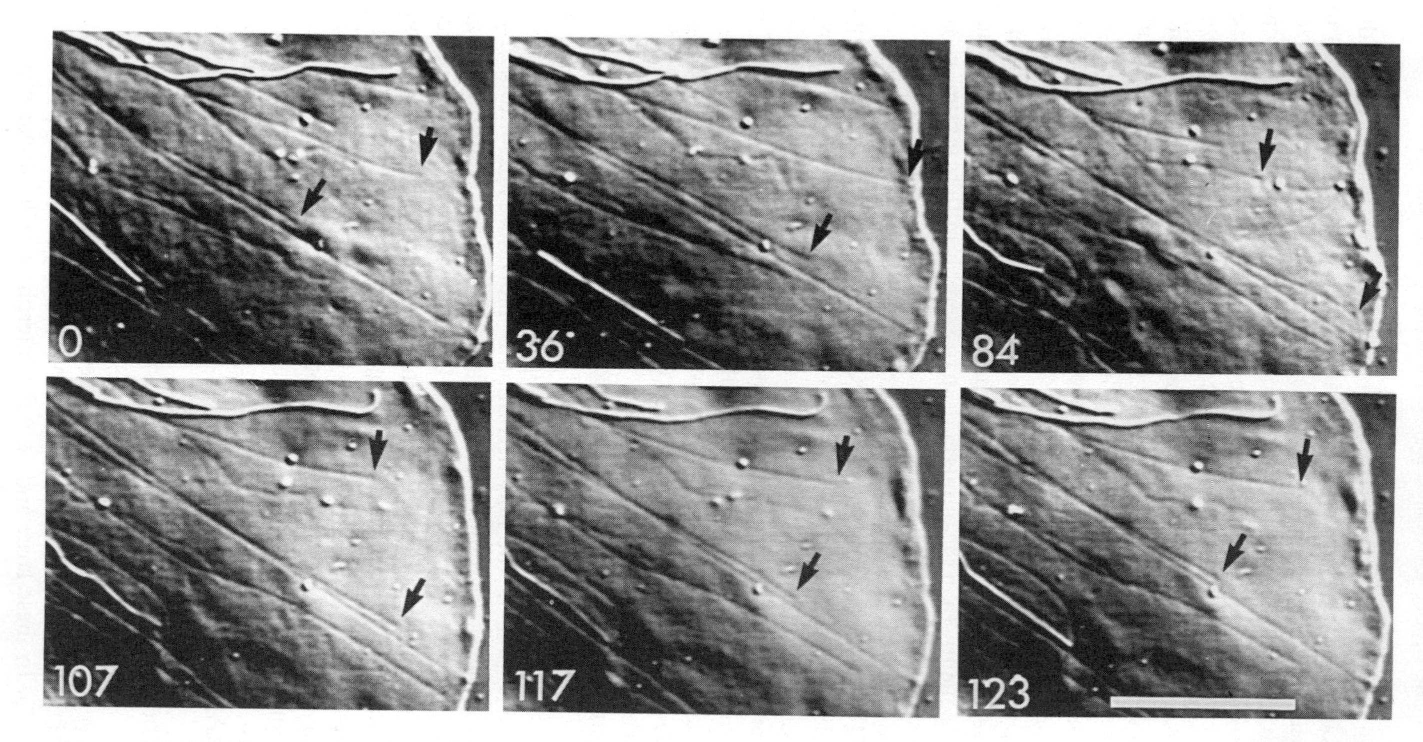

Figure 4.8. Real-time observations of the dynamic instability of individual microtubules in the CMTC of interphase newt cells. Micrographs were taken from a video recording which was made using video enhanced DIC microscopy and digital image processing to reveal the microtubules in the cytoplasm. Arrows mark the tips of two dynamic microtubules. Time in seconds is given on each frame. Bar = 20 μm. (From Cassimeris *et al.*, 1988b.)

coextensive with microtubules were seen to elongate at a constant velocity of about 7 μm/min for an average of 71 ± 61 s and rapidly shorten at a constant velocity of about 17 μm/min for an average duration of 22 ± 15 s. At least 70% of the rapid shortening phases were rescued, showing that rescue occurs frequently for microtubules of the CMTC. Direct observations of individual microtubule dynamic instability in interphase mammalian cells has also been obtained from sequential fluorescent images of cells microinjected with fluorescently labelled tubulin (Sammak and Borisy, 1988; Schulze and Kirschner, 1988).

Fluorescent or biotinylated analogues of tubulin have provided substantial information about the pathways of tubulin assembly in living cells. Tubulin, purified from mammalian brain, has been covalently coupled to a fluorophore (DTAF, FITC, and rhodamine) (Keith *et al.*, 1981; Leslie *et al.*, 1984; Wadsworth and Salmon, 1986c; Gorbsky *et al.*, 1987, 1988; Sammak *et al.*, 1987) or biotin (Mitchison *et al.*, 1986; Kristofferson *et al.*, 1986; Schulze and Kirschner, 1986, 1987) in a way that preserves *in vitro* reassembly capabilities, then microinjected into living cells to serve as a labelled tracer in the cell's tubulin pool. The distribution of fluorescent tubulin can be recorded in cells using fluorescence microscopy, low-light-level video cameras or CCD detectors coupled to digital image processors (Salmon and Wadsworth, 1986). Higher-resolution light and electron microscopy images have been obtained by lysing and fixing the cells. The hapten-labelled subunits were located within the microtubules using antibodies against the hapten (Soltys and Borisy, 1985; Schulze and Kirschner, 1986, 1987; Mitchison and Kirschner, 1986).

Forward incorporation studies have shown that tubulin incorporation occurs at the plus ends of polar microtubules in the CMTC and the mitotic spindle and by nucleated elongation from the centrosome or spindle poles (Soltys and Borisy, 1985; Schulze and Kirschner, 1986, 1987; Mitchison and Kirschner, 1986). There is no evidence that tubulin exchanges at sites along the lengths of microtubules in either of these arrays. In addition, about 80% of the microtubules in the CMTC become fully labelled with a half-time of about 5–10 min depending on cell type (Saxton *et al.*, 1984; Schulze and Kirschner, 1986; Sammak *et al.*, 1987). The non-kinetochore microtubules of mammalian spindles are fully labelled within 1 min after labelled tubulin injection (Saxton *et al.*, 1984; Mitchison *et al.*, 1986).

These forward incorporation rates correlate well with the rates of tubulin turnover within microtubules determined by measurement of fluorescence redistribution after photobleaching (FRAP) (Salmon *et al.*, 1984a; Saxton *et al.*, 1984; Salmon and Wadsworth, 1986; McIntosh *et al.*, 1986; Wadsworth and Salmon, 1986a,b). In this approach, cells are microinjected with fluorescently labelled tubulin and allowed to equilibrate so that the fluorescent tubulin becomes uniformly distributed within the cellular tubulin pool and

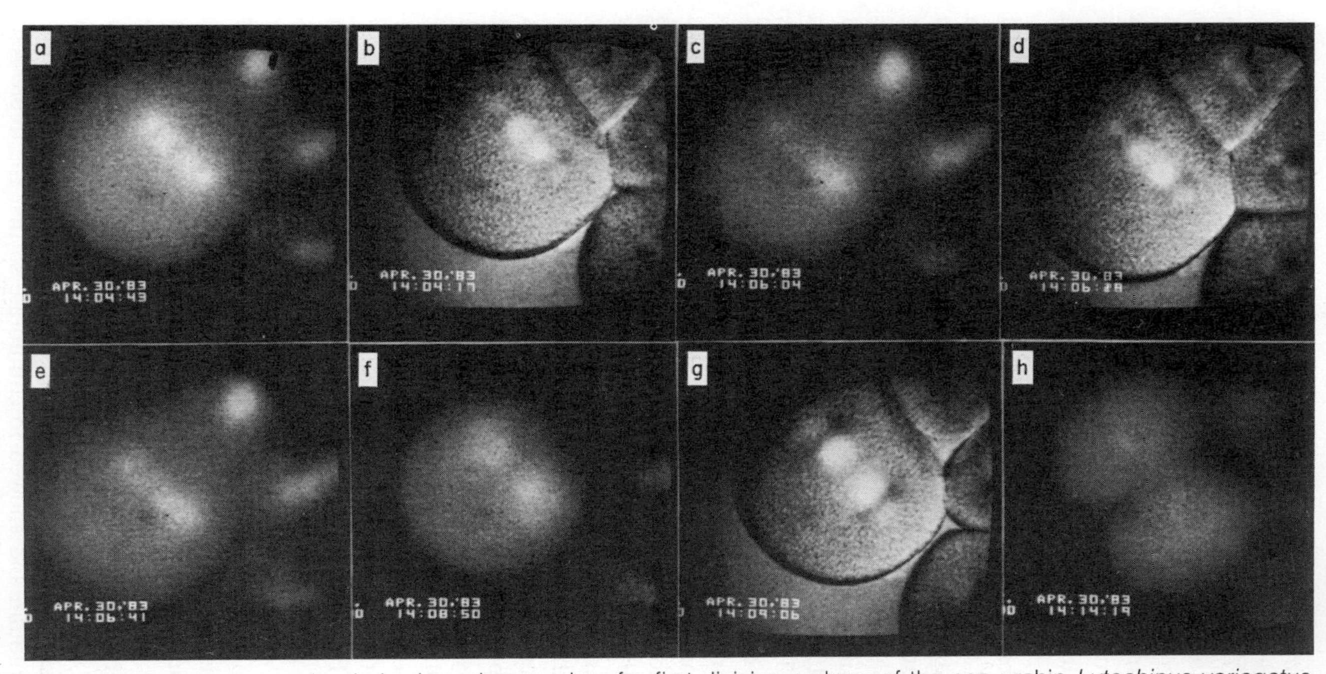

Figure 4.9. Fluorescence and polarization micrographs of a first-division embryo of the sea urchin *Lytechinus variegatus* demonstrating that fluorescence redistribution after photobleaching (FRAP) is rapid and that photobleaching does not significantly alter the normal amount and distribution of spindle and astral fibre birefringence, the morphological changes in assembly during mitosis, or the timing of mitotic events. The cell was microinjected with DTAF-tubulin about 20 min before mitosis. Frames (a), (c), (e), (f) and (h) are fluorescence micrographs and frames (b), (d) and (g) are polarization micrographs. The contrast seen in the spindle in the polarization micrographs is produced by the birefringence of the parallel array of microtubules. Time in hours: minutes: seconds is given in the lower left-hand corner of each frame. Bleaching for 8 seconds with a 12 μm diameter argon laser microbeam ended at 14:06:03. (From Salmon *et al.*, 1984a.)

incorporated throughout the length of microtubules (Fig. 4.9). A brief (0.1 s) pulse from a focused laser beam is then used to rapidly bleach a fraction (30–50%) of the fluorophores in a local region of the microtubule array. The rate and pattern of FRAP is measured using either photomultiplier spot methods (Fig. 4.10) or by analysis of digitized images (Figs. 4.9 and 4.11). FRAP studies initially showed that fluorescence recovery was rapid, extensive and uniform throughout the bleached region (Salmon *et al.*, 1984a; Saxton *et al.*, 1984). Although photobleaching fluorescently labelled microtubules can cause microtubule fragmentation (Vigers *et al.*, 1988), we found that extensive photobleaching one half of a metaphase spindle in sea-urchin embryos did not alter the birefringence of the spindle fibres (Fig. 4.9) (Salmon *et al.*, 1984a). This result demonstrated that photobleaching destroyed the fluorophores attached to the tubulins within the spindle fibre microtubules *in vivo*, but did not cut microtubules as occurs with UV irradiation (Forer, 1965, 1966; Wilson and Forer, 1988). Thus, fluorescence recovery was a measure of the steady-state turnover of tubulin within the dynamic spindle fibre microtubules. Fluorescence recovery within a bleached region occurred in three phases, as shown in Fig. 4.10. There is an initial very rapid phase produced by the diffusion of unpolymerized tubulin. The second phase is termed the rapid incorporation phase. Fluorescence recovery during this phase is exponential

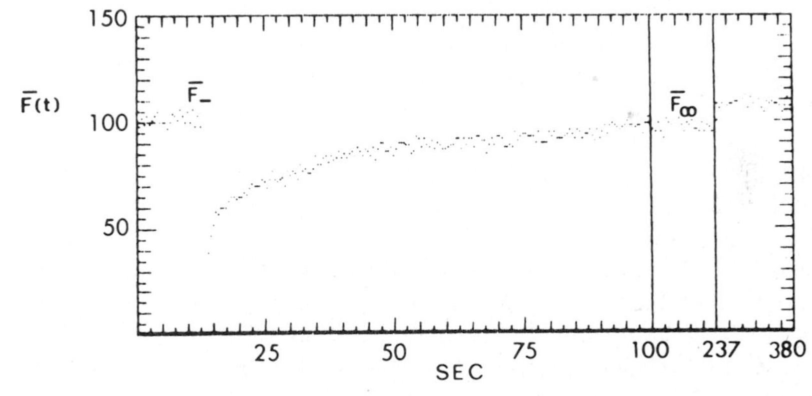

Figure 4.10. Computer records of spindle FRAP in a sea-urchin metaphase spindle obtained by photometric techniques. Fluorescence excitation and bleaching were produced by a laser microbeam (488 nm, 4.5 μm diameter) focused in a central half-spindle of first-mitotic *Lytechnius variegatus* embryos that had been microinjected before mitosis with DTAF–tubulin. Spindle fluorescence was bleached by 50–60% by removing an attenuating filter (5×10^{-4}) from the path of the laser beam. Photon counts were acquired every second using a photomultiplier. The first five fluorescence values were used to normalize the data. A total of 382 s of normalized data were acquired, and data for 0–100, 227–237, and 367–382 seconds are plotted. (From Wadsworth and Salmon, 1986b.)

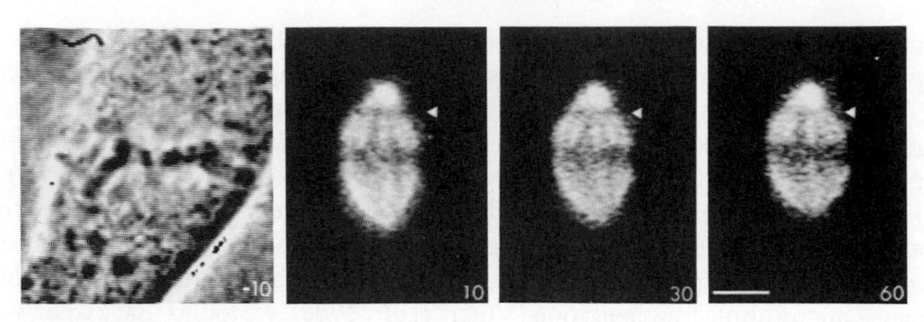

Figure 4.11. VIDEO-FRAP of a metaphase PtK_1 cell at 23–24°C. The cell was microinjected with DTAF-tubulin and a narrow bar pattern was photobleached (100 ms laser pulse) across the upper half-spindle. A phase-contrast image of the cell is shown on the left. It was taken 10 s prior to photobleaching. Fluorescent images during fluorescence recovery after bleaching are shown to the right of the phase-contrast image. Each image was obtained by summing frames for 1 s using a SIT camera. Time in seconds after photobleaching is given on each frame. Before bleaching, the distribution of fluorescence in the spindle was symmetric about the metaphase plate. The lower half-spindle tilted out of focus during the recovery period. Bar = 5 μm. (From Cassimeris *et al.*, 1988a.)

with a half-time of 16–60 s depending on cell type (Wadsworth and Salmon, 1986b) and represents typically 70–80% of the non-diffusible bleached fluorescence. In the third phase, the remaining 20–30% of the bleached fluorescence recovers more slowly and spindles usually looked fully recovered after about 10 min. These results show that tubulins throughout 70–80% of the microtubules in the spindle—most probably the labile polar, non-kinetochore microtubules—exchange with tubulins in the cellular pool with a half-time of tens of seconds. The slow phase of fluorescence recovery is probably due to the slower turnover of tubulin within the differentially stable kinetochore microtubules, as discussed below.

These results provide strong support for the dynamic instability of polar microtubule assembly. The rapid rate of fluorescence recovery and forward incorporation is several orders of magnitude too fast to be produced by simple equilibrium diffusional exchange of tubulin at microtubule ends (Salmon *et al.*, 1984a).

There is also no pattern to the recovery of fluorescence during the rapid incorporation phase (Salmon *et al.*, 1984a; Saxton *et al.*, 1984; Wadsworth and Salmon, 1986a,b; McIntosh *et al.*, 1986; McIntosh and Vigers, 1987; Cassimeris *et al.*, 1988a). As shown in Fig. 4.11, bar patterns bleached across the middle of a half-spindle in metaphase PtK_1 cells, recover fluorescence without movement of the pattern. This type of experiment showed that plus to minus treadmilling (Margolis *et al.*, 1978; Margolis and Wilson, 1981) did not produce the fluorescence recovery. Since tubulin does not appear to exchange at sites along the lengths of microtubules, whole microtubules must be rapidly

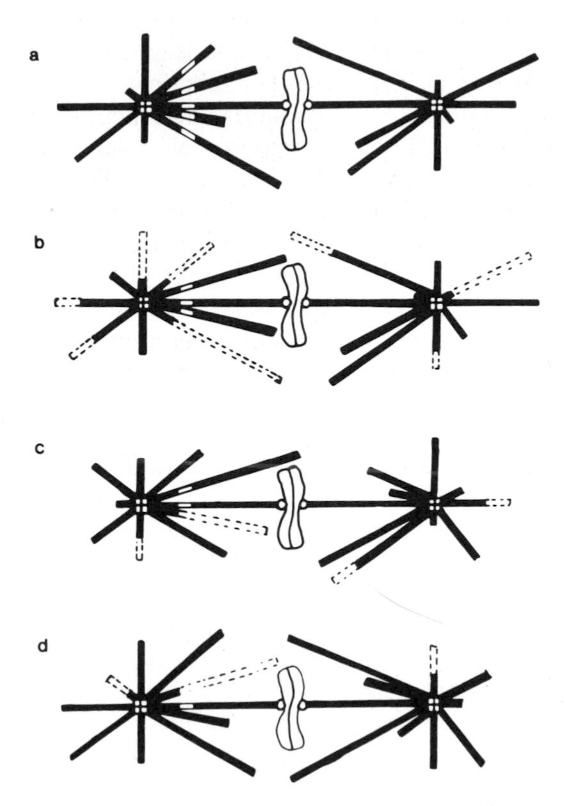

Figure 4.12. Sketch of the fluorescence recovery in a metaphase spindle by microtubule dynamic instability. For simplicity, few microtubules and one chromosome are drawn. (a) A bar pattern is photobleached across the left half-spindle. Solid black, fluorescent microtubules; white, photobleached regions. (b–d) As microtubules elongate, and rapidly shorten asynchronously, fluorescence recovers in the bar pattern without translocation of the bleach. Broken lines, original length of rapidly shortening microtubules. (From Cassimeris *et al.*, 1987a.)

polymerizing and rapidly depolymerizng completely (Fig. 4.12) in order to achieve the rapid rates and uniform pattern observed of fluorescence recovery as illustrated in Figs. 4.9, 4.10 and 4.11.

Another way to look at the dynamics of microtubules is to determine how fast they depolymerize when elongation is abruptly blocked. Colchicine and other drugs, like colcemid and nocodazole, that bind to the colchicine-binding site on tubulin have been used to abruptly block microtubule elongation in cells (Salmon *et al.*, 1984b; DeBrabander *et al.*, 1986). These drugs act by binding to a site on tubulin not exposed when tubulin is assembled into microtubules (Margolis and Wilson, 1977; Margolis *et al.*, 1980; Bergen and Borisy, 1983). It is the tubulin–drug complex which inhibits microtubule elongation and blocks nucleation. This can be demonstrated by microinjecting tubulin–colchicine complex into cells. The tubulin–colchicine complex is very stable and produces similar rates of spindle microtubule depolymerization to those produced by high concentrations of colchicine or nocodazole (Wadsworth and Salmon, 1986b; McIntosh *et al.*, 1986). In the presence of high concentrations of colchicine, colcemid, or nocodazole, non-kinetochore, polar spindle microtubules depolymerize with a half-time of 6–10 s depending on cell type (Fig. 4.4) (Salmon *et al.*, 1984b). The rates of microtubule depolymerization have been estimated to be of the order of 500–1000 dimers/s (Salmon *et al.*, 1984b) as predicted for rapid shortening during microtubule dynamic instability (Mitchison and Kirschner, 1984b; Horio and Hotani, 1986; Walker *et al.*, 1988).

Although the real-time behaviour of individual microtubules within the spindle has yet to be observed, the above observations leave little doubt that dynamic instability is the dominant mechanism of polar (non-kinetochore) microtubule assembly in the spindle as well as the CMTC. Thus, consideration of mechanisms that may produce or regulate the assembly of the mitotic spindle (Harris, 1978; Vandré *et al.*, 1984) must now be considered in terms of frequencies of nucleation, catastrophe and rescue as well as the rate constants of elongation and rapid shortening (Cassimeris *et al.*, 1987a). This is somewhat complicated. The amount of microtubule polymer can be increased by increasing the frequency of nucleation and/or the mean length of microtubules. Microtubule length can be increased by increasing the rate of elongation or the frequency of rescue and/or by decreasing the frequency of catastrophe or the rate of rapid shortening. Comprehensive information about all these parameters is not yet available. The mean elongation time for a polar spindle microtubule in newt cells is about 100 s based on a measured half-time of FRAP of 87 s (Wadsworth and Salmon, 1986a). The mean length of elongation before catastrophe is likely to be in the range of 10 μm in these spindles, whose interpolar length is 40–50 μm. These values are similar to the mean time of elongation and the mean length of elongation measured for the microtubules in the interphase newt CMTC (Cassimeris *et al.*, 1988b). It now appears that the velocity of elongation, the frequency of catastrophe and probably the rate of rapid shortening are similar during mitosis and interphase in the newt. This suggests that, during mitosis, the short mean length and life-time of spindle microtubules appears to be produced by changes in cell physiology that substantially reduce the frequency of rescue in comparison to the rate occurring during interphase. These arguments (see also Kirschner and Mitchison, 1986a) predict that polar spindle microtubules in general elongate at several micrometres per minute for about a minute before rapidly shortening and disappearing. This cycle would be repeating continuously, but asycronously, among the population of polar microtubules, so that at all times there are growing microtubule plus ends throughout the spindle (Fig. 4.12).

IV. Chromosome Fibre Formation by Microtubule Attachment to Kinetochores

A. Kinetochore capture of microtubules

Initially, it was thought that kinetochore microtubules predominantly formed by nucleated microtubule assembly and that chromosomal fibres resulted from the interdigitation of kinetochore and non-kinetochore microtubules (McIntosh *et al.*, 1969; Nicklas, 1971; McIntosh *et al.*, 1975). In lysed cell preparations, kinetochores as well as centrosomes were shown to initiate the assembly of microtubules (Snyder and McIntosh, 1975; Pepper and Brinkley, 1979; Telzer and Rosenbaum, 1979; Bergen *et al.*, 1980; Telzer *et al.*, 1980; Brinkley *et al.*, 1981; Brinkley, 1986; Mitchison and Kirschner, 1985a,b; McIntosh, 1985). However, nucleation of microtubules from kinetochores *in vitro* required much higher tubulin concentrations than required for centrosomes. Nucleation of microtubules from kinetochores *in vivo* was also demonstrated immediately following reversal of drug treatments that had induced complete spindle microtubule disassembly and high free tubulin concentrations (Witt *et al.*, 1980; DeBrabander *et al.*, 1981a,b,c, 1986). Once the spindle was reassembled and the free tubulin concentration was lowered, it was difficult to tell whether these nucleated microtubules persisted. Nevertheless, the above evidence makes it clear that kinetochores can initiate the growth of microtubules at elevated tubulin concentrations.

However, as initially stressed by Pickett-Heaps *et al.* (1982) and Rieder (1982), attachment, not nucleation, now appears to be the major mechanism of kinetochore microtubule formation during mitosis (Brinkley *et al.*, this volume). When chromosomes initially form connections to a spindle pole and begin moving towards that pole (orientation), microtubules are seen extending from the kinetochore polewards (Nicklas and Kubai, 1985; Nicklas, 1987a). As seen in Fig. 4.13(a), mono-oriented chromosomes have kinetochore microtubules only at the kinetochore facing the pole. The sister kinetochore faces away from the pole and is devoid of microtubules. This occurs for mono-oriented chromosomes close to or many micrometres away from the pole in either monopolar or bipolar spindles (Rieder *et al.*, 1986). If kinetochore microtubules normally formed by nucleation at kinetochores, microtubules should be evident at the kinetochores on mono-oriented chromosomes facing away from the pole. The concentration of tubulin is not expected to be much different between adjacent sister kinetochores. When a mono-oriented chromosome begins moving toward the opposite pole (when it bi-orients), microtubules are then seen extending towards that pole from the sister kinetochore facing that pole (Fig. 4.13(b)).

There are several other observations consistent with the capture concept (see also Pickett-Heaps *et al.*, 1982; Rieder, 1982). The plus ends of kinetochore

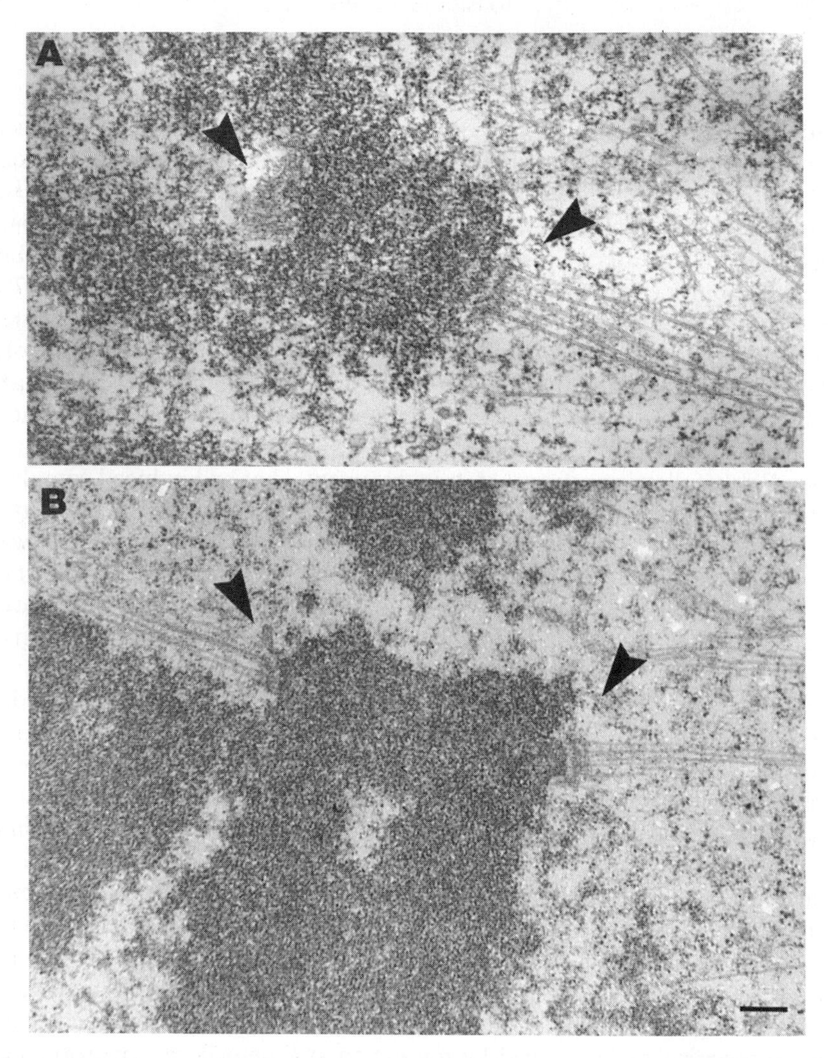

Figure 4.13. Electron micrographs comparing the attachment of microtubules to kinetochores of (A) mono-oriented and (B) bi-oriented newt chromosomes. In (A), the pole is to the lower right. There are no microtubules on the kinetochore facing away from the pole. Bar = 0.25 μm. (Original micrographs provided by Lynne Cassimeris and Conly Rieder.)

microtubules are proximal to the kinetochores (McIntosh and Euteneuer, 1984; McIntosh, 1984, 1985). Many kinetochore microtubules have been found to extend all the way between the kinetochore and the pole (Fig. 4.6). The dynamic instability of polar microtubule assembly can produce a continuous supply flux of microtubule ends in the vicinity of the kinetochores for efficient capture (Hill, 1985). The further a chromosome is from a pole, the longer it usually takes before it forms connections to that pole (personal observation in mitotic newt cells). This result is predicted, since the frequency of capture will depend on microtubule density, which decreases with distance from the pole. Finally, kinetochore capture of microtubule ends *in vitro* has recently been demonstrated for purified preparations of chromosomes and stabilized microtubules (Mitchison and Kirschner, 1985a,b; Huitorel and Kirschner, 1988; Koshland and Kirschner, 1988). Capture occurs efficiently at concentrations of microtubule ends similar to the values expected in the cell (Huitorel and Kirschner, 1988). Huitorel and Kirschner (1988) have shown that kinetochores can capture either plus or minus microtubule ends, but that the minus ends detach much more quickly than the plus ends.

In the living cell, Nicklas, Begg and co-workers (reviewed by Ellis and Begg, 1981) have shown that a kinetochore can be detached reversibly from microtubules using micromanipulation techniques. In first meiotic grasshopper spermatocytes, chromosomes can be broken free of a pole by hard jerks with a microneedle. The break does not occur within the chromosomal fibre, but by the detachment of microtubules from the kinetochore (Nicklas *et al.*, 1982; Nicklas and Kubai, 1985). The kinetochores are not damaged by the detachment process. Kinetochore microtubules reform, and the chromosome moves poleward and segregates normally in anaphase (see also Nicklas, 1987a, 1988a).

B. Differential stability of kinetochore microtubules

In contrast to the polar, non-kinetochore microtubules, kinetochore microtubules are differentially stable. Kinetochore microtubules have been seen to persist much longer than non-kinetochore microtubules following cooling (Fig. 4.5) (Brinkley and Cartwright, 1975; Lambert and Bajer, 1977; Salmon and Begg, 1980; Rieder, 1981), heating (Rieder and Bajer, 1977), pressurization (Salmon *et al.*, 1976), cell lysis (McIntosh *et al.*, 1975) or treatment with colchicine-like drugs (Brinkley *et al.*, 1967; Salmon *et al.*, 1984b). When biotinylated tubulin was microinjected into metaphase BSC1 cells, Mitchison *et al.* (1986) found that the non-kinetochore microtubules were fully labelled within about 1 min, while the kinetochore microtubules were still not fully labelled 10 min after microinjection.

The differential stability of kinetochore microtubules observed *in vivo* is most likely produced in part by the persistent attachment of their plus ends to kinetochores (Salmon *et al.*, 1976; Mitchison and Kirschner, 1985a,b). In support of this concept, Cassimeris *et al.* (1988) have measured the half-life of microtubule attachment to kinetochores in PtK_1 cells at 23°C following treatment of the cells with 10 μg/ml nocodazole to abruptly block microtubule elongation. The non-kinetochore microtubules disappear within seconds, while the half-life of kinetochore microtubules is 5–7 min; no microtubules are seen after 20 min. Mitchison and Kirschner (1985a,b) have shown in isolated preparations that the attachment of kinetochores to the distal ends of microtubules nucleated from centrosomes produces microtubules stable to tubulin dilution. A half-life of 17 min has been measured for the attachment of plus microtubule ends to kinetochores on isolated chromosomes *in vitro* (Huitorel and Kirschner, 1988).

Persistent attachment to kinetochores may be the primary mode of stabilization of kinetochore microtubules, but they could also be stabilized subsequently by complexes along their lengths. Membranes and proteins like calmodulin have been shown to be selectively localized to chromosomal fibres (Welsh, Sweet and Hepler, this volume). In the cold-stable chromosomal fibre, there is a high density of amorphous material that surrounds the bundles of microtubules (Fig. 4.14) (Brinkley and Cartwright, 1975; Rieder, 1981). The nature of this material is unknown. The persistent kinetochore fibre micro-

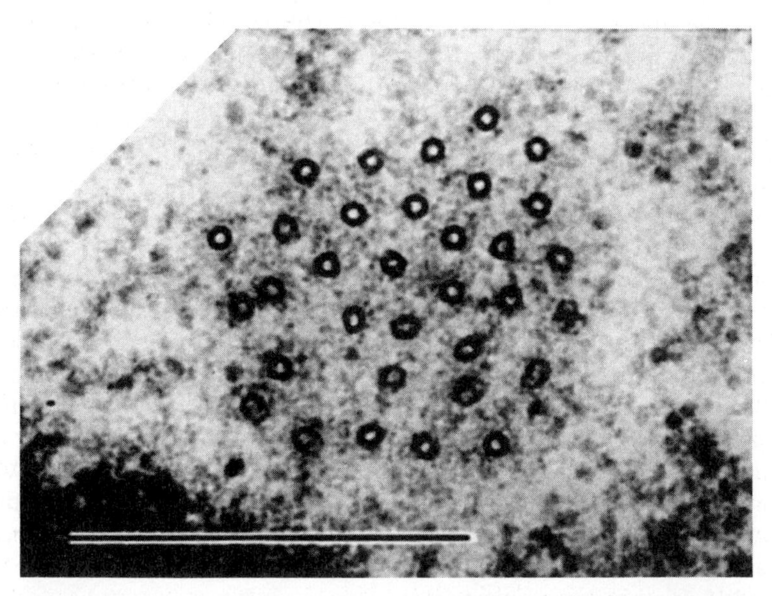

Figure 4.14. Electron micrograph of a cross-section through a cold-stable kinetochore fibre near the chromosomes. Note the electron-dense material that surrounds the microtubules. (From Rieder, 1981.)

tubules in cells depleted of non-kinetochore microtubules are held together into tight bundles (Fig. 4.5) by some form of unknown cross-linking (Rieder and Bajer, 1977; Brinkley and Cartwright, 1975; Lambert and Bajer, 1977; Witt *et al.*, 1981). These lateral associations between microtubules may be a general property of spindle microtubules as discussed later (reviewed in McIntosh, 1984, 1985).

V. Tubulin Association and Dissociation Occur At Kinetochores

Although attachment at the kinetochore may produce persistent kinetochore microtubules, attachment does not block tubulin exchange at the kinetochore. Studies using labelled tubulins and photobleaching techniques indicate that, during prometaphase and metaphase, tubulin can incorporate into kinetochore microtubules proximal to the kinetochore; in anaphase, tubulin can dissociate from the kinetochore microtubules also proximal to the kinetochore.

Mitchison *et al.* (1986) initially reached these conclusions on the basis of forward incorporation studies. They microinjected biotinylated tubulin into metaphase BSC1 cells, fixed them at various times after microinjection, labelled the biotinylated tubulin with gold-conjugated antibodies and analysed the preparations by electron microscopy. They found that kinetochore microtubules contained biotinylated tubulin within 1 min after injection (Fig. 4.15(a)). The kinetochore microtubules were labelled only within 0.5 μm proximal to the kinetochore; the remaining lengths of the kinetochore microtubules toward the poles were unlabelled. Using DTAF–tubulin, Wise *et al.* (1986) have confirmed that tubulin can incorporate into kinetochore microtubules at the kinetochore at metaphase (Fig. 4.18).

To see where tubulin left the kinetochore microtubules in anaphase, Mitchison and Kirschner (1986a) microinjected cells with biotinylated tubulin in late metaphase and fixed the cells at various times in anaphase. No label was found proximal to the kinetochores in mid anaphase, indicating that tubulin dissociates from kinetochore microtubules at the kinetochores as the kinetochores move polewards.

Gorbsky *et al.* (1987, 1988) have analysed the sites of tubulin dissociation from kinetochore microtubules in anaphase. Cells were microinjected with fluorescently labelled tubulin to uniformly label the spindle microtubules before anaphase. After the onset of anaphase, a narrow bar pattern was photobleached across a half-spindle between the kinetochores and a pole. The chromosomes were seen to move to and then through the persistent bleach pattern (Fig. 4.15(b) and Fig. 4.16). Little change occurred in the position of the bleach pattern with respect to the pole. Assuming that the unrecovered fluorescence in the bleach region represented stationary kinetochore micro-

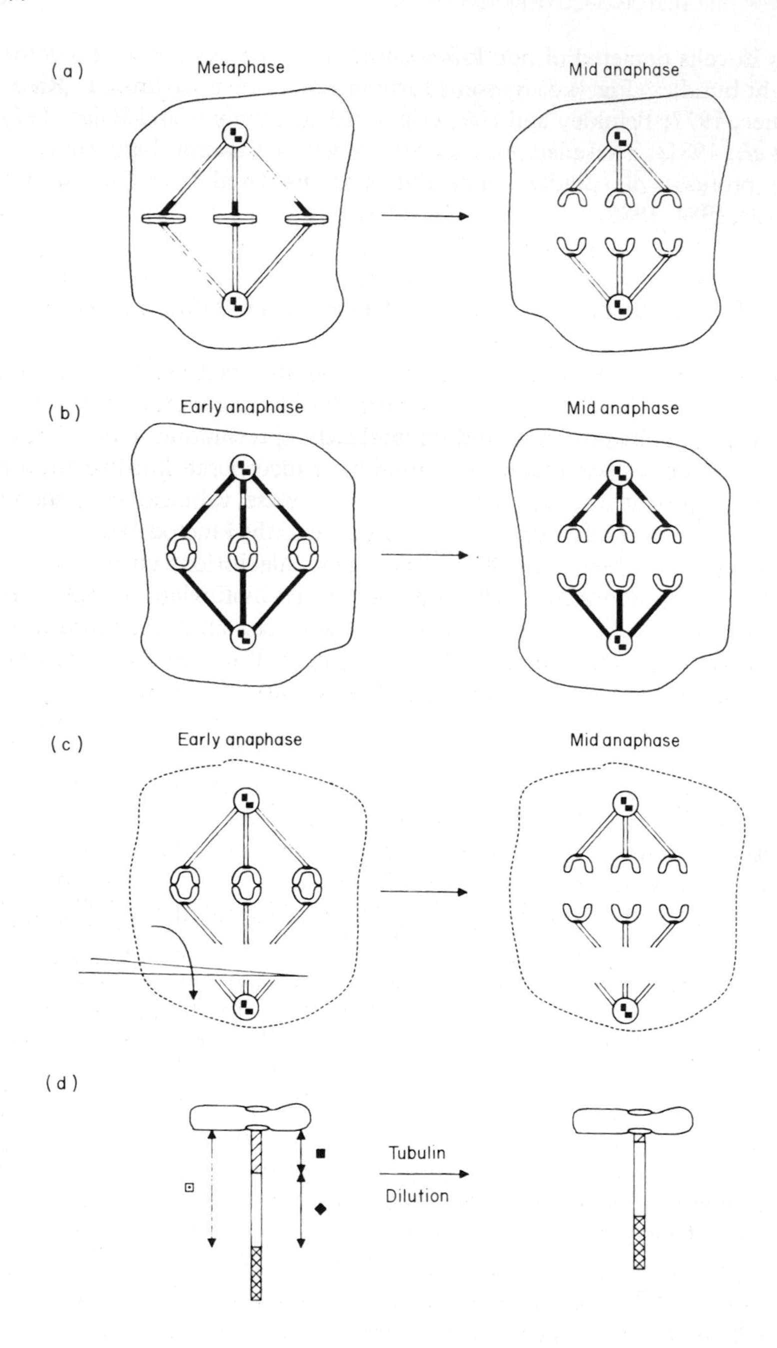
(a)
Metaphase
Mid anaphase
(b)
Early anaphase
Mid anaphase
(c)
Early anaphase
Mid anaphase
(d)
Tubulin
Dilution

Figure 4.15.

tubules, these results provide added support to the initial findings of Mitchison *et al.* (1986) that kinetochore microtubules shorten during anaphase poleward movement by losing subunits proximal to the kinetochore.

Nicklas (1987b) has provided additional support for this conclusion. He used a microneedle to cut the microtubules across the middle of the spindle between the chromosome and the pole. These experiments were done after the onset of anaphase in partially lysed preparations of first meiotic grasshopper spermatocytes. The chromosomes moved to the ends of the microtubules at the edge of the cut site; the chromosomal fibre microtubules did not appear to translate through the cut site (Fig. 4.15c).

Koshland *et al.* (1988) have approached this question using *in vitro* reconstituted preparations. They made short segments of microtubules labelled with biotinylated tubulin and stabilized these with a cross-linker that did not impair their ability to nucleate microtubule growth. They then assembled tubulin onto these segments using buffer conditions that produced preferential growth at the plus ends. These complexes were incubated with isolated chromosomes at concentrations that produced kinetochore capture of microtubule plus ends. When the concentration of tubulin was diluted, the number of microtubules attached to a kinetochore decreased. In the remaining attached microtubules, the labelled ends persisted and the distance between the label and the kinetochore became shorter over time (Fig. 4.15(d) and Fig. 4.17). They

---

Figure 4.15. Summary diagrams of the evidence for sites of tubulin association and dissociation in kinetochore microtubules. (a) Mitchison *et al.* (1986) found that when biotinylated tubulin was microinjected into metaphase BSC1 cells, tubulin incorporated into kinetochore microtubules only proximal to the kinetochores. When cells injected just before metaphase went into anaphase before they were fixed and stained for biotinylated tubulin, no label was seen proximal to the kinetochores. (b) Gorbsky *et al.* (1987, 1988) allowed fluorescently labelled tubulin to incorporate throughout the spindle microtubules then used photobleaching techniques to label a region of the kinetochore microtubules between the chromosomes and the poles during anaphase. They found that the kinetochores moved polewards towards the bleached zone, but the bleached zone remained stationary with respect to its near pole as seen in Fig. 4.16. (c) Nicklas (1987b) used a fine microneedle to sever the microtubules between the kinetochores and the poles during anaphase in partially lysed preparations of first meiotic grasshopper spermatocyte cells. The kinetochores moved to the edge of the cut zone as their chromosomal fibres shortened, but did not penetrate into the cut zone. (d) Koshland *et al.* (1988) assembled microtubules *in vitro* from stable, biotinylated labelled, microtubule fragments in a way that produced growth primarily from the plus ends. These segmented microtubules were bound to kinetochores on isolated mammalian chromosomes. When the concentration of free tubulin was reduced by dilution, the lengths of the unlabelled segments of microtubules that remained bound to the kinetochore were found to be shorter, indicating that depolymerization occurred at the kinetochore. Symbols in (d) define the lengths plotted in Fig. 4.17.

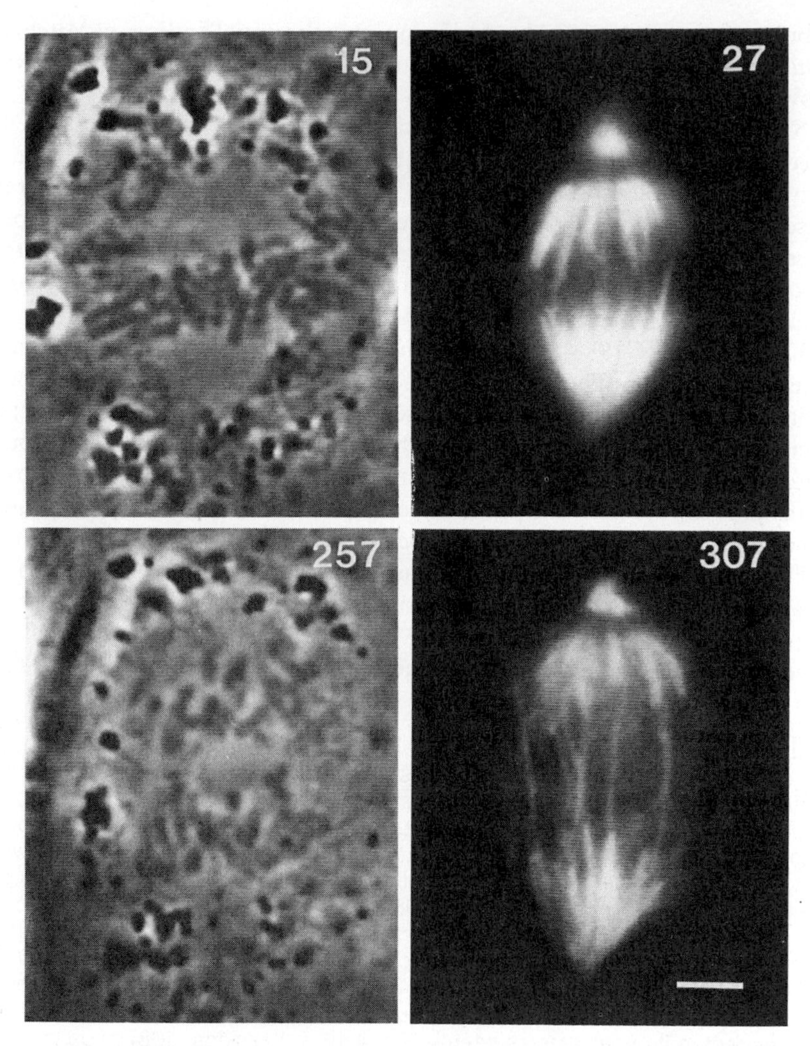

Figure 4.16. VIDEO-FRAP of a LLC-PK cell during anaphase demonstrating that kinetochores move polewards along stationary kinetochore microtubules. Cells were microinjected with rhodamine–tubulin to label the spindle microtubules. Time in seconds after photobleaching a bar pattern across the upper half-spindle is indicated on each frame. Corresponding phase-contrast and fluorescence video micrographs were recorded shortly after photobleaching and later after the chromosome-to-pole distance had shortened. The bleach zone remained stationary with respect to the near pole and accompanied it during pole–pole separation. Bar = 5 μm. (Original micrographs from Gary Gorbsky.)

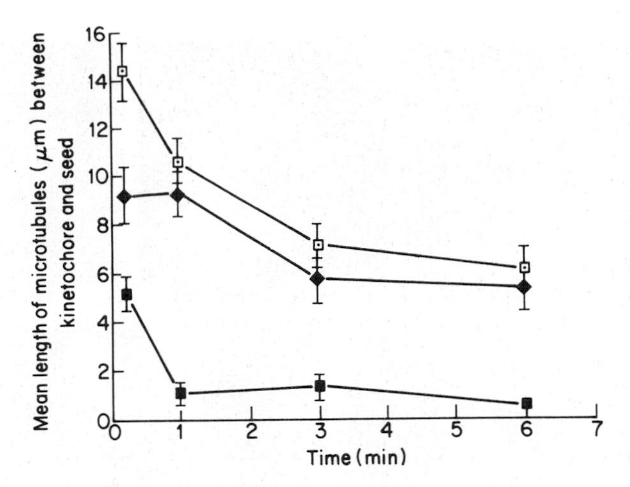

Figure 4.17. The shortening of kinetochore microtubules *in vitro* by tubulin dissociation at the kinetochore. The lengths represented by the different symbols are defined in Fig 4.15d. (From Koshland *et al.*, 1988.)

concluded from these results that microtubule shortening occurred by tubulin dissociation at the kinetochore.

In summary, the above experimental results show that chromosomes can be attached to their spindle poles while tubulin exchange occurs at their kinetochores. In addition, the shortening of kinetochore microtubules during poleward movements of chromosomes in anaphase and in reconstituted preparations *in vitro* involves tubulin dissociation at the kinetochore.

VI. Kinetochore Microtubule Dynamics at Metaphase may be Complex

An interesting puzzle concerns the assembly dynamics of kinetochore microtubules during metaphase (and prometaphase). Kinetochore microtubules in cultured mammalian cells at metaphase appear to become fully labelled with fluorescent or biotinylated tubulin throughout their length by about 10–15 min after microinjection (Saxton *et al.*, 1984; Mitchison *et al.*, 1986). At short times after injection, incorporation is proximal to the kinetochore (Mitchison *et al.*, 1986; Mitchison, 1988; Wise *et al.*, 1986). Over time, the extent of incorporation away from the kinetochore increases, but the extent of incorporation is different between sister kinetochore microtubules and between neighbouring chromosomes (Fig. 4.18) (Wise *et al.*, 1986; Mitchison, 1988).

There are several mechanisms which may contribute to the turnover of tubulin within kinetochore microtubules (Fig. 4.19). At first glance, this pattern of incorporation suggests that tubulin subunits flux poleward within the

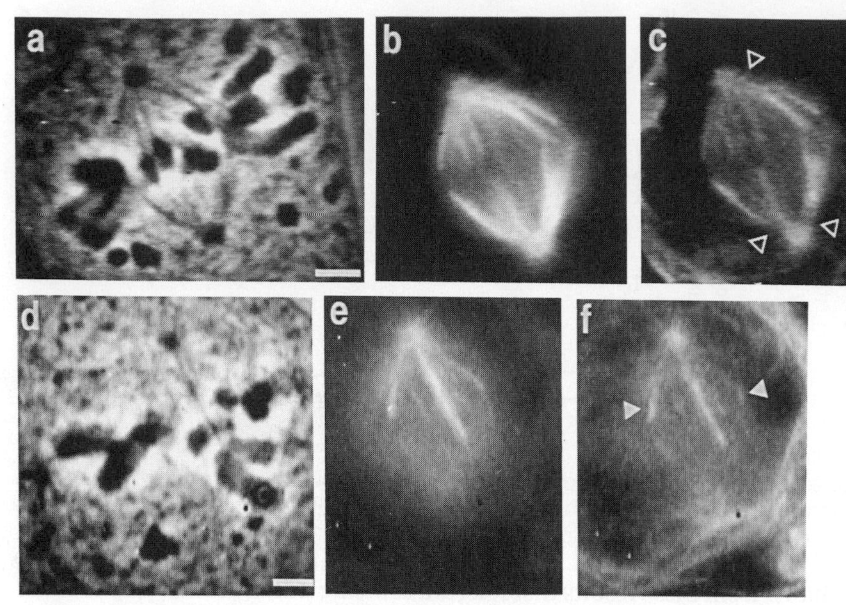

Figure 4.18. The asymmetrical incorporation of tubulin into kinetochore microtubules at metaphase in PtK_1 cells at 22–23°C. Five minutes after the cells were microinjected with DTAF–tubulin, they were lysed in a buffer to induce depolymerization of non-kinetochore microtubules, then fixed for immunofluorescence microscopy. The results for two cells are shown in (a)–(c) and (d)–(f). Only the lower surface of the upper half-spindle is in focus in (d)–(f). For each cell, the micrographs correspond to the same optical section. (a) and (d) are phase-contrast micrographs; (b) and (e) are micrographs of antitubulin immunofluorescence to show the extent of the kinetochore fibres; (c) and (f) are micrographs of antiDTAF–tubulin immunofluorescence to show DTAF–tubulin incorporation into the kinetochore microtubules. In (c) and (f) the secondary antibody had some non-specific staining for the intermediate filaments that surround the spindle region. In each cell, tubulin incorporation was proximal to the kinetochores, not the poles, and the extent of incorporation towards the pole was different for different kinetochores. Open arrows mark regions of non-incorporation while the solid arrows mark regions of incorporation. Bar = 2 μm. (From Wise *et al.*, 1986.)

lattice of kinetochore microtubules (Mitchison *et al.*, 1986; Mitchison, 1988), with the rate of treadmilling different for different kinetochores (Fig. 4.19b). This result is consistent with earlier observations of poleward movements of areas of reduced birefringence produced in chromosomal fibres by UV microbeam irradiation (Forer, 1965; Schaap and Forer, 1984a,b; Wilson and Forer, 1988) and the poleward transport of particle states along chromosomal fibres (Hard and Allen, 1977). There is one report that FRAP patterns move poleward in spindle fibres of a marine embryo (Hamaguchi *et al.*, 1987), but photobleaching studies in metaphase have not yet revealed any poleward flux

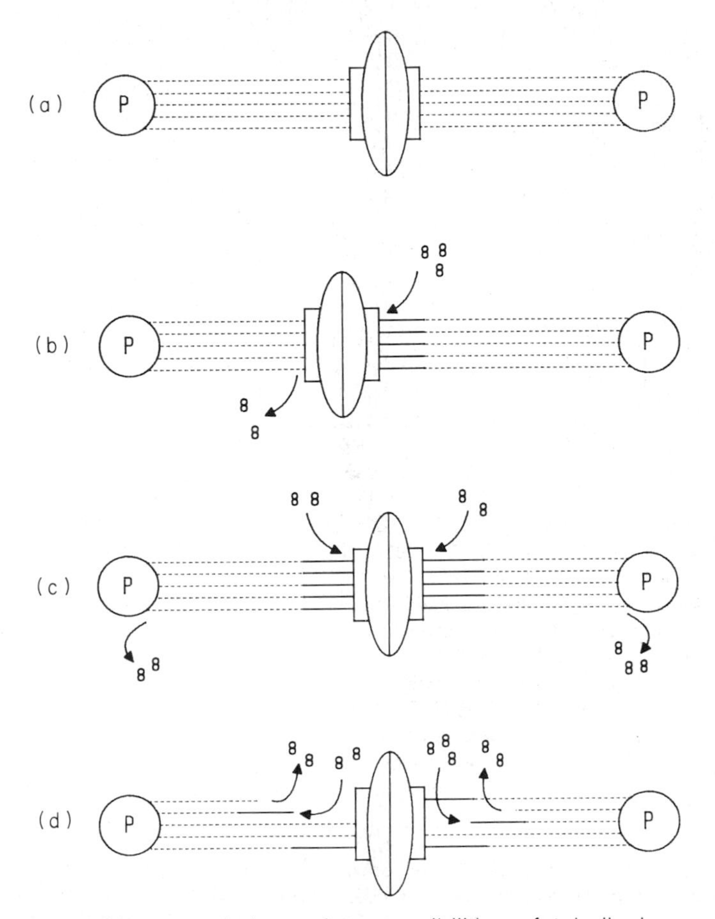

Figure 4.19. Diagram of the various possibilities of tubulin incorporation into kinetochore microtubules before anaphase. (a) A chromosome at the metaphase plate with four kinetochore microtubules attached to each sister kinetochore. Dotted lines represent kinetochore microtubules extending from poles, P, to kinetochores. (b) The chromosome has moved towards one pole during its oscillations back and forth at the metaphase plate. The kinetochore microtubules that shorten lose subunits at their kinetochore and the kinetochore microtubules that elongate gain subunits at their kinetochore. The dark lines represent newly added tubulin subunits. (c) Both sister kinetochore microtubules continuously incorporate subunits at the kinetochores and lose subunits at the poles, producing a treadmilling of subunits poleward. (d) Tubulin incorporation into kinetochore microtubules by the dynamic microtubules at the kinetochore. When a microtubule is released, it rapidly shortens for a distance before rescue, regrowth and recapture. Fully labelled polar microtubules could also be recaptured by the kinetochore.

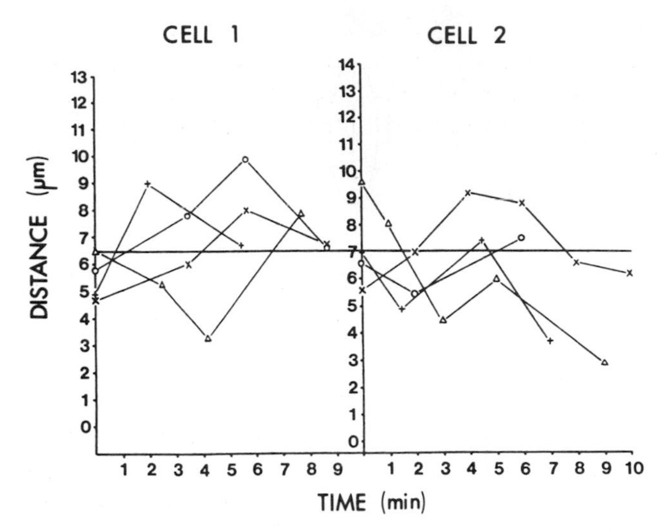

Figure 4.20. Plot of the oscillatory movements of bi-oriented chromosomes back and forth between the spindle poles during prometaphase and metaphase in PtK_1 cells at room temperature. Kinetic plots of the position of the centromere regions of chromosomes for two cells are shown. Four chromosomes were followed in each cell by time-lapse video phase microscopy. (Modified from Wise *et al.*, 1986.)

of tubulin in the chromosomal fibres of newt, BSC1 or PtK_1 cells (Fig. 4.9). Treadmilling in kinetochore microtubules could occur in some cell types, but not in others. The difference between sister kinetochores in the extent of tubulin incorporation proximal to the kinetochore suggests that some incorporation may be driven by the oscillatory movements of the chromosomes that occur back and forth between the spindle poles (Fig. 4.19(c)). As a chromosome oscillates away from a pole (Fig. 4.20), the kinetochore microtubules may elongate by tubulin incorporation at the kinetochore. Repeated back-and-forth oscillation would produce incorporation into both sister kinetochore microtubules, but the extent of incorporation toward the pole would be limited to the extent of the oscillations. Oscillations do not appear to occur far enough towards the poles to account for the complete exchange of tubulin throughout the length of kinetochore microtubules (Wise *et al.*, 1986). Since the lifetime of microtubule attachment at the kinetochore is of the order of several minutes, kinetochore microtubules could also become labelled by cycles of detachment, rapid shortening, rescue and reattachment (Fig. 4.15(d)) (McIntosh and Vigers, 1987; Cassimeris *et al.*, 1988a). Transient lateral associations between tightly clustered microtubules within the chromosomal fibre (see below) could promote rescue at a much higher frequency than occurs for the majority of polar microtubules within the spindle. Over time, the incorporation of labelled tubulin would increase in the kinetochore microtubules because more and more of the original kinetochore microtubules would be replaced with completely labelled polar microtubules. Unfortunately, there is insufficient evidence at this time to determine accurately whether one or all of the above processes are responsible for tubulin subunit turnover within kinetochore microtubules at metaphase.

VII. Chromosomal Fibres are Dynamic Clusters of Kinetochore and Non-kinetochore Microtubules

A. Transient lateral interactions

It is important to distinguish between chromosomal fibres and kinetochore microtubules. Chromosomal fibres are also frequently termed kinetochore fibres. They are a cluster of both kinetochore and non-kinetochore polar microtubules (McIntosh *et al.*, 1975; Inoué, 1981; Nicklas, 1985, 1987a; McDonald, this volume). Chromosomal fibres are distinctly visible in fixed spindles by antitubulin immunofluorescence microscopy (Brinkley *et al.*, 1980), thick-section electron microscopy (McIntosh *et al.*, 1975) and in living cells using fluorescent analogues of tubulin (Figs. 4.11 and Fig. 4.16). They can also be seen in living cells using polarization microscopy, as illustrated in Fig. 4.21.

Polarization microscopy has the advantage that dynamic aspects of microtubule assembly and alignment can be recorded in real time. Parallel bundles of microtubules are form-birefringent (Sato *et al.*, 1975) and they can be seen in either bright or dark contrast with a sensitive polarization microscope, depending on the orientation of a compensator. In the newt spindles shown in Fig. 4.21, the birefringent chromosomal fibres of mono-oriented and bi-oriented chromosomes appear to be about 1 μm in diameter and they change in length concurrently with the 1 μm/min oscillations of kinetochores towards and away from the poles (Bajer, 1982; Rieder *et al.*, 1986).

The kinetochores on newt chromosomes are about 0.25 μm in diameter and have attached to them about 25 kinetochore microtubules (Rieder *et al.*, 1986). This 0.25 μm diameter bundle of kinetochore microtubules is not visible at low resolution, but can be distinguished using the high-resolution, high-sensitivity polarization microscope of Inoué (1986) in large, flat mitotic cells such as newt lung epithelial cells (Cassimeris *et al.*, 1988a). This instrument has a depth of field of about 150 nm when a 1.4 NA objective is fully and uniformly illuminated by the condenser lens. As seen in Fig. 4.22, 0.25 μm diameter cables or bundles of kinetochore microtubules can be seen attached to kinetochores. The kinetochore microtubule bundles extend variable distances

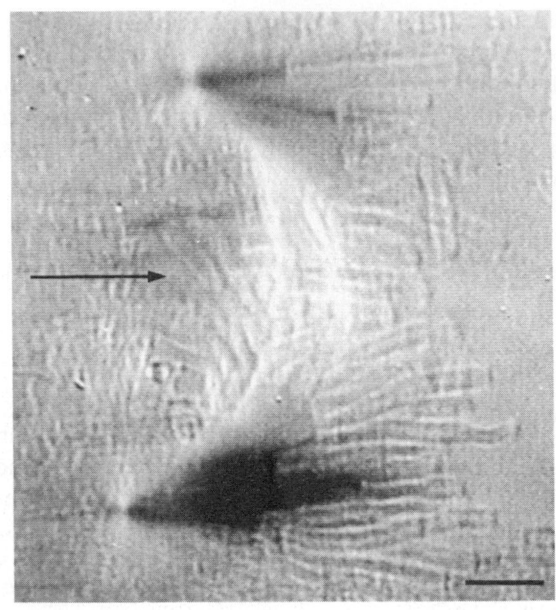

Figure 4.21. Polarization micrograph from a video sequence of very early prometaphase in newt cells. Many of the chromosomes are mono-oriented to either the upper or lower pole. The arrow indicates the position of the spindle equator. The contrast of the birefringent spindle fibres depends on their orientation with respect to the polarizer–analyser directions in the polarization microscope. Here they appear in bright contrast when oriented vertically and in dark contrast when oriented horizontally. Note that the chromosomal fibres only extend from kinetochores oriented to poles and that they appear to be about 1 μm in diameter. Images were obtained using a 40 × /NA = 0.75 Nikon rectified objective. Bar = 10 μm. (Original micrograph provided by Lynne Cassimeris and Shinya Inoué.)

Figure 4.22. High-resolution polarization micrographs from video sequences of mitosis in the newt. These optical sections are approximately 150 nm thick. Photographs are not all of the same cell. (a) and (b) are prometaphase while (c) and (d) are anaphase. Contrast of the spindle fibres varies with orientation as described in Fig. 4.21. Large black and white arrows point to the bundles or cables of kinetochore microtubules. Small black arrows point to fine fibular strands and rods that may be clusters of a few microtubules. (From Cassimeris *et al.*, 1988a.)

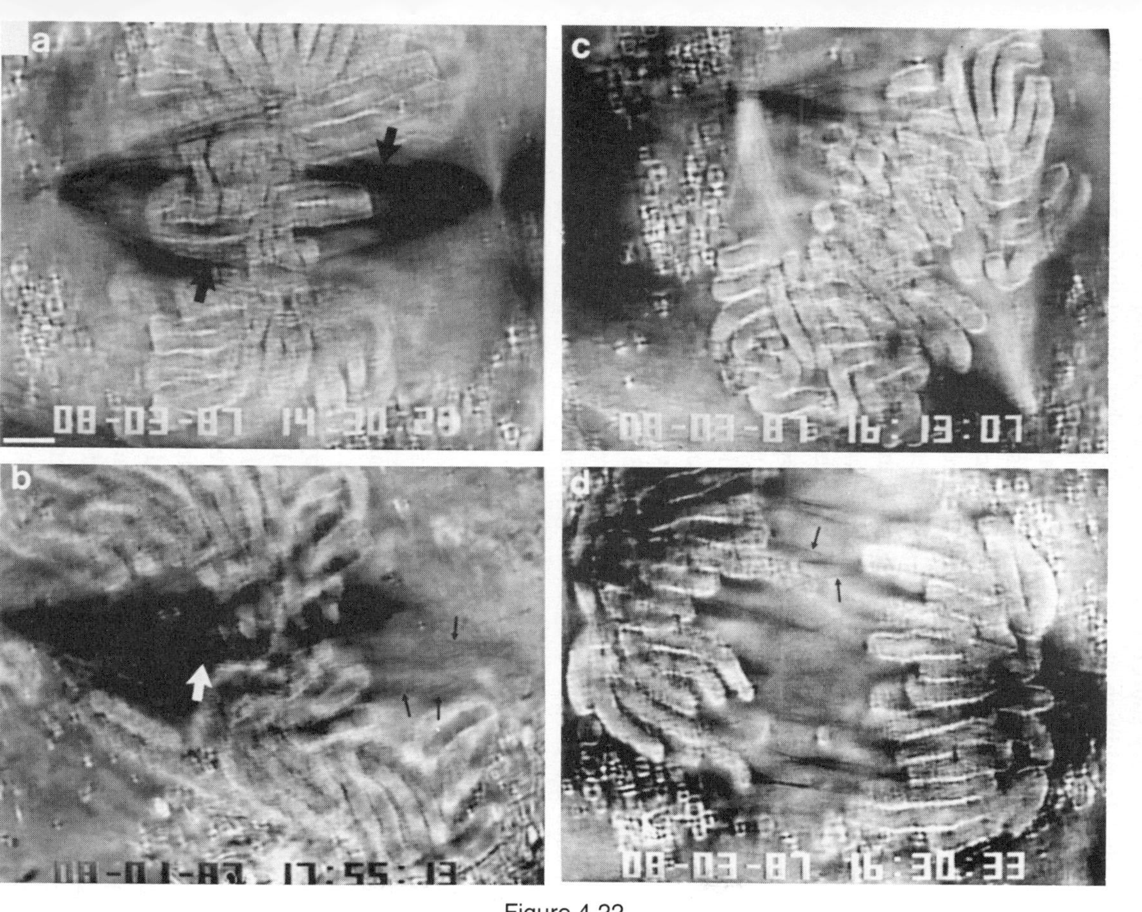

Figure 4.22.

towards the poles before splaying. Time-lapse recording of living cells has shown that the length of these bundles increases and decreases with time, becoming very short during full metaphase and anaphase (Cassimeris *et al.*, 1988a). In addition, it was also apparent that microtubules were transiently forming clusters, several micrometres in length, throughout the spindle by lateral interactions. As the microtubule density increased in the spindle towards metaphase and anaphase, transient lateral interactions between non-kinetochore microtubules and kinetochore microtubules appeared to pull apart the bundle of kinetochore microtubules. Cooling or heating cells to induce the disappearance of non-kinetochore microtubules produces tight bundles of microtubules extending from the kinetochores to the poles (Lambert and Bajer, 1977; Rieder and Bajer, 1977).

The above observations indicate that chromosomal fibres are dynamic clusters of kinetochore microtubules and non-kinetochore polar microtubules produced in part by the persistence of kinetochore microtubules and transient, weak, lateral interactions between kinetochore microtubules and non-kinetochore microtubules (Cassimeris *et al.*, 1988a). Lateral associations and cross-bridges between microtubules have been seen in electron micrographs of spindles (reviewed by Bajer and Mole-Bajer, 1975 and McIntosh, 1984, 1985). An important possibility is that the transient lateral associations of microtubules are produced by force generators like kinesin or dynein, but this important issue is not yet resolved.

B. Mechanical properties

The mechanical nature of chromosomal fibres has been studied in detail by centrifugation and micromanipulation methods. Results of many of these studies have been reviewed in depth by Ellis and Begg (1981) and Nicklas (1971, 1975, 1977, 1985, 1987a, 1988a, 1988b). Chromosomal fibres are much stiffer than chromatin (Nicklas, 1971, 1983). Chromosome arms can be stretched radially away from a pole over distances several times longer than their resting length without changing the length of its chromosomal fibre to the pole. Chromosomes can easily be pulled in an equatorial direction and rotated about a pole. Mono-oriented chromosomes, such as occur in early prometaphase and anaphase, are also easily pushed towards the pole. These observations characterize the chromosomal fibre as strong in tension but weak in compression, the characteristics of a thread made of strong, stiff elements. As discussed previously, the weakest link in the connection between a chromosome and the pole is the junction of the kinetochore and the kinetochore microtubules (Nicklas *et al.*, 1982; Nicklas and Kubai, 1985; Nicklas, 1985, 1987a).

Before discussing possible mechanisms of poleward-force generation, chromosome congression and spindle morphogenesis, I will first summarize several important characteristics associated with chromosome movement.

VIII. There Are Several Major Features of Chromosome Poleward Movement

A. Chromosomes are pulled polewards by their kinetochores

It is well established that, after attachment, chromosome arms are pulled towards the spindle poles throughout mitosis by forces at their kinetochores (Schrader, 1953; Mazia, 1961; Nicklas, 1971). The chromatin connecting kinetochores to the bulk of the chromosome arms can be seen to be stretched towards the poles in time-lapse records of living cells and in fixed preparations. Destroying the kinetochore blocks poleward movement of its chromosome (McNeil and Berns, 1981; Hays and Salmon, 1985; Rieder *et al.*, 1986).

B. Movement is tightly coupled to changes in chromosome fibre length

There is a tight coupling between the movement of chromosomes and the lengthening and shortening of their chromosomal fibres (reviewed by Ellis and Begg, 1981). This behaviour occurs for both mono-oriented and bi-oriented chromosomes naturally in prometaphase and anaphase and during experimental manipulations that do not produce detachment of kinetochores from their fibres. In prometaphase, the movement of a chromosome towards one pole is accompanied by the shortening of the chromosome fibre attached to that pole and the concurrent lengthing of the chromosomal fibre attached to the opposite pole (if it exists) (Nicklas, 1977; Ellis and Begg, 1981). In anaphase, kinetochores move polewards as their chromosomal fibres shorten (Fig. 4.16).

C. The resultant force on a chromosome determines whether a chromosomal fibre lengthens or shortens

Östergren (1945, 1951) initially proposed that chromosomes move to a position between the spindle poles at which opposing forces are balanced. At metaphase in a bipolar spindle, a chromosome duplex usually has chromosome fibres extending to opposite poles and poleward forces directed towards opposite poles. Since these chromosomes spontaneously move to an equatorial position between the poles, this would be the position at which opposing forces are balanced. There are several different kinds of evidence in support of this force-balance concept. At the onset of anaphase, these opposing forces are uncoupled and the poleward force at each kinetochore pulls its chromatid polewards. This same result occurs at metaphase when the opposing poleward forces are uncoupled. McNeil and Berns (1981) and Hays and Salmon (1985) have shown that when one kinetochore of a metaphase chromosome is completely destroyed, the chromosome moves from the metaphase plate toward the pole attached to the undamaged kinetochore. Conly Rieder and I have used a laser microbeam to sever the mechanical connection between sister chromatids in metaphase newt bipolar spindles. The separated chromatids move towards their attached poles at metaphase with kinetics typical of anaphase (unpublished observations).

A clear demonstration of how the resultant force on a chromosome directs movement has been produced by Nicklas (1977) in bipolar spindles of grasshopper spermatocytes near first meiotic metaphase. At this stage, the chromosomes have formed chromosomal fibres to both poles and they have congressed to near the spindle equator. By snagging the end of a chromosome arm with the tip of a fine microneedle, Nicklas stretched the arm towards one pole. The added poleward force was proportional to the amount of stretch of the elastic chromatin (Nicklas, 1983). This stretching resulted in the slow movement of the chromosome from an equatorial position between the poles towards the pole in the direction of stretch. If the stretch was maintained, the chromosome slowly moved all the way to the pole. When the microneedle was removed, the chromosome moved back to the spindle equator. At all times, the chromosome remained tethered to both poles by its chromosomal fibres. One fibre shortened, while the other elongated, at a velocity typical of chromosome movements during prometaphase during stretching and after release.

D. Tubulin kinetics probably govern velocity

Chromosome velocity with respect to the spindle poles is very slow compared to velocities associated with the unhindered translocation produced by kinesin, dynein or myosin (Nicklas, 1984; Vale *et al.*, 1985a; Vale, 1987; Paschal *et al.*, 1987; Lye *et al.*, 1987). Chromosome movements occur at velocities in the range of 0.1–5 μm/min depending on the type of cell and the temperature. Saltatory particle transport, both towards and away from the centrosome occurs at 1–10 μm/s, equivalent to the rates of kinesin and dynein motility measured *in vitro*. For comparison, if you read down this page at the speed a chromosome moves, it would take you about 200 days to get to the foot!

Chromosome velocity is load-independent for magnitudes of opposing force about two orders of magnitude greater than estimated for the normal drag force on the chromosome. The drag force on a chromosome moving at about 1 μm/min has been estimated at 1×10^{-7} dyne (Nicklas, 1963, 1965). Using

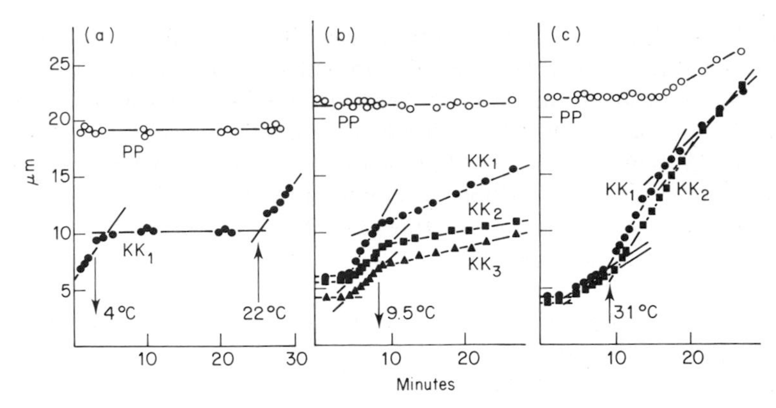

Figure 4.23. Abrupt shifts in temperature produce abrupt changes in the velocity of the poleward movement of chromosomes during anaphase in first-meiotic crane-fly spermatocytes. KK is the separation between kinetochores and PP is the interpolar distance. (From Salmon and Begg, 1980.)

micromanipulation methods, Nicklas (1983, 1988b) has found that the amount of force needed to stall chromosome poleward movement in anaphase of grasshopper spermatocytes is in the range of $2–5 \times 10^{-5}$ dyne. Thus, the rate of translocation of the kinetochore polewards is usually not limited by opposing force, but by the intrinsic rate of translocation of the poleward force-generation mechanism.

The velocity of chromosome movement is very sensitive to temperature (Mazia, 1961; Inoué, 1964; Fuseler, 1975; Lambert and Bajer, 1977; Salmon and Begg, 1980). In anaphase, abrupt changes in temperature have been shown to produce abrupt shifts in the velocity with which chromosomes move to the poles (Salmon and Begg, 1980). Figure 4.23 shows that different chromosomes can have different velocities, but that each shifts by the same relative amount when temperature is abruptly changed. Ahrrenius plots of velocity verses reciprocal temperature for crane fly spermatocytes produce straight lines and the Q_{10} is 2.8, typical of many enzymatic processes.

There does not appear to be a positive correlation between the number of kinetochore microtubules and the velocity of chromosome movement (Salmon, 1975c, 1976). The fastest rates of chromosome movement are typically seen in early prometaphase when chromosomes are initially forming attachments to the spindle and moving towards the poles.

The above observations suggest that the mechanism that governs chromosome velocity is probably the mechanism that regulates the kinetics of tubulin exchange with kinetochore microtubules (Salmon, 1975c, 1976; Salmon and Begg, 1980; Nicklas, 1975, 1987a). Although the direction of the resultant force on a chromosome appears to regulate the direction of chromosome movement

(Nicklas, 1977; Hays *et al.*, 1982), there is a rate-limiting step in the reactions that produce chromosomal fibre lengthening and shortening. In anaphase, these reactions probably occur at the kinetochore for two reasons. First, tubulin appears predominantly to leave kinetochore microtubules at the kinetochore (Mitchison *et al.*, 1986; Gorbsky *et al.*, 1987, 1988). Second, different chromosomes can exhibit different constant velocities at the same temperature (Fig. 4.21) (Salmon and Begg, 1980), a result consistent with velocity regulation at the kinetochore.

E. Poleward force at a kinetochore depends on kinetochore microtubule number

Thus far, techniques have not been developed to measure directly the relation between the number of kinetochore microtubules and the poleward force at the kinetochore. Hays and co-workers (Hays *et al.*, 1982; Hays and Salmon, 1985) measured this relation indirectly in the following ways using chromosome equilibrium positions within the spindle as an indicator of poleward force at a kinetochore. In the first study (Hays *et al.*, 1982), multivalent chromosomes were generated from the fusion of chromosome fragments produced by γ-irradiation of grasshopper nymphs. These multivalent chromosomes had one to four functional kinetochore complexes in first meiotic spermatocytes. Hays *et al.* (1982) found that multivalent chromosomes congressed to a position between the spindle poles closer to the pole that was attached to the greater number of kinetochore complexes. Östergren observed similar behaviour earlier for naturally occurring multivalents (1949, 1951). Increasing the number of chromosomal fibres to a pole and, hence, the number of kinetochore microtubules, increased the poleward force on the chromosome. In a second study, Hays and Salmon (1985; Hays, 1986) used Bern's laser microbeam methods (McNeil and Berns, 1981) to reduce the number of kinetochore microtubules on bivalent metaphase chromosomes in grasshopper spermatocytes at first meiotic metaphase. Electron microscopy showed that opposing kinetochores on chromosomes at the metaphase plate had similar number kinetochore microtubules. Damaging a kinetochore resulted in the movement of the chromosome complex closer to the pole attached to the undamaged kinetochore (Fig. 4.24). After the damaged chromosome achieved a new metaphase equilibrium position, the cells were fixed and processed for electron microscopy and the number of kinetochore microtubules at the damaged and undamaged kinetochores was measured by serial-section analysis. Undamaged kinetochores had about 38 kinetochore microtubules, independently of the lengths of their kinetochore fibres. Damaged kinetochores had fewer microtubules in proportion to the degree of damage. There was a direct correlation between the difference in number of kinetochore

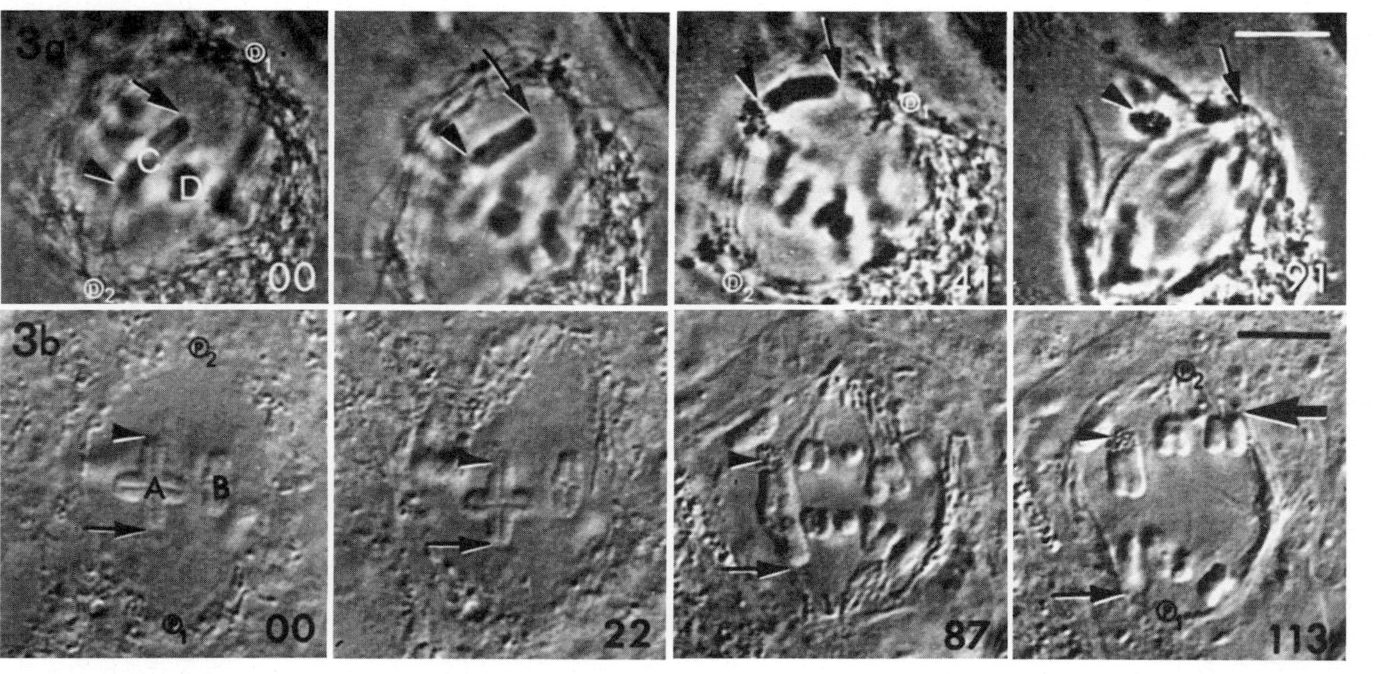

Figure 4.24. Laser microsurgery of a kinetochore induces a shift in the position of its chromosome between the spindle poles at metaphase in grasshopper first meiotic spermatocytes. Video micrographs for two experiments are shown. (a) The total face of the lower kinetochore (arrowhead) of bivalent (C) was irradiated (at 00 min). The chromosome moved towards the upper pole, P_1, to which the unirradiated kinetochore (arrow) was attached (11 and 41 min). At anaphase (91 min), the irradiated kinetochore did not segregate properly to the lower pole, P_2, but remained in the upper half-spindle (arrowhead). Unirradiated bivalents, e.g. chromosome D, divided normally. (b) Microbeaming the part of the upper kinetochore (arrowhead) of bivalent A (at 00 min) resulted in a partial shift in the bivalent towards the lower spindle pole, P_1 (22 min). At anaphase (87 min), both kinetochores and their homologues were segregated properly to opposite poles (113 min). (From Hays and Salmon, 1985.)

microtubules at undamaged and damaged kinetochores on a chromosome complex and the change in position of the chromosome between the spindle poles. The fewer the number of microtubules on the irradiated kinetochore, the further the chromosome moved away from the pole faced by the irradiated kinetochore and toward the pole faced by the undamaged kinetochore (Fig. 4.25). Thus, based on the force balance behaviour of chromosome congression,

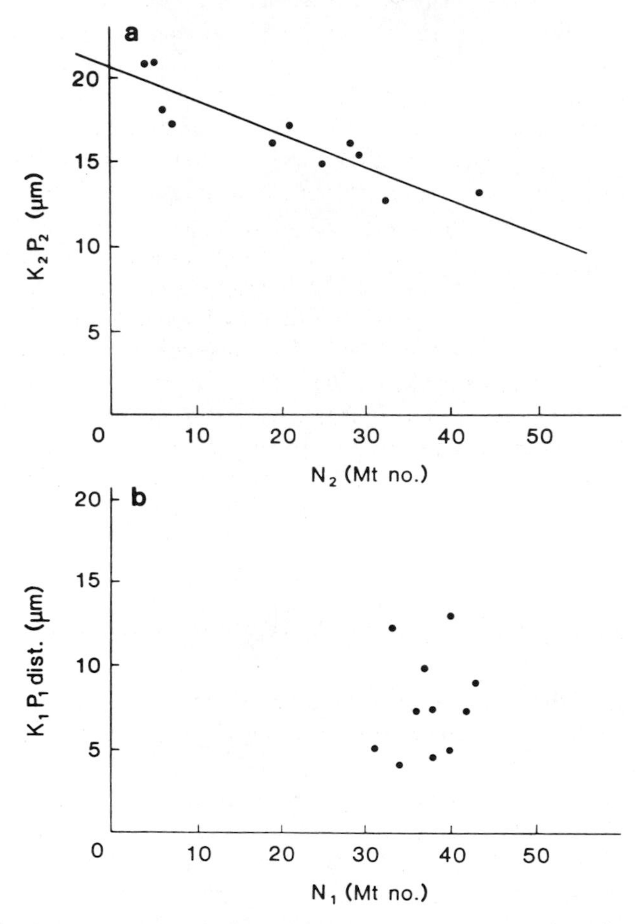

Figure 4.25. Plots of chromosomal fibre length versus kinetochore microtubule number for (a) irradiated and (b) unirradiated homologous kinetochores. The values for the lengths of the fibres, L_1 and L_2 from kinetochores K_1 and K_2 to poles P_1 and P_2, respectively, are plotted against the number of kinetochore microtubules, N_1 and N_2. The values of fibre length and kinetochore microtubule number were determined from serial thin sections of irradiated cells as described below Fig. 4.24, fixed after the irradiated chromosome had reached a new congression equilibrium position between the poles. (From Hays, 1986.)

poleward force at a kinetochore is reduced by reducing the size of a kinetochore and the number of kinetochore microtubules.

F. Poleward force at the kinetochore can oscillate in strength

Bajer was the first to stress the importance of considering the oscillations of chromosomes in analysing the mechanisms that produce and govern chromosome movements in the spindle. In newt (Bajer, 1982; Rieder *et al.*, 1986) and PtK_1 (Fig. 4.20) spindles, chromosome oscillations are pronounced during mitosis. Mono-oriented chromosomes in both monopolar and bipolar spindles stochastically move several micrometres towards and then away from their poles. Bi-oriented chromosomes at the spindle equator also oscillate to and fro between the spindle poles (Fig. 4.22). Oscillations in chromosome position could be produced by variations in either the strength of the poleward force at the kinetochore or the outward force on the chromosome arms produced by polar ejection forces (see below). Since the degree of stretch of the chromatin at the centromere region increases with poleward movement and decreases with movement away from the pole (Bajer, 1982; Rieder *et al.*, 1986), it is probably a variation in the poleward force at the kinetochore that makes the major contribution to frequent oscillations in chromosome position. Oscillation in the strength of the poleward force at the kinetochore could be due to changes in the number of kinetochore microtubules or the chemistry of the kinetochore.

IX. Poleward Force may be Generated at the Kinetochore

The mechanism of poleward force-generation is unresolved at this time. I present here several possible schemes that involve only microtubule-dependent force-generating mechanisms, because evidence for a function of actin or myosin in poleward chromosome movement is negative and chromosome movement requires microtubules attached to the kinetochore (reviewed by Inoué, 1981b). This latter fact excludes models that propose that the poleward force at the kinetochore is generated independently of kinetochore microtubules by contractile or elastic complexes extending between the kinetochore and the pole. The variety of microtubule-dependent models proposed for generating poleward force at the kinetochore can be divided into two major classes: "traction fibre models" in which the kinetochore microtubules pull the kinetochore polewards, and "kinetochore motor" models in which the kinetochore pulls the chromosome polewards along stationary kinetochore microtubules (Fig. 4.26).

A. Traction fibre models

Traction fibre models treat the kinetochore solely as a handle on the chromosome for attaching microtubules, and poleward force is generated within the chromosomal fibre. This has been the working hypothesis in the field of mitosis since the early 1950s. In Fig. 4.26, I have drawn two different schemes which could generate poleward force by traction fibre mechanisms that differ in the mechanism of force generation.

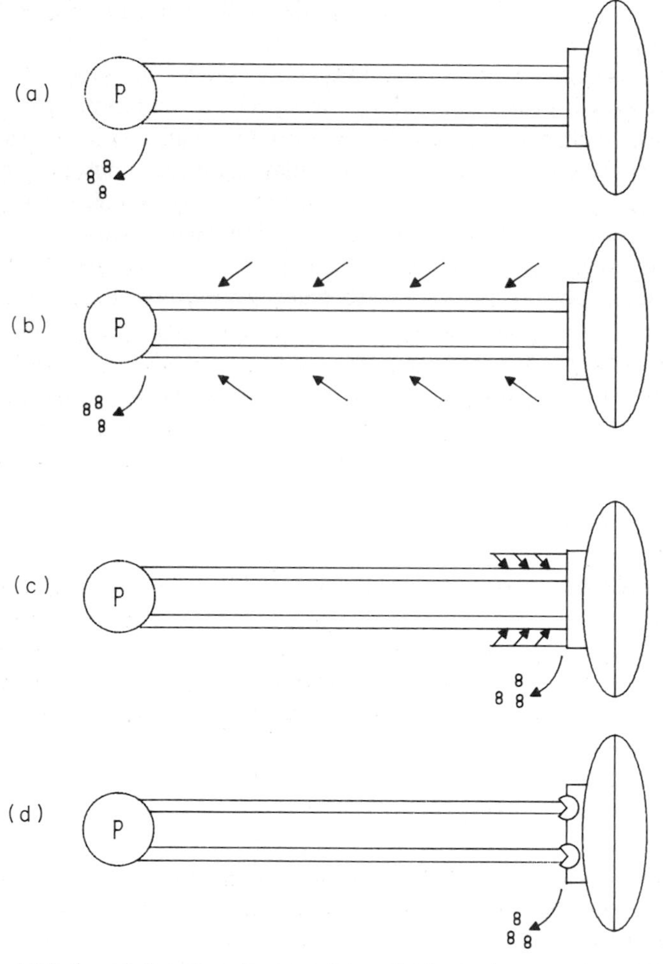

Figure 4.26. (a, b) Traction fibre and (c, d) kinetochore motor models for force production. Details of the models are given in the text. In all models, tubulin incorporation occurs at the kinetochore; only proposed sites for dissociation are shown.

The first traction fibre mechanism (Fig. 4.26(a)) involves a balance between tubulin incorporation at the kinetochore and dissociation near the pole, or treadmilling of tubulin along the kinetochore microtubules (derived from Inoué and Sato, 1967; Inoué, 1981b). When incorporation at the kinetochore matches dissociation at the pole, the chromosomal fibre has a constant steady-state length. If incorporation is blocked at the kinetochore, then the kinetochore moves polewards at the rate of poleward disassembly. The energetics of the disassembly reaction drives the poleward movement and poleward force could be proportional to the number of kinetochore microtubules.

The second traction fibre mechanism (Fig. 4.26(b)) involves treadmilling of tubulin as described above but also proposes force producers that actively push kinetochore microtubules polewards (derived from McIntosh *et al.*, 1969; McIntosh *et al.*, 1975; Margolis *et al.*, 1978; Margolis and Wilson, 1981; McIntosh, 1984, 1985). Kinesin is a good possibility for this type of force producer since experiments *in vitro* have demonstrated that kinesin actively moves microtubules towards their minus ends (Vale *et al.*, 1985b, 1987; Pryer *et al.*, 1986). In sea-urchin embryos, Scholey *et al.* (1985) have shown that kinesin is localized within the spindle fibre microtubules. The inactive end of kinesin could be bound to other polar microtubules or to some other lattice (Leslie *et al.*, 1987) or membranous components within the spindle (Hepler, this volume). In this model, poleward force at a kinetochore would be a function of the length of kinetochore microtubules as well as their number. Since many kinetochore microtubules extend all the way to the pole, the compression of these microtubules could take up much of the force generated along their lengths and only a fraction of the generated force would be applied to the kinetochore. At the pole these microtubules would be in compression, while at the kinetochore they would be in tension. Theoretically, compression can promote depolymerization and tension can promote polymerization (Hill and Kirschner, 1982), so that the stresses generated by the translocators along the kinetochore microtubules would be predicted to produce poleward flux of tubulin subunits.

B. Kinetochore motor models

The kinetochore motor models differ from the traction fibre models in that the kinetochore complex is involved in poleward force generation as well as microtubule attachment.

One possibility (Fig. 4.26(c)) (derived from Pickett-Heaps *et al.*, 1982) is that there are filaments that extend away from the kinetochore and interdigitate with the kinetochore microtubules. Attached to these filaments could be force producers that pull the kinetochore microtubules toward their attachment sites at the kinetochore. Initially, kinetochores would slide polewards along the walls of microtubules until their ends were pulled into the attachment sites at the kinetochore. Then poleward movement would depend on the tubulin dynamics at the attachment sites. There is structural evidence in a plant cell for filaments extending several micrometres from the kinetochore, but their existence in other cell types is uncertain (Schibler and Pickett-Heaps, 1980; Pickett-Heaps *et al.*, 1982). Associated with the outer plate of the kinetochore in mammalian cells is a fine filamentous network termed the corona (Rieder, 1982; Brinkley *et al.*, this volume), but its function is unknown. These filaments could produce poleward force in conjunction with cytoplasmic forms of dynein. *In vitro*, these translocators actively move microtubules toward their plus ends as required by this model (Sale and Satir, 1977; Paschal *et al.* 1987). In this scheme, the lengthening and shortening of kinetochore microtubules can occur solely at ends attached to the kinetochore, without requiring the complexity of subunit exchange at the spindle poles. The kinetochore microtubules that extend to the poles will be in tension for regions close to the pole and in compression or tension for regions close to the kinetochore. In anaphase, the region proximal to the kinetochore will be in compression, which would promote tubulin dissociation at the kinetochore. During metaphase, tension in microtubule regions near a kinetochore would occur if the poleward force at the opposing kinetochore were greater. This tension could promote tubulin association and elongation, while the compression in the opposing kinetochore micro tubules could promote tubulin dissociation and shortening in those microtubules. If the association and dissociation reactions are rate-limiting steps, then only a small fraction of the forces generated along the kinetochore microtubules would be transmitted to the centromere chromatin between opposing kinetochores. In this model, poleward force would depend on the length of the kinetochore microtubules overlapping with the filaments extending from the kinetochore and may not be proportional to the number of kinetochore microtubules.

The second version of the kinetochore motor model we term the "Pac-man" model (Fig. 4.26(d)) (derived from Mitchison and Kirschner, 1985b; Mitchison *et al.*, 1986; Kirschner and Mitchison, 1986a,b; Cassimeris *et al.*, 1987a; Huitorel and Kirschner, 1988; Koshland *et al.*, 1988; Gorbsky *et al.*, 1987, 1988). This model differs from the first version only by the localization of the poleward force producers within or near the kinetochore attachment site rather than on some filamentous complex extending from the kinetochore. The kinetochore grabs the ends of polar microtubules and tries to pull itself poleward while depolymerization and shortening occur proximal to the kinetochore. Each attachment site is viewed as a unit force producer. The strength of the poleward force would be proportional to the number of kinetochore microtubules and could be generated by distinctly different types of force producers, as described next.

C. Recent evidence for a kinetochore motor

There is now strong support for some type of kinetochore motor. The traction fibre and kinetochore motor models in Fig. 4.26 make specific predictions about the sites of tubulin association and dissociation within kinetochore microtubules. The traction fibre models predict tubulin flux and dissociation towards the pole whereas the kinetochore motor models are consistent with tubulin association and dissociation occurring only at the kinetochore. As discussed previously, the available data indicate that kinetochores move polewards in anaphase mammalian cells along stationary kinetochore microtubules, indicating that tubulin dissociation occurs predominantly at the kinetochore (Figs. 4.17, 4.18 and 4.19). Removing the spindle pole does not appear to block the movement of the chromosomes or the shortening of the persistent chromosomal fibres (Hiromoto and Shoji, 1982; Hiromoto and Nakano, 1988; Nicklas, 1987b).

Before anaphase, evidence clearly supports tubulin incorporation into kinetochore microtubules at the kinetochore (Fig. 4.17 and 4.20), but the sites of tubulin dissociation are uncertain. We have been unable to see any poleward movement of bleach patterns within chromosomal fibres in our FRAP studies on PtK_1 (Fig. 4.11) and newt cells (Wadsworth and Salmon, 1986a; Cassimeris *et al.*, 1988a), indicating that treadmilling of tubulin poleward does not take place at detectable rates in these cell types at metaphase. This issue will not be resolved completely until techniques are developed to detect the sites of dissociation independently from complications induced by the dynamic non-kinetochore microtubules and cycles of microtubule detachment and reattachment at the kinetochore. Mitchison *et al.* (1986) have suggested that a kinetochore motor could be turned on in anaphase, but is inactive or less active in metaphase. However, when sister chromatids in newt spindles are separated in metaphase by laser microsurgery, the kinetics of poleward movement are nearly identical to those in normal anaphase (Salmon and Rieder, unpublished observation). This suggests that the motors that move chromosomes polewards in anaphase also operate before anaphase.

There may be molecular complexes that bind along kinetochore microtubules in higher concentrations toward the poles. A kinetochore motor could use this concentration gradient to regulate the strength of poleward force-generation. Thus, poleward force would be greater the further the kinetochore is from the pole, which is the required vectorial relationship to produce spontaneous alignment of chromosomes at the spindle equator, as discussed below.

Koshland *et al.* (1988) have provided evidence that the disassembly of microtubules attached to kinetochores *in vitro* can occur by tubulin dissociation at the kinetochore without microtubule detachment (Figs. 4.15d and 4.17). The rate of shortening of these kinetochore microtubules following dilution of the tubulin concentration was the same in the presence and absence of ATP.

From this they suggest that poleward force at the kinetochore may be generated by some form of energy stored in the microtubule lattice (perhaps a conformational change at the microtubule end generated by the hydrolysis of the GTP bound to tubulin) or by a diffusion motor as proposed by Hill (1985). Inoué initially proposed in his dynamic equilibrium model of spindle assembly and function that tubulin insertion into kinetochore microtubules could produce pushing forces, while tubulin dissociation would produce pulling forces (Inoué, 1964, 1976, 1981b; Inoué and Sato, 1967; Inoué and Ritter, 1975). It has always been difficult to comprehend how microtubules could shorten and still maintain their mechanical integrity. The kinetochore appears to resolve this issue.

In the cell, there is evidence that anaphase poleward movement requires protein phosphorylation, but can occur in the absence of ATP (Spurck and Pickett-Heaps, 1987). However, ATP dependence is controversial (Hepler and Palevitz, 1985) and the identity of the poleward force producer(s) in the cell is not yet resolved. Since it does not take much force to drag a chromosome through the cytoplasm (Nicklas, 1965), an ATPase, such as dynein, could be attached to the kinetochore attachment sites *in vivo* in very low numbers and contribute significantly to poleward force production at very low concentrations of ATP (Nicklas, 1984). Nevertheless, the kinetochore is clearly not just a handle on the chromosome, but appears to be an organelle capable of transducing tubulin association and dissociation reactions and external forces into translocation along microtubules.

X. Polar Microtubule Arrays Can Push Chromosomes Away From the Pole

A. Chromosome congression in monopolar spindles

In addition to the poleward forces at kinetochores, chromosomes in animal spindles are also pushed outwards away from the spindle poles by a second class of forces. These polar ejection forces are generated by interactions between the polar spindle microtubule arrays and the chromosome arms (Mole-Bajer *et al.*, 1975; Bajer, 1982; Rieder *et al.*, 1986).

The actions of these two different forces are very apparent in monopolar spindles. Monopolar spindles in newt mitotic cells have provided much useful information. Figure 4.4 shows a monopolar and a bipolar spindle printed at the same magnification. In the monopolar spindles, a chromosome is pulled toward the pole by poleward directed forces at the kinetochore that faces the pole. Only kinetochores facing the pole form chromosomal fibres as seen in the polarization micrographs in Fig. 4.21. Ultrastructural analysis has shown that the number of kinetochore microtubules in monopolar spindles is similar to the number in bipolar spindles (Cassimeris and Rieder, unpublished observations). Kinetochores that face away from the pole have no kinetochore microtubules (Fig. 4.13) and these kinetochores have been shown by microsurgery studies to be inactive (Rieder *et al.*, 1986). Thus, in monopolar spindles, there is one poleward force on the chromosome.

Attached chromosomes in monopolar spindles do not usually move all the way to the pole. Instead, they achieve average distances from the pole that are similar in magnitude to the distance between the metaphase plate and the poles of a normal bipolar spindle (Fig. 4.4, Fig. 4.15). Why are chromosomes in monopolar spindles not pulled by the poleward force at their kinetochores all the way to the pole?

Rieder *et al.* (1986) have shown in newt cells that forces associated with the polar microtubule arrays push the chromosome arms outward away from the pole. Chromosome arms extend radially outward in monopolar microtubule arrays. When chromosome arms are cut free of the kinetochore complex by laser microsurgery, they are transported at 1–2 μm/min radially outwards (Fig. 4.27). Small severed fragments of the chromosome containing the kinetochore complex move up close to the pole (Fig. 4.28) (Salmon and Rieder, unpublished observation). When polar microtubules are depolymerized by cold or nocodazole, centric chromosomes will move up close to the pole (Rieder *et al.*, 1985; Cassimeris *et al.*, 1987b). Upon rewarming and reassembly of the polar microtubule arrays, chromosomes are pushed away from the pole. The lifetime and mean length of microtubules in monopolar spindles of the newt appear similar to the respective values in bipolar spindles (Cassimeris *et al.*, 1987b).

B. Origins of polar ejection forces

These studies have shown that chromosomes are pushed outwards away from a pole by ejection forces associated with the polar array of microtubules, which act at sites all along the length of the chromosome (Rieder *et al.*, 1986). The molecular origins of the polar ejection force are unknown. It may be produced by the pushing of elongating polar microtubules, which occurs continuously because of their dynamic instability assembly. Bajer *et al.* (1982) have shown that taxol-induced polymerization of polar microtubules can severely stretch chromosome arms away from the pole. The outward force could also be generated by translocator molecules located on the surface of the chromosome that have the transport polarity of kinesin. In either case, the magnitude of this polar ejection force at a region of the chromosome is likely to depend on the density of polar microtubules in that region and may depend on the orientation of the chromosome surface with respect to the orientation of the microtubules.

Microtubule density increases closer to the pole, so that the strength of the outward force on a chromosome will increase as the chromosome moves closer to the pole.

C. Chromosome congression in a bipolar spindle

The net orientation and magnitude of the resultant force on a chromosome depends on the geometry of the spindle. In monopolar spindles, chromosome position is produced by the resultant action of one poleward force and one outward force, which are oriented in opposite radial directions (Fig. 4.27). In

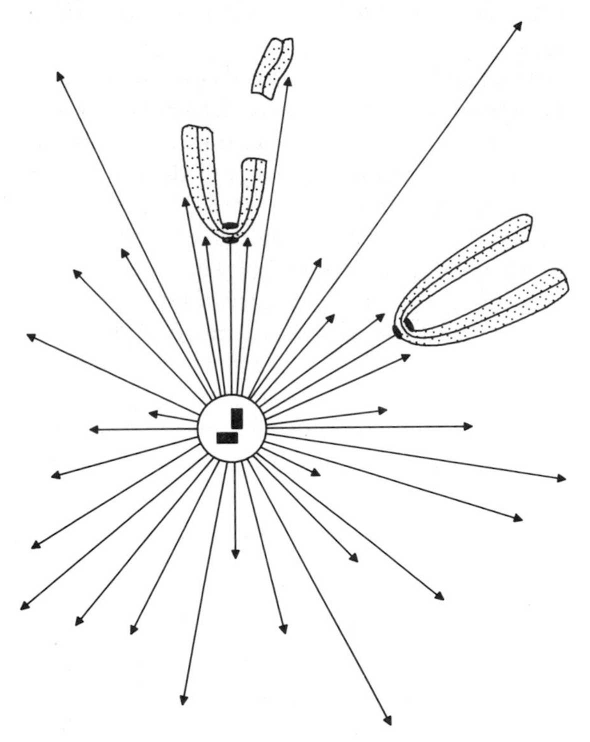

Figure 4.27. Schematic drawing of chromosome alignment and orientation in monopolar astral spindles. Chromosomes are pulled polewards by forces at the kinetochore attached to the pole and they are pushed outward away from the pole by forces acting between the polar kinetochore and the chromatin. The outward forces on the chromosome are revealed by the radial orientation of the chromosome arms and by the outward translocation of chromosome arms that occurs when they are severed from the centromere by laser microsurgery. In the newt, severed arms move outwards at 1 to 2 μm/min. Chromosomes may achieve a position away from the pole where the strength of the poleward force at the kinetochore is balanced by the outward pushing forces on the chromosome arms.

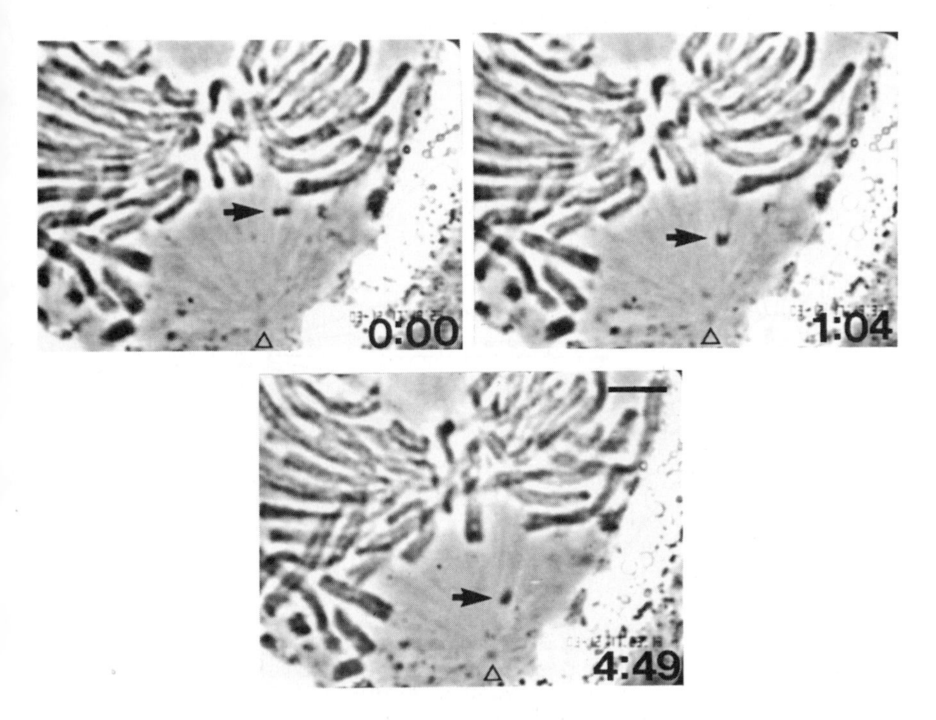

Figure 4.28. Laser microsurgery of the centromere region of a mono-oriented chromosome in a newt spindle. A small region of the chromosome including the centromere was severed from the bulk of the chromosome arms by laser microbeam irradiation. The centric fragment (arrows) moves to the pole (open triangle), while the bulk of the chromosome is pushed away from the pole. Time in minutes:seconds is given on each video frame. (Salmon and Rieder, original micrographs.)

bipolar spindles, chromosome position is produced by one poleward force if the chromosome is mono-oriented or two oppositely oriented poleward forces and two oppositely oriented outward forces if the chromosome is bi-oriented (Fig. 4.29). The resultant of the opposing outward forces on the chromosome has a vectorial component directed along the spindle interpolar axis and a vectorial component directed normal to the interpolar axis. This normal component pushes the chromosomes to the periphery of the central spindle and orients their arms towards a direction perpendicular to the spindle interpolar axis. This phenomenon is commonly observed in animal spindles (Rieder *et al.*, 1986).

Bi-oriented chromosomes usually move to a position near the equator of bipolar spindles. Östergren (1945, 1951) initially proposed that the poleward force at a kinetochore increases with distance away from the pole and that

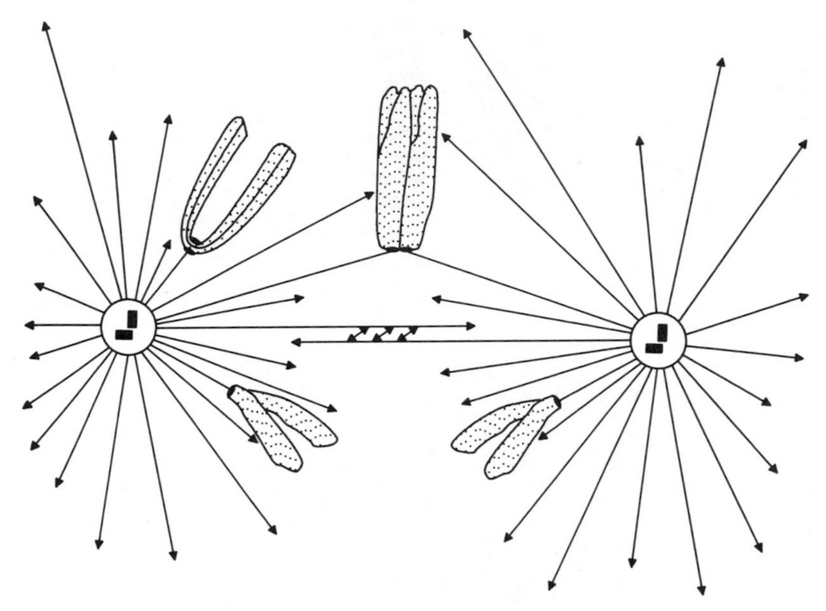

Figure 4.29. Schematic drawing of chromosome alignment and orientation in a bi-polar spindle. Chromosome position and the orientation of its arms depend on whether it is mono- or bi-oriented and the strength of the polar ejection forces, which probably depends on the density of microtubules. Each polar microtubule array pushes the chromosome arms away from the poles. The orientation of the chromosome arms depends on the position of the chromosome in the spindle. At the spindle equator, the arms are pushed outward in a direction normal to the spindle interpolar axis because that is the resultant direction of the opposing polar ejection forces.

chromosomes move in response to the difference in the forces towards opposite poles (Fig. 4.30(a)). The equator is the position in a bipolar spindle where the distance, and hence the net force toward opposite poles is equal.

We (Rieder *et al.*, 1986; Cassimeris *et al.*, 1987a; Salmon, 1988) have modified Östergren's concept to include the effects of polar ejection forces (Fig. 4.30(b)) in order to explain the positioning of mono-oriented chromosomes in monopolar and bipolar spindles. If the only force on a mono-oriented chromosome were the poleward force at the kinetochore, then these chromosomes would always move all the way to the pole—and they do not. In our model, the poleward force at a kinetochore depends primarily on the number of kinetochore microtubules and hence may not depend directly on distance from the pole. The net force on a chromosome towards a pole increases with distance from the pole because the strength of the polar ejection forces decreases with distance from the pole as the density of microtubules decreases.

(a)

P1
kF1
kF2
P2
L1
L2
RF

(b)

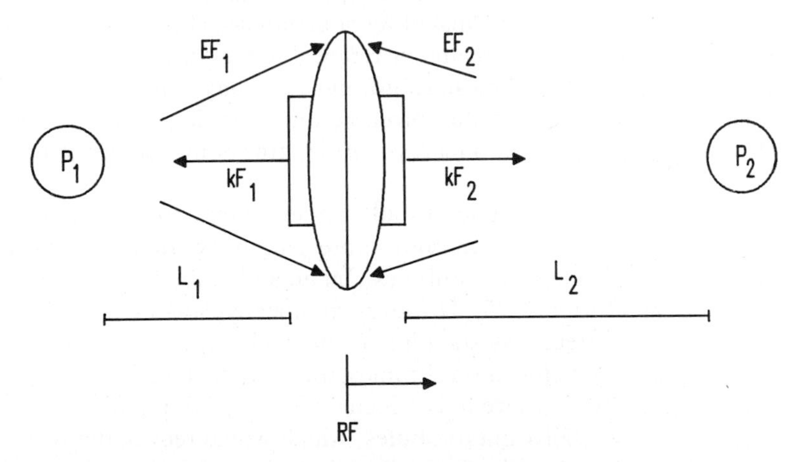

Figure 4.30. Models of chromosome congression in a bipolar spindle. (a) Östergren's model: poleward force at the kinetochore, KF, increases with the distance from the pole. Chromosomes move and the lengths of their chromosomal fibres change until $(KF_1 - KF_2) = RF = 0$. (b) Kinetochore motor, polar ejection model: poleward force at the kinetochore is independent of the distance from the pole, but the strength of the polar ejection forces increase closer to their associated poles. Chromosomes move to a position where $(KF_1 - KF_2) + (EF_2 - EF_1) = RF = 0$.

Thus, congression to the metaphase plate in bipolar spindles occurs by the balancing of four forces on a chromosome: two opposing poleward forces and two opposing polar ejection forces.

D. Metaphase spindle shortening upon microtubule depolymerization

In early prometaphase, before much chromosome attachment, the spindle poles usually separate until few of their microtubules overlap. As more chromosomes bi-orient on the spindle, the poles get closer together. Metaphase spindles are usually noticeably shorter than early prometaphase spindles. After chromosome separation in anaphase, the spindle re-elongates. This behaviour indicates that pushing forces are generated between overlapping polar arrays of microtubules that oppose the pulling forces associated with bi-oriented chromosomes in prometaphase and metaphase spindles (LaFountain, 1972).

The length of metaphase spindles is a function of microtubule assembly. Inoué and co-workers observed that decreases and increases in spindle microtubule assembly at metaphase produced concurrent changes in the lengths of the interpolar spindle and the chromosomal fibres (Inoué, 1952, 1964, 1976; Inoué and Sato, 1967; Inoué and Ritter, 1975; Salmon, 1975a,b,c, 1976). Moderate cooling, pressure or concentrations of colchicine-like drugs not only reduced the number of microtubules in a spindle, but also induced a shortening of the lengths of the chromosomal fibres at metaphase (Fig. 4.7). Rewarming, returning to atmospheric pressure, or removal of the drug produces increases in microtubule number and mean length and a re-elongation of the spindle and chromosomal fibres. Agents that promote microtubule polymerization, like D_2O, have also been shown to increase the lengths of the chromosomal fibres (Inoué and Sato, 1967).

Agents that affect polar microtubule assembly may induce the observed changes in spindle and chromosome fibre length by direct effects on the assembly of kinetochore microtubules (Inoué and Sato, 1967; Inoué, 1976, 1981a; Inoué and Ritter, 1975). However, another explanation is also possible, based upon the differential stability of kinetochore microtubules and the effects of polar ejection forces on chromosome position in monopolar spindles. Cooling, hydrostatic pressure or colchicine-like drugs all rapidly decrease the density of the labile polar microtubules, which would reduce the strength of the polar ejection forces. The strength of the poleward forces at kinetochores may be more stable in proportion to the differential stability of the kinetochore microtubules. Thus, shortening of the chromosomal fibres and spindle occurs until the chromosomes move close enough to the spindle poles where the density of polar microtubules is sufficient to produce ejection forces strong enough to balance the poleward forces at the kinetochores. Promotion of microtubule assembly increases the density of polar microtubules further away from the poles. This would increase the strength of the polar ejection forces and push the chromosomes further away from the poles until the strength of the polar ejection forces matched that of the poleward forces at the kinetochores.

The above discussion provides much support for Mazia's concept that a bipolar spindle should be considered as functionally equivalent to two monopolar spindles held together by the attachment of chromosomes to microtubules extending from opposite poles (Mazia *et al.*, 1981).

XI. A "Pac-Man, Polar Ejection" Working Hypothesis of Chromosome Movement

Assuming the kinetochore is both the site of attachment of the ends of polar microtubules and the site of energy transduction for poleward force-generation, in combination with the dynamic instability of polar microtubules and polar ejection forces, produces a very simple view of the way chromosomes are aligned and segregated by the mitotic spindle (Fig. 4.29).

In preparation for mitosis, the mitotic centre or centrosome duplicates and the kinetochores on the duplicated chromosomes become oriented to face in opposite directions. The duplicated mitotic centres move apart to establish the bipolarity required for segregation. Microtubules in each polar array are oriented with their plus ends distal from their poles. At the breakdown of the nuclear envelope, the nucleation activity of the centres is enhanced to increase the number of polar microtubules growing out toward the kinetochores and the frequency of rescue is greatly reduced from the value in interphase, so that the half-life of a polar microtubule is brief. Kinetochores attach to poles by capturing the plus distal growing ends of the polar microtubules. The high density of polar microtubules and their frequent turnover by dynamic instability ensure a statistically significant number of ends to make capture efficient. Once attached, kinetochores try to move polewards, toward the minus ends of the kinetochore microtubules.

Initially, one kinetochore on a chromosome captures a microtubule emanating from one pole. The kinetochore motor pulls the chromosome polewards along this microtubule. This orients the kinetochore to the pole and prevents the opposing kinetochore from making microtubule attachments to the same pole by making it face away from the pole. When the opposing kinetochore captures a microtubule from the other pole, the chromosome begins moving towards that pole.

The position of a chromosome in the spindle is determined by a force-balance mechanism, with each duplex chromosome experiencing four possible forces (Figs. 4.29 and 4.30(b)): two poleward forces at oppositely oriented kinetochores whose strength may depend only on the number of kinetochore microtubules; and two polar ejection forces which push the chromosome out, away from the poles and whose strength depends on the density of elongating microtubules, which decreases away from a pole. Chromosomes align at the spindle equator because this is the position where the sum of the polar ejection forces and poleward force on the chromosome produced in association with one polar array of microtubules is equivalent but opposite in direction to the resultant force produced in association with the opposite polar array. Kinetochore microtubules and chromosome fibres change length in response to imbalances in these forces.

At the onset of anaphase, sibling chromatids become mechanically separated. Their kinetochore motors pull them toward opposite poles as the density of microtubules and the corresponding polar ejection forces decrease. Concurrent with the decreased density of polar spindle microtubules, the frequency of rescue increases, and microtubules grow to the long lengths chracteristic of the CMTC.

XII. Future Considerations

The above description of the mechanism of mitosis is only a working hypothesis. Several different mechanisms may have evolved for force generation to ensure successful segregation of the chromosomes (see reviews by Kubai, 1975; Bajer and Mole-Bajer, 1975; Goode, 1981; Pickett-Heaps *et al.*, 1982; McIntosh, 1984, 1985; Fuge *et al.*, 1985; Pickett-Heaps, 1986; Cohn *et al.*, 1986; Nicklas, 1985, 1987, 1988a; Mitchison, 1988). One mechanism may be more dominant in one organism than in another. For animal spindles, we have stressed a significant contribution for the polar ejection forces in the mechanism of chromosome congression. In plants, the chromosome arms do not appear to align normal to the spindle axis as is the case for many animal spindles. Polar ejection forces in plant spindles may not be significant or there may be other ways chromosome arms are pulled polewards that are absent in animal cells.

The kinetochore has become of great interest, because it appears to be able to pull itself polewards along kinetochore microtubules while permitting tubulin dissociation without frequent detachment. As Koshland *et al.* (1988) have suggested, the kinetochore may simply use the energy stored in the tubulin lattice of a captured microtubule, generated previously from the hydrolysis of GTP bound to tubulin during elongation, to produce the poleward force at the kinetochore. On the other hand, other force producers such as dynein may be involved with poleward chromosome movement *in vivo*, an interesting issue not yet resolved. The kinetochore may also have other capabilities. It may sense the concentration of molecular complexes bound to kinetochore microtubules and adjust the strength of poleward force-generation as discussed above. The kinetochore appears to use tension to regulate attachment. Nicklas has shown that when two kinetochore complexes on a meiotic duplex are attached to the same pole, one kinetochore will detach and reorient towards the opposite pole unless a force away from the pole is applied to the chromosome complex (Nicklas, 1971, 1985, 1987a). Attachment of both kinetochores to one pole is stable only if there is constant tension. Kinetochores also appear capable of moving by an ATP-dependent mechanism along microtubules towards their plus ends *in vitro* (Mitchison and Kirschner, 1985b), a direction away from the spindle poles *in vivo*. Outward translocation of kinetochores along spindle microtubules has been reported for one species of diatom, but it has not been reported yet for higher eukaryotic cells (Pickett-Heaps *et al.*, 1982). In several cell types, chromosomes segregate non-randomly. Non-random segregation appears to depend on unknown interactions between kinetochores (Swedak and Forer, 1987). Evidently, the kinetochore is a fascinating organelle and it will clearly be the major focus of mitosis research in the near future.

Acknowledgements

I want to thank Bill Brinkley and Jeremy Hyams for their remarkable sense of humour, patience and dedication to elucidating the mechanisms of mitosis. I also want to thank Leah Haimo and Bruce Telzer for providing me a dry spot in Woods Hole where I could finish this paper; Shinya Inoué for critically reading earlier versions of this manuscript; and Conly Rieder, Lynne Cassimeris and Rich Walker for putting together many of the illustrations.

REFERENCES

Allen, R. D., Allen, N. S. and Travis, J. L. (1981). *Cell Motil.* **1**, 291–302.
Bajer, A. S. (1982). *J. Cell Biol.* **93**, 33–48.
Bajer, A. S. and Mole-Bajer, J. (1971). In *Advances in Cell and Molecular Biology*, Vol. 1 (ed. E. J. Dupraw), pp. 213.
Bajer, A. S. and Mole-Bajer, J. S. (1972). *Int. Rev. Cytol. Suppl.* **3**, 1–271.
Bajer, A. and Mole-Bajer, J. (1975). In *Molecules and Cell Movement* (ed. S. Inoué and R. E. Stephens), pp. 77–96. Raven Press, New York.
Bajer, S. A., Cypher, C., Mole-Bajer, J. and Howard, H. M. (1982). *Proc. Natl. Acad. Sci. USA* **79**, 6569–6573.
Bergen, L. G. and Borisy, G. G. (1980). *J. Cell Biol.* **84**, 141–150.
Bergen, L. G. and Borisy, G. G. (1983). *J. Biol. Chem.* **258**, 4190–4194.
Bergen, L. G., Kuriyama, R. and Borisy, G. G. (1980). *J. Cell Biol.* **84**, 151–159.
Bloom, G. S., Luca, F. C., Collins, C. A. and Vallee, R. B. (1986). *Proc. Ann. NY Acad. Sci.* **466**, 328–339.

Borisy, G. G. and Gould, R. R. (1977). In *Mitosis Facts and Questions* (ed. M. Little, N. Paweletz, C. Petzelt, H. Ponstingl, D. Schroeter and H.-P. Zimmermann), pp. 78–87, Springer-Verlag, Berlin.
Brinkley, B. R. (1986). *Ann. Rev. Cell Biol.* **1**, 145–172.
Brinkley, B. R. and Cartwright, J. (1975). *Ann. NY Acad. Sci.* **253**, 428–439.
Brinkley, B. R., Stubblefield, E. and Hsu, T. C. (1967). *J. Ultrastruct. Res.* **19**, 1–18.
Brinkley, B. R., Fistel, S. H., Marcum, J. M. and Pardue, R. L. (1980). *Int. Rev. Cytol.* **63**, 59–95.
Brinkley, B. R., Cox, S. M., Pepper, D. A., Wible, L., Brenner, S. C. and Pardue, R. L. (1981). *J. Cell Biol.* **90**, 554–562.
Browne, C. L., Lockwood, A. H., Su, J.-L., Beavo, J. and Steiner, A. L. (1980). *J. Cell Biol.* **87**, 336–345.
Caplow, M. (1986). *Ann. NY Acad. Sci.* **466**, 510–518.
Carlier, M. F., Hill, T. L. and Chen, Y. D. (1984). *Proc. Natl. Acad. Sci. USA* **51**, 771–775.
Cassimeris, L. U. (1988). Microtubule assembly dynamics. PhD thesis, submitted to University of North Carolina.
Cassimeris, L. U., Wadsworth, P. and Salmon, E. D. (1986). *J. Cell Biol.* **102**, 2023–2032.
Cassimeris, L. U., Walker, R. A., Pryer, N. K. and Salmon, E. D. (1987a). *Bioassays* **7**, 149–154.
Cassimeris, L., Rieder, C. L. and Salmon, E. D. (1987b). *J. Cell Biol.* **105**, 205a.
Cassimeris, L., Inoué, S. and Salmon, E. D. (1988a). *Cell Motil. Cytoskel.* **10**, 185–196.
Cassimeris, L., Pryer, N. K. and Salmon, E. D. (1988b). *J. Cell Biol.* **107**, 2223–2231.
Cohn, S. A., Tippit, D. H. and Spurck, T. P. (1986). *J. Theor. Biol.* **122**, 277–301.
DeBrabander, M., Geuens, G., Nuydens, R., Willebrords, R. and DeMey, J. (1980). In *Microtubules and Microtubule Inhibitors* (eds. M. DeBrabander and J. DeMey), pp. 57–70. Elsevier/North-Holland Biomedical Press, Amsterdam.
DeBrabander, M., Geuens, G., DeMey, J. and Joniau, M. (1981a). *Cell Motil.* **1**, 469–483.
DeBrabander, M., Geuens, M. G., Nuydens, R., Willebrords, R. and DeMey, J. (1981b). *Proc. Natl. Acad. Sci. USA* **78**, 5608–5612.
DeBrabander, M., Geuens, G., Nuydens, R., Willebrords, R. and DeMey, J. (1981c). *Cell Biol. Int. Rep.* **5**(9), 913–920.
DeBrabander, M., Geuens, G., Nuydens, R., Willebrords, R., Aerts, F., DeMey, J. and McIntosh, J. R. (1986). *Int. Rev. Cytol.* **101**, 215–274.
Dinsmore, J. H. and Sloboda, R. D. (1988). *Cell* **53**, 769–780.
Ellis, G. W. and Begg, D. A. (1981). In *Mitosis/Cytokinesis* (ed. A. Forer, and A. M. Zimmerman), pp. 155–179. Academic Press, New York.
Evans, L., Mitchison, T. and Kirschner, M. (1985). *J. Cell Biol.* **100**, 1185–1191.
Forer, A. (1965). *J. Cell Biol.* **25**, 95–117.
Forer, A. (1966). *Chromosoma* **19**, 44–98.
Fuge, H., Bastmeyer, M. and Steffen, W. (1985). *J. Theor. Biol.* **115**, 391–399.
Fuseler, J. W. (1975). *J. Cell Biol.* **67**, 789–800.
Goode, D. (1973). *J. Mol. Biol.* **80**, 531–538.
Goode, D. (1981). *Biosystems* **14**, 271.
Gorbsky, G. J., Sammak, P. J. and Borisy, G. G. (1987). *J. Cell Biol.* **104**(1), 9–18.
Gorbsky, G. J., Sammak, P. J. and Borisy, G. G. (1988). *J. Cell Biol.* **106**(1), 1185–1192.

Hamaguchi, Y. (1975). *Develop. Growth Differ.* **17**(2), 111–117.
Hamaguchi, Y., Toriyama, Y. M., Sakai, H. and Hiromoto, Y. (1987). *Cell Struct. Funct.* **12**, 43–52.
Hard, R. and Allen, R. D. (1977). *J. Cell Sci.* **27**, 47–56.
Harris, P. (1978). In *Cell Cycle Regulation* (ed. E. D. Bretow, I. L. Cameron and G. M. Padilla), pp. 75–104, Academic Press, New York.
Hays, T. S. (1986). Ph.D. thesis. University of North Carolina, Chapel Hill.
Hays, T. S. and Salmon, E. D. (1985). *J. Cell Biol.* **101**, 6a.
Hays, T. S., Wise, D. and Salmon, E. D. (1982). *J. Cell Biol.* **93**, 374–382.
Heidemann, S. R. and McIntosh, J. R. (1980). *Nature* **286**(5772), 517–519.
Hepler, P. K. and Palevitz, B. A. (1985). *J. Cell Biol.* **102**, 1995–2005.
Hill, T. L. (1985). *Proc. Natl. Acad. Sci. USA* **82**, 4404–4408.
Hill, T. L. and Kirschner, M. W. (1982). *Int. Rev. Cytol.* **78**, 1–125.
Hiromoto, Y. and Shoji, Y. (1982). In *Biological Functions of Microtubules and Related Structures* (ed. H. Sakai, H. Mohri and G. G. Borisy), pp. 247–259. Academic Press, New York.
Hiromoto, Y. and Nakano, Y. (1988). *Cell Motil. Cytoskel.* **10**, 172–184.
Horio, T. and Hotani, H. (1986). *Nature* **321**, 605–607.
Huitorel, P. and Kirschner, M. W. (1988). *J. Cell Biol.* **106**, 151–159.
Inoué, S. (1952). *Exp. Cell Res. Suppl.* **2**, 305–318.
Inoué, S. (1964). In *Primitive Motile Systems in Cell Biology* (ed. R. H. Allen and N. Kamiya), pp. 549–598. Academic Press, New York.
Inoué, S. (1976). In *Cell Motility*. Cold Spring Harbor Conferences on Cell Proliferation, Vol. 3 (ed. R. Goldman, T. Pollard and J. Rosenbaum), pp. 1317–1328.
Inoué, S. (1981a). *J. Cell Biol.* **89**, 346–356.
Inoué, S. (1981b). *J. Cell Biol.* **91**(3, Pt. 2), 131s–147s.
Inoué, S. (1986). *Video Microscopy*. Plenum Press, New York.
Inoué, S. and Ritter, H., Jr. (1975). In *Molecules and Cell Movement* (ed. S. Inoué and R. E. Stephens), pp. 3–30. Raven Press, New York.
Inoué, S. and Sato, H. (1967). *J. Gen. Physiol.* **50**, 259–292.
Inoué, S., Fuseler, J., Salmon, E. D. and Ellis, G. W. (1975). *Biophys. J.* **15**, 725–744.
Izant, J. C. (1983). *Chromosoma* **88**, 1–10.
Kane, R. E. (1962). *J. Cell Biol.* **12**(1), 47–55.
Keith, C. H., Feramisco, J. R. and Shelanski, M. (1981). *J. Cell Biol.* **88**, 234–240.
Kiehart, D. (1981). *J. Cell Biol.* **88**, 604–617.
Kirschner, M. W. (1979). *Int. Rev. Cytol.* **54**, 1–71.
Kirschner, M. and Mitchison, T. (1986a). *Cell* **45**, 329–342.
Kirschner, M. and Mitchison, T. (1986b). *Nature* **324**, 621.
Koshland, D. E., Mitchison, T. J. and Kirschner, M. W. (1988). *Nature* **331**, 499–504.
Kristofferson, D., Mitchison, T. and Kirschner, M. (1986). *J. Cell Biol.* **102**, 1007–1019.
Kubai, D. (1975). *Int. Rev. Cytol.* **43**, 167.
LaFountain, J. R., Jr. (1972). *J. Cell Sci.* **10**, 79–93.
Lambert, A.-M. and Bajer, A. S. (1977). *Cytobiologie* **15**, 1–23.
Leslie, R. J. and Pickett-Heaps, J. D. (1983). *J. Cell Biol.* **96**, 548–561.
Leslie, R. J., Saxton, W. M., Mitchison, T. J., Neighbors, B., Salmon, E. D. and McIntosh, J. R. (1984). *J. Cell Biol.* **99**, 2146–2156.
Leslie, R. J., Hird, R. B., Wilson, L., McIntosh, J. R. and Scholey, J. (1987). *Proc. Natl. Acad. Sci. USA* **84**, 2771–2775.
Lye, R. J., Porter, M. E., Scholey, J. M. and McIntosh, J. R. (1987). *Cell* **51**, 309–318.

Luykx, P. (1970). *Int. Rev. Cytol. Suppl.* **2**, 1–173.
Mandelkow, M.-E. and Mandelkow, E. (1985). *J. Mol. Biol.* **181**, 123–135.
Mandelkow, M.-E., Schultheiss, R., Rapp, R., Muller, M. and Mandelkow, E. (1986). *Cell* **102**, 1067–1073.
Margolis, R. L. and Wilson, L. (1977). *Proc. Natl. Acad. Sci. USA* **74**(8), 3466–3470.
Margolis, R. L. and Wilson, L. (1981). *Nature* **293**, 705–711.
Margolis, R. L., Wilson, L. and Kiefer, B. I. (1978). *Nature* **272**, 450–452.
Margolis, R. L., Rauch, C. T. and Wilson, L. (1980). *Biochemistry* **19**, 5550–5557.
Marsland, D. (1970). In *High Pressure Effects on Cellular Processes* (ed. A. M. Zimmerman), pp. 260–306.
Mazia, D. (1961). In *The Cell* (ed. J. Brachet and A. E. Mirsky), Vol. 3, pp. 77–412. Academic Press, New York.
Mazia, D. and Dan, K. (1952). *Proc. Natl. Acad. Sci. USA* **38**, 826–838.
Mazia, D. (1984). *Exp. Cell Res.* **153**, 1–15.
Mazia, D. (1987). *Int. Rev. Cytol.* **100**, 49–92.
Mazia, D., Paweletz, N., Sluder, G. and Finze, E. (1981). *Proc. Natl. Acad. Sci. USA* **78**(1), 377–381.
McIntosh, J. R. (1981). In *International Cell Biology 1980–1981* (ed. H. G. Schweiger), pp. 359–368.
McIntosh, J. R. (1983). In *Spatial Organization of Eukaryotic Cells* (ed. J. R. McIntosh), pp. 115–142. Alan R. Liss, New York.
McIntosh, J. R. (1984). *Trends Biochem. Sci.* **9**, 195–198.
McIntosh, J. R. (1985). In *Aneuploidy* (ed. V. L. Dellarco, P. E. Voytek and A. Hollander), pp. 197–229. Plenum Press, New York.
McIntosh, J. R. and Euteneur, U. (1984). *J. Cell Biol.* **98**, 525–533.
McIntosh, J. R. and Vigers, G. P. A. (1987). In *Proceedings of the 45th Annual Meeting of the Electron Microscopy Society of America* (ed. G. W. Bailey), pp. 794–797. San Francisco Press, San Francisco.
McIntosh, J. R., Hepler, P. K. and Van Wie, D. G. (1969). *Nature* **224**, 659–663.
McIntosh, J. R., Cande, W. Z. and Snyder, J. A. (1975). In *Molecules and Cell Movement* (ed. S. Inoué and R. E. Stephens), pp. 31–76. Raven Press, New York.
J. R., Saxton, W. M., Stemple, D. L., Leslie, R. J. and Welsh, M. J. (1986). *Ann. NY Acad. Sci.* **466**, 566–579.
McNeil, P. A. and Berns, M. W. (1981). *J. Cell Biol.* **88**, 543–553.
Mitchison, T. J. (1986). *J. Cell Sci. Suppl.* **5**, 121–128.
Mitchison, T. J. (1988). *Ann. Rev. Cell Biol.* **4**, 527–549.
Mitchison, T. J. and Kirschner, M. (1984a). In *Molecular Biology of the Cytoskeleton* (ed. G. G. Borisy, D. W. Cleveland and D. B. Murphy), pp. 27–44, Cold Spring Harbor Laboratory, Cold Spring Harbor.
Mitchison, T. J. and Kirschner, M. (1984b). *Nature* **312**, 232–237,
Mitchison, T. J. and Kirschner, M. (1984c). *Nature* **312**, 237–242.
Mitchison, T. J. and Kirschner, M. W. (1985a). *J. Cell Biol.* **101**, 755–765.
Mitchison, T. J. and Kirschner, M. W. (1985b). *J. Cell Biol.* **101**, 766–777.
Mitchison, T. J., Evans, L., Schulze, E. and Kirschner, M. (1986). *Cell* **45**, 515–527.
Mole-Bajer, J., Bajer, A. S. and Owczarzak, A. (1975). *Cytobios* **13**, 45–65.
Nicklas, R. B. (1963). *Chromosoma* **14**, 276–295.
Nicklas, R. B. (1965). *J. Cell Biol.* **25**, 119.
Nicklas, R. B. (1971). In *Advances in Cell Biology*, Vol. 2 (ed. D. M. Prescott, L. Goldstein and E. McConkey), pp. 225–297, Plenum Press, New York.

Nicklas, R. B. (1975). In *Molecules and Cell Movement* (ed. S. Inoué and R. E. Stephends), pp. 97–118, Raven Press, New York.
Nicklas, R. B. (1977). In *Mitosis Facts and Questions* (ed. M. Little, N. Paweletz, C. Petzelt, H. Ponstingl, D. Schroeter and H.-P. Zimmermann), pp. 150–155, Springer-Verlag, Berlin.
Nicklas, R. B. (1983). *J. Cell Biol.* **97**, 542–548.
Nicklas, R. B. (1984). *Cell Motil.* **4**, 1–5.
Nicklas, R. B. (1985). In *Aneuploidy* (ed. V. L. Dellarco, P. E. Voytek and A. Hollander), pp. 183–195. Plenum Press, New York.
Nicklas, R. B. (1987a). In *Chromosome Structure and Function: The Impact of New Concepts* (ed. J. P. Gustafson and R. J. Kaufman), pp. 53–74. Plenum Publishing, New York.
Nicklas, R. B. (1987b). *J. Cell Biol.* **105**, 176A.
Nicklas, R. B. (1988a). *J. Cell Sci.* **89**, 283–285.
Nicklas, R. B. (1988b). *Ann. Rev. Biophys. Chem.* **17**, 431–449.
Nicklas, R. B. and Kubai, D. F. (1985). *Chromosoma* **92**, 313–324.
Nicklas, R. B., Kusai, D. F. and Hays, T. S. (1982). *J. Cell Biol.* **95**, 91–104.
Östergren, G. (1945). *Hereditas* **31**, 49.
Östergren, G. (1951). *Hereditas* **37**, 85–156.
Östergren, G., Mole-Bajer, J. and Bajer, A. (1960). *Proc. NY Acad. Sci.* **90**, 381–408.
Pantaloni, D. and Carlier, M.-F. (1986). *Ann. NY Acad. Sci.* **466**, 496–509.
Paschal, B. M., Shpetner, H. S. and Vallee, R. B. (1987). *J. Cell Biol.* **105**, 1273–1282.
Pepper, D. A. and Brinkley, B. R. (1979). *J. Cell Biol.* **82**, 585–591.
Pickett-Heaps, J. (1986). *TIBS* **44**, 504–507.
Pickett-Heaps, J. D., Tippit, D. H. and Porter, K. R. (1982). *Cell* **29**, 729–744.
Pratt, M. M., Otter, T. and Salmon, E. D. (1980). *J. Cell Biol.* **86**, 738–745.
Pryer, N. K., Wadsworth, P. and Salmon, E. D. (1986). *Cell Motil. Cytoskel.* **6**, 537–548.
Purich, D. L. and Kristofferson, D. (1984). *Adv. Protein Chem.* **36**, 133–211.
Rappaport, R. (1988). *J. Exp. Zool.* **246**, 253–257.
Rebhun, L., I., Jemiolo, D., Ivy, N., Mellon, M. and Nath, J. (1975). *Ann. NY Acad. Sci.* **253**, 362–377.
Rebhun, L. I. and Palazzo, R. E. (1988). *Cell Motil. Cytoskel.* **10**, 197–209.
Rieder, C. L. (1981). *Chromosoma* **84**, 145–158.
Rieder, C. L. (1982). *Int. Rev. Cytol.* **79**, 1–57.
Rieder, C. L. and Bajer, A. S. (1977). *J. Cell Biol.* **74**, 717.
Rieder, C. L., Davison, E. A., Jensen, L. C. W. and Salmon, E. D. (1985). In *Microtubules and Microtubule Inhibitors* (ed. M. DeBrabander and J. DeMey), pp. 253–260, Elsevier, Amsterdam.
Rieder, C. L., Davison, E. A., Jensen, L. C. W., Cassimeris, L. and Salmon, E. D. (1986). *J. Cell Biol.* **103**, 581–591.
Sale, W. and Satir, P. (1977). *Proc. Natl. Acad. Sci. USA* **74**, 2045–2049.
Sakai, H. (1978). *Int. Rev. Cytol.* **55**, 23–48.
Salmon, E. D. (1975a). *J. Cell Biol.* **65**, 603–614.
Salmon, E. D. (1975b). *J. Cell Biol.* **66**, 114–127.
Salmon, E. D. (1975c). *Ann. NY Acad. Sci.* **253**, 383–406.
Salmon, E. D. (1976). In *Cell Motility*. Cold Spring Harbor Conferences on Cell Proliferation, Vol. 3 (ed. R. Goldman, T. Pollard and J. Rosenbaum), pp. 1329–1342, Cold Spring Harbor Laboratory, Cold Spring Harbor.

Salmon, E. D. (1982). *Methods Cell Biol.* **25**, 69–105.
Salmon, E. D. (1988). In *Cell Movement, Vol. 2: Kinesin, Dynein and Microtubule Dynamics* (ed. F. D. Warner and J. R. McIntosh), pp. 431–440. Alan R. Liss, New York.
Salmon, E. D. and Begg, D. A. (1980). *J. Cell Biol.* **85**, 853–865.
Salmon, E. D. and Segall, R. R. (1980). *J. Cell Biol.* **86**, 355–365.
Salmon, E. D. and Wadsworth, P. (1986). In *Applications of Fluorescence in the Biomedical Sciences* (ed. D. L. Taylor, A. S. Waggoner, F. Lanni, R. F. Murphy and R. R. Birge), pp. 377–403, Alan R. Liss, New York.
Salmon, E. D., Goode, D., Maugel, T. K. and Bonar, D. B. (1976). *J. Cell Biol.* **69**, 443–454.
Salmon, E. D., Saxton, W. M., Leslie, R., Karow, M. L. and McIntosh, J. R. (1984a). *J. Cell Biol.* **99**, 2165–2174.
Salmon, E. D., McKeel, M. and Hays, T. (1984b). *J. Cell Biol.* **99**, 1066–1075.
Salmon, E. D., Walker, R. A., O'Brien, E. T., Pryer, N. K., Voter, W. A. and Erickson, H. P. (1987). *Proceedings of the 45th Annual Meeting of the Electron Microscopy Society of America* (ed. G. W. Bailey), pp. 636–637. San Francisco Press, San Francisco.
Sammak, P. J. and Borisy, G. G. (1988). *Nature* **332**, 724–726.
Sammak, P. J., Gorbsky, G. J. and Borisy, G. G. (1987). *J. Cell Biol.* **104**(3), 395–406.
Sanger, J. W. (1977). In *Mitosis Facts and Questions* (ed. M. Little, N. Paweletz, C. Petzelt, H. Ponstingl, D. Schroeter and H.-P. Zimmermann), pp. 98–120, Springer-Verlag, Berlin.
Sato, H., Ellis, G. W. and Inoué, S. (1975). *J. Cell Biol.* **67**, 501–517.
Saxton, W. M., Stemple, D. L., Leslie, R. J., Salmon, E. D. and McIntosh, J. R. (1984). *J. Cell Biol.* **99**, 2175–2186.
Saxton, W. M. and McIntosh, J. R. (1987). *J. Cell Biol.* **105**, 875–886.
Schaap, C. J. and Forer, A. (1984a). *J. Cell Sci.* **65**, 21–40.
Schaap, C. J. and Forer, A. (1984b). *J. Cell Sci.* **65**, 41–60.
Schibler, M. J. and Pickett-Heaps, J. D. (1980). *Eur. J. Cell Biol.* **22**, 687–698.
Scholey, J. M., Porter, M. E., Grissom, P. M. and McIntosh, J. R. (1985). *Nature* **318**(6045), 483–486.
Schrader, F. (1953). *Mitosis. The Movements of Chromosomes in Cell Division.* Columbia University Press, New York.
Schroeder, T. E. (1987). *Dev. Biol.* **124**, 9–22.
Schulze, E. and Kirschner, M. (1986). *J. Cell Biol.* **102**, 1020–1031.
Schulze, E. and Kirschner, M. (1987). *J. Cell Biol.* **104**(2), 277–288.
Schulze, E. and Kirschner, M. (1988). *Nature* **334**, 356–359.
Silver, R. B. (1986). *Proc. Natl. Acad. Sci. USA* **83**, 4302–4306.
Sluder, G. (1976). *J. Cell Biol.* **70**, 75–85.
Sluder, G. and Rieder, C. L. (1985). *J. Cell Biol.* **100**, 897–903.
Sluder, G., Miller, F. J. and Rieder, C. L. (1986). *J. Cell Biol.* **103**, 1873–1881.
Snyder, J. A. and McIntosh, J. R. (1975). *J. Cell Biol.* **67**, 744–760.
Soltys, B. J. and Borisy, G. G. (1985). *J. Cell Biol.* **100**, 1682–1689.
Spurck, T. P. and Pickett-Heaps, J. D. (1987). *J. Cell Biol.* **105**, 1691–1705.
Spurck, T. P., Pickett-Heaps, J. D. and Klymkowsky, M. W. (1986a). *Protoplasma* **131**, 47–59.
Spurck, T. P., Pickett-Heaps, J. D. and Klymkowsky, M. W. (1986b). *Protoplasma* **131**, 60–74.
Swedak, J. A. M. and Forer, A. (1987). *J. Cell Sci.* **88**, 441–452.

Tao, W., Walter, R. J. and Berns, M. (1988). *J. Cell Biol.* **107**, 1025–1035.
Telzer, B. R. and Haimo, L. T. (1981). *J. Cell Biol.* **89**, 373–378.
Telzer, B. R. and Rosenbaum, J. L. (1979). *J. Cell Biol.* **81**, 484–488.
Telzer, B. R., Moses, M. J. and Rosenbaum, J. L. (1980). *Proc. Natl. Acad. Sci. USA* **72**(10), 4023–4027.
Vale, R. D. (1987). *Ann. Rev. Cell Biol.* **3**, 347–378.
Vale, R. D., Reese, T. S. and Sheetz, M. P. (1985a). *Cell* **42**, 39–50.
Vale, R. D., Schnapp, B. J., Mitchison, T., Steuer, E., Reese, T. S. and Sheetz, M. P. (1985b). *Cell* **43**, 623–632.
Vandré, D. D., Kronebusch, P. and Borisy, G. G. (1984). In *Molecular Biology of the Cytoskeleton* (ed. Borisy, G. G., Cleveland, D. W. and Murphy, D. B.), pp. 1–16.
Vigers, G. P. A., Coue, M. and McIntosh, J. R. (1988). *J. Cell Biol.* **107**, 1011–1024.
Wadsworth, P. and Salmon, E. D. (1986a). *J. Cell Biol.* **102**, 1032–1038.
Wadsworth, P. and Salmon, E. D. (1986b). *Ann. NY Acad. Sci.* **466**, 580–592.
Wadsworth, P. and Salmon, E. D. (1986c). *Methods Enzymol.* **134**, 519–528.
Walker, R. A., Pryer, N. K., Cassimeris, L. U., Soboeiro, M. and Salmon, E. D. (1986). *J. Cell Biol.* **103**, 432a.
Walker, R. A., O'Brien, E. T., Pryer, N. K., Soboeiro, M. F., Voter, W. A., Erickson, H. P. and Salmon, E. D. (1988). *J. Cell Biol.* **107**, 1437–1448.
Weisenberg, R. C. (1972). *Science* **177**, 1104–1105.
White, J. G. and Borisy, G. G. (1983). *J. Theor. Biol.* **101**, 289–316.
Wilson, E. B. (1925). *The Cell in Development and Heredity*. MacMillan, New York.
Wilson, P. J. and Forer, A. (1988). *J. Cell Sci.* **91**, 455–468.
Wise, D., Cassimeris, L. U., Rieder, C. L., Wadsworth, P. and Salmon, E. D. (1986). *J. Cell Biol.* **103**, 412a.
Witt, P. L., Ris, H. and Borisy, G. G. (1980). *Chromosoma* **81**, 483–505.
Witt, P. L., Ris, H. and Borisy, G. G. (1981). *Chromosoma* **83**, 523–540.
Wolniak, S. M. (1988). *Biochem. Cell Biol.* **66**, 490–514.
Zieve, G. and Solomon, F. (1982). *Cell* **28**, 233–242.

CHAPTER 5

Microtubule-associated Proteins in the Sea-urchin Egg Mitotic Spindle

GEORGE S. BLOOM[1] and RICHARD B. VALLEE[2]

Cell Biology Group, Worcester Foundation for Experimental Biology, Shrewsbury, Massachusetts, USA

I. Introduction

Much research effort has been invested in studying the behaviour and organization of the mitotic spindle. However, despite the remarkable advances that have been made in the past few decades in characterizing and understanding the molecular basis of cellular behaviour in general, relatively little has been learned about the molecular mechanism of mitosis. Microtubules have long been known to be prominent structures within the mitotic spindle, and it seems certain that they are largely responsible for its organization. Tubulin is the major protein constituent of microtubules, and it is probable that at least some aspects of mitosis may be explained on the basis of the polymerization and depolymerization properties of this protein alone. However, it also seems clear that other molecules must be involved in mitosis to organize the spindle microtubules into a coherent array, and to mediate their complex reorganizational behaviour during mitosis.

Several proteins have been implicated as spindle components in the course of investigating cytoplasmic microtubules obtained from brain tissue and cultured mammalian cells. Most of these proteins were originally identified by their co-purification with tubulin and are known collectively as microtubule-

---

[1]For present address see list of contributors at the front of this volume.

[2]Author to whom correspondence and reprint requests should be sent.

MITOSIS: Molecules and Mechanisms
ISBN 0-12-363420-2

associated proteins, or MAPs. A number of the MAPs were found by immunofluorescence microscopy to be present on mitotic as well as interphase microtubules (Bulinski and Borisy, 1980; Bloom *et al.*, 1984a).

The MAPs represent a variety of protein species of different size and with different structural features. The cellular distribution of the MAPs is not fully characterized. It seems evident that at least some species vary extremely in concentration among cell types (Miller *et al.*, 1982; Vallee, 1982; Caceres *et al.*, 1984; De Camilli *et al.*, 1984; Huber and Matus, 1984; Parysek *et al.*, 1984a,b; Binder *et al.*, 1985), while others are more uniformly distributed (Bloom *et al.*, 1984a,b). Thus, some fundamental questions regarding the function of the MAPs during mitosis remain unanswered. Do they serve a common function that can be performed by any one MAP in a given cell type? Alternatively, are they all involved in mitosis, but required at different levels depending on the particular cell type? So far, these questions have proved to be difficult to address.

In part, this dilemma stems from the fact that most of the biochemical characterization of cytoplasmic microtubules has been performed with material obtained from non-mitotic tissues and cells, principally brain tissue. This chapter describes the purification and characterization of microtubules from a system—the sea-urchin egg—in which the microtubules are devoted primarily to a role in mitosis. It is hoped that this will allow for the comprehensive analysis of the component proteins of mitotic microtubules obtained from a single, homogenous cellular source.

II. Advantages of the Sea-urchin Egg

Sea-urchin eggs, as well as eggs from many other organisms, are primed to undergo multiple rapid mitotic divisions upon fertilization. A considerable body of literature indicates that much of the RNA and protein that will be required during early embryogenesis is stockpiled in the egg. In particular, it has been found that tubulin levels show little change during early embryogenesis in sea urchins and other organisms (Raff and Kaumeyer, 1973; Raff *et al.*, 1975). This is despite the abrupt transition from a state in the unfertilized egg in which there is no detectable microtubule polymer, to the appearance of numerous cytoplasmic microtubules following fertilization. This suggests that other component proteins of the mitotic spindle may be stockpiled in the egg, and work carried out by us has indicated this to be true for the MAPs (Vallee and Bloom, 1983; Bloom *et al.*, 1985b; and see below).

The sea-urchin egg is a relatively accessible system for biochemical investigation. Very large quantities of homogeneous cells arrested at a synchronous stage of development can be obtained, and these contain a high concentration of tubulin and MAPs (see below). We obtain 0.4–0.5 mg of microtubules per millilitre of packed, de-jellied eggs; thus, yields of up to 25–50 mg of protein can be obtained in the course of a few hours' work.

This is not to say that the sea-urchin egg is without its own biochemical challenges. For instance, it contains a high level of proteolytic enzymes, which are involved in the fertilization process. However, these are well characterized, and can be blocked with specific inhibitors. A second problem is that actin self-assembly and gelation occur very readily, and can present a serious contamination problem (Kane, 1975). This can also be avoided by conducting most of the procedure in the cold.

Most serious is the potential complication presented by the formation of cilia at the blastula stage of development. The egg may stockpile precursors for these microtubule-containing structures (Auclair and Siegel, 1966), as well as for the mitotic spindle. The extent to which this occurs is not yet certain, nor is it known how the egg might prevent ciliary precursors from interacting with the mitotic spindle. These problems are of considerable interest in their own right.

III. Purification of Microtubules versus Mitotic Spindles as a Biochemical Strategy

One of the advantages of the sea-urchin egg as a source for mitotic proteins is the ease with which mitotic spindles can be isolated and purified for direct analysis. However, isolated spindles are exceedingly complex both biochemically and structurally. As a consequence, they have so far served only a limited role in illuminating the molecular basis of mitosis.

At both the light-microscopic and electron-microscopic levels, the organization of microtubules and chromosomes in isolated spindles appears to be well preserved (Kane, 1962; Salmon and Segall, 1980; Silver *et al.*, 1980). It is not certain, however, how accurately their protein composition reflects that of the working components of the spindle. Isolated spindles contain a considerable variety of proteins. Tubulin, for example, has been found to be enriched in these preparations, but accounts for only 20% of the total protein at most (Salmon, 1982). The remaining proteins are diverse and, for the most part, uncharacterized. The complexity of these preparations is probably partly a reflection of the structural complexity of the spindle, which, in addition to microtubules and chromosomes, contains membranous organelles, centrosomes, and a variety of less well characterized structures. However, it also reflects the limited specificity inherent in differential centrifugation, which is the basis for spindle-isolation techniques. Thus, it is not known how many of the proteins present in isolated spindle preparations are specific spindle components rather than adventitious contaminants, nor is it clear how this issue can be resolved conclusively.

Further complicating the analysis of isolated spindles is the presence of high levels of protease activity in sea-urchin eggs, as discussed above. Proteolysis can be reasonably well controlled with appropriate inhibitors when cells are lysed into aqueous solutions that are iso-osmotic with the protease-containing vesicles (Vallee and Bloom, 1983; Bloom *et al.*, 1985b; and see below). However, most spindle-isolation techniques utilize detergents or glycols, which are likely to permeabilize vesicles and thereby aggravate the proteolysis problem.

For these reasons, it seems useful to attempt to purify individual spindle components that are more conducive to biochemical manipulation. Microtubule purification, like spindle isolation, involves differential centrifugation techniques. However, the sedimentation properties of the microtubule proteins can be altered drastically at will, allowing for their specific isolation and the elimination of non-specific contaminants. In addition, the use of detergents or glycols is unnecessary and egg lysis can readily be achieved using iso-osmotic conditions. This, along with the use of appropriate inhibitors, satisfactorily controls proteolysis (Vallee and Bloom, 1983; Bloom *et al.*, 1985b).

Clearly, the spindle may require more working parts than its microtubules alone. It is worth noting in this regard that important proteins may fail to co-purify with the egg microtubules for any of a variety of reasons. Nonetheless, given the importance of microtubules in spindle organization and function, their purification and characterization are bound to contribute to a clearer understanding of how mitosis is accomplished.

IV. History of the Sea-urchin Egg as a System for the Biochemical Characterization of Microtubule Proteins

While brain tissue has been the favoured source for the biochemical investigation of cytoplasmic microtubules, the sea-urchin egg has been recognized for many years as a biochemical system. However, it has provided a number of interesting pitfalls not experienced with brain tissue.

Bryan *et al.* (1975) attempted to assemble microtubules from the cytoplasm of the sea-urchin egg and found that, in contrast to brain tissue, self-assembly would not occur. However, purified egg tubulin was found to assemble even more readily than mammalian brain tubulin (Kuriyama, 1977; Detrich and Wilson, 1983; Suprenant and Rebhun, 1983). Together, these observations suggested that the sea-urchin egg contains factors that inhibit microtubule assembly *in vitro*, and possibly *in vivo* as well (Bryan *et al.*, 1975; Naruse and Sakai, 1981).

Kane (1975) attempted to purify microtubules from sea-urchin eggs using conditions designed for reversible assembly purification of brain microtubules. Microtubules failed to self-assemble. Instead, a large amount of actin was obtained, apparently due both to self-assembly and to the formation of a gel by specific cross-linking factors. This effect itself was of considerable interest. However, from the point of view of the microtubule biochemist, it revealed that contamination with actin and its associated proteins would be a serious problem in microtubule purification, even if microtubule assembly could be achieved.

Keller and Rebhun (1980) attempted to purify microtubules using isolated mitotic spindles as starting material rather than whole-egg cytosol. This procedure was successful, and resulted in the purification of microtubules that were capable of reversible self-assembly. In addition to tubulin, they identified a second protein (M_r 80 000) that persisted along with tubulin through multiple reversible assembly purification cycles, suggesting that it was a *bona fide* microtubule component.

Our laboratory has applied two additional approaches to identifying and purifying microtubule proteins in the egg, a taxol-based purification procedure and the use of anti-MAP monoclonal antibodies. As discussed in the subsequent sections of this chapter, these approaches have been successful in identifying a variety of sea-urchin MAPs and showing them to be associated with the mitotic spindle *in vivo*.

V. Use of Taxol to Promote Microtubule Assembly

This laboratory has been experimenting for several years with the use of taxol for purifying microtubules. Taxol was first identified as an antimitotic agent (Wani *et al.*, 1971), and was later shown to exert its effects through a direct and specific interaction with tubulin (Parness and Horwitz, 1981). Unlike other tubulin-specific drugs, all of which inhibit microtubule assembly, taxol has the unique effect of promoting assembly.

This suggested that the compound could be useful in purifying microtubules from a large number of experimental systems in which self-assembly does not occur (Vallee, 1982). Because microtubules assembled in the presence of taxol will no longer depolymerize readily, the reversible assembly purification procedure that had been successful with brain tissue and a few other systems could not be applied. Instead, a new strategy for microtubule purification was devised. One of the key elements to this strategy was the finding that the MAPs could be dissociated from microtubules by exposure to conditions of elevated ionic strength (Vallee, 1982). This allowed for the specific solubilization of these proteins, and for their further biochemical manipulation.

As a first step in developing this procedure, microtubules were purified from brain tissue and HeLa cells (Vallee, 1982), systems that had already been extensively characterized using reversible assembly purification as a starting point. It was found that the composition of the microtubules purified with the aid of taxol was identical to that obtained by the reversible assembly procedure. No additional protein species were observed in the taxol-containing preparations, nor were any of the known MAPs diminished in amount. This indicated that the drug did not interfere with the normal binding of MAPs or affect the specificity of binding.

As an added bonus, it appeared that the yield of tubulin and MAPs was greater than with the self-assembly procedure. The effect on tubulin yield was not unexpected, in view of the decrease in the critical concentration for assembly in the presence of the drug (Schiff *et al.*, 1979). The effect on MAP yield was probably due to the increased availability of tubulin binding sites for the MAPs. In fact, in brain tissue we have identified a major high-molecular-weight (HMW) MAP using taxol–MAP 1B (Bloom *et al.*, 1985a) — that had gone unnoticed in traditional microtubule preparations, apparently owing to competition for limited microtubule-binding sites.

VI. Biochemical and Ultrastructural Identification of MAPs in the Sea-urchin Egg Using the Taxol Procedure

On the basis of the suitability of the taxol method for obtaining microtubule proteins, we applied the technique, with a number of modifications, to cytosolic extracts prepared from unfertilized sea-urchin eggs. Analysis by SDS PAGE of the stages from a typical preparation is shown in Fig. 5.1A. Taxol was added to a cytosolic extract (lane E) to promote microtubule assembly. Centrifugation yielded a microtubule-depleted supernate (lane S1) and a microtubule pellet (lane P1) in which tubulin was the major component. In addition, numerous other proteins spanning a broad range of molecular weight were present in the pellet. Most prominent among these was a species of 77 kDa. This may correspond to the 80 kDa protein found by Keller and Rebhun to co-assemble with tubulin extracted from isolated sea-urchin spindles (Keller and Rebhun, 1980). Among the other major non-tubulin bands were a 100 kDa protein and several HMW proteins. The electrophoretic mobilities of the latter were similar to, but not identical to those of the HMW

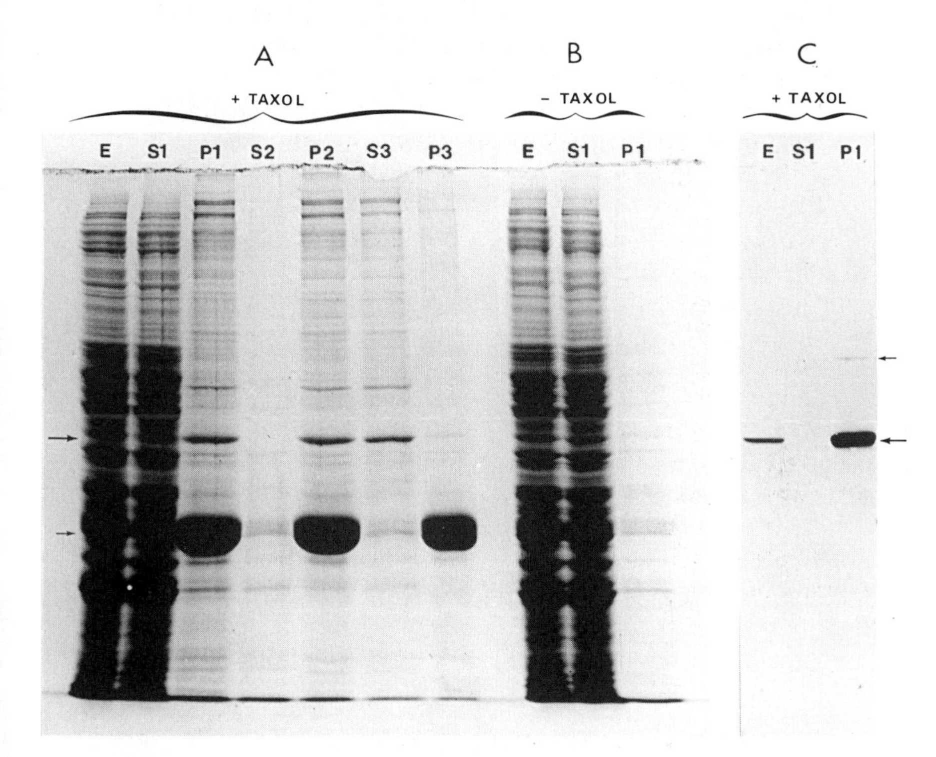

Figure 5.1. Electrophoretic and immunoblot analysis of microtubules purified from unfertilized eggs of *Strongylocentrotus purpuratus*. A 7% polyacrylamide SDS gel stained with Coomassie blue compares the purification procedure performed in the presence (A) and absence (B) of 20 μM taxol. (A) Microtubules polymerized when taxol was added to the cytosolic extract (lane E). Centrifugation of the extract yielded a microtubule-depleted supernate (lane S1) and a microtubule pellet (lane P1). After resuspension of the pellet in taxol-containing buffer, the microtubules were centrifuged again. Very little protein was solubilized (lane S2) and the new pellet (lane P2) was resuspended into taxol-containing buffer supplemented with 0.4 M NaCl. Centrifugation under these conditions yielded a supernate containing virtually all of the non-tubulin proteins (lane S3) and a microtubule pellet composed of nearly pure tubulin (lane P3). The positions of tubulin and the most prominent non-tubulin protein, a 77 kDa species, are indicated on the left by the small and large arrows, respectively. (B) When taxol was not used, microtubule proteins failed to polymerize, as indicated by their absence in what otherwise would have been the first microtubule pellet (lane P1). (C) Immunoblot of the first three steps in microtubule purification with taxol using a monoclonal antibody to the 77 kDa protein. The positions of that protein and a minor immunoreactive band are indicated by the large and small arrows, respectively. (From Bloom *et al.*, 1985b.)

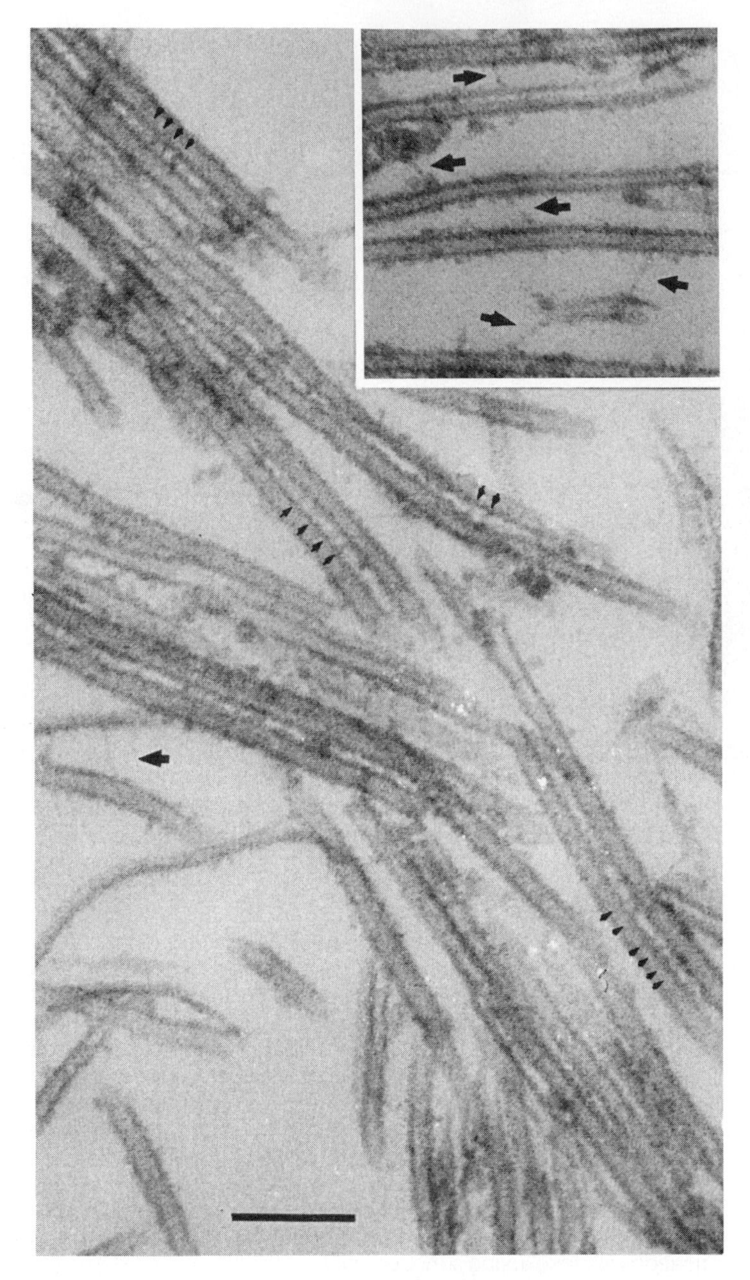

Figure 5.2. Electron microscopy of microtubules purified from unfertilized *Lytechinus variegatus* eggs. Microtubules often appeared to be cross-linked into bundles by periodic short projections (small arrows) and longer, but less abundant fibres (large arrows). Bar = 100 nm. (From Vallee and Bloom, 1983.)

brain MAPs. A limited amount of actin appeared in the microtubule pellet, but contamination by this protein was nearly eliminated in later stages of the preparation.

When the microtubule pellet was resuspended and centrifuged again, much of the actin was solubilized (lane S2), but the other non-tubulin proteins remained bound to the microtubules (lane P2). These proteins could be dissociated from tubulin by resuspension of the microtubules in a high-salt buffer containing taxol and a subsequent centrifugation step. The non-tubulin proteins were recovered in the supernate (lane S3), while most of the tubulin was pelleted in the form of MAP-free microtubules (lane P3).

Neither tubulin nor any of the non-tubulin proteins sedimented when the cytosolic extract was centrifuged in the absence of taxol (Fig. 5.1B). Thus, the non-tubulin proteins, with the exception of actin, behaved as observed for the known MAPs in brain and HeLa cells (Vallee, 1982).

By thin-section electron microscopy, the purified microtubules were also found to contain associated structures (Fig. 5.2). Rows of short projections were observed, and these appeared to cross-link microtubules into bundles at selected locations. In addition, longer but less-abundant filaments were seen to span the gaps between adjacent microtubules. Both classes of projections are reminiscent of structures that have been observed in association with spindle microtubules (Wilson, 1969; Brinkley and Cartwright, 1971; McIntosh, 1974; Salmon and Segall, 1980; Witt *et al.*, 1981; McDonald, this volume), an important indication that the non-tubulin proteins co-purifying with taxol-stabilized microtubules represent components of the spindle.

VII. Association of MAPs with the Mitotic Spindle

To investigate further the possibility that the non-tubulin proteins were indeed spindle MAPs, we began a program of monoclonal antibody production. Two strategies were used. First, we immunized mice with total non-tubulin proteins obtained by salt extraction of taxol-purified microtubules (as in Fig. 5.1A, lane S3). This approach had the advantage of not presupposing which of the non-tubulin proteins were, in fact, *bona fide* MAPs. It introduced the potential complication of obtaining antibodies to tubulin, which often contaminates the MAPs fraction to a detectable extent (Vallee and Bloom, 1983).

Screening was performed by both immunoblotting and immunofluorescence microscopy. Immunoblotting revealed that most hybridoma wells were producing antibodies to multiple polypeptides, including tubulin in some cases. We were delighted to find that more than half of the total wells (57 of 91) reacted with the mitotic spindle by immunofluorescence microscopy.

To focus upon only those positive hybridoma colonies most likely to be producing antibodies to MAPs, most wells showing reaction with tubulin by immunoblotting were rejected. A total of 12 wells were cloned using immunoblot analysis as the primary method of screening. Antibodies to four distinct non-tubulin proteins were obtained.

A second and more direct strategy was also employed. The 77 kDa species was the major putative MAP and had apparently been identified as a microtubule protein by two different approaches (Keller and Rebhun, 1980; Vallee and Bloom, 1983). We felt that it was worth producing antibodies to this protein alone, since it seemed probable that it would prove to be of biological interest. The protein was isolated from taxol-purified microtubules by preparative SDS gel electrophoresis and was used in this form for immunization. In the subsequent hybridoma fusion, all 39 wells tested reacted on immunoblots with the immunogen. Five separate antibodies were ultimately produced from independent fusion wells, all reacting with the same 77 kDa protein species.

The antibodies have proved to be invaluable as a means of monitoring the behaviour of individual proteins during the early stages of microtubule purification, when they cannot be assayed by other means. Figure 5.1C shows the results of immunoblot analysis of the early microtubule purification stages using one of five monoclonal antibodies specific for the 77 kDa protein. It may be seen that the antibody detected the protein in the cytosolic extract (E), but not in the supernate (S1) produced by centrifugation of microtubules. The pellet (P1), which contained microtubules enriched approximately 12-fold relative to the extract, can be seen to be comparably enriched in the immunoreactive protein. Thus, the entire detectable pool of the 77 kDa protein co-purified with the microtubules directly out of the cytosolic extract. Comparable results have been obtained using the other four monoclonal antibodies specific for this protein. Similar data are shown in Fig. 5.3 for proteins of molecular weights 37, 150, 205 and 235 kDa. By the biochemical criterion of efficient co-purification with microtubules, therefore, a total of five distinct proteins were found to behave as genuine MAPs in the sea-urchin egg.

On Coomassie blue-stained gels the 77 kDa, 205 kDa and 235 kDa proteins were among the most prominent non-tubulin components of the purified microtubules. In contrast, the 150 kDa and 37 kDa bands were barely visible by Coomassie blue staining of SDS gels (Fig. 5.3A). We do not understand the basis for this diversity in abundance. One interesting possibility is that, as in the axoneme, the spindle microtubules are very complex in composition, containing a variety of associated structures (see Fig. 5.2). Each of these might be composed of multiple subunits, as in the case of dynein (Bell *et al.*, 1979; Piperno and Luck, 1979; Pfister *et al.*, 1982; Paschal *et al.*, 1987) or the radial spokes (Piperno *et al.*, 1981). This would imply that the MAPs might co-purify in groups. We have not yet been able to test this interesting possibility.

Proteolysis does not seem to be responsible for generating the several MAP

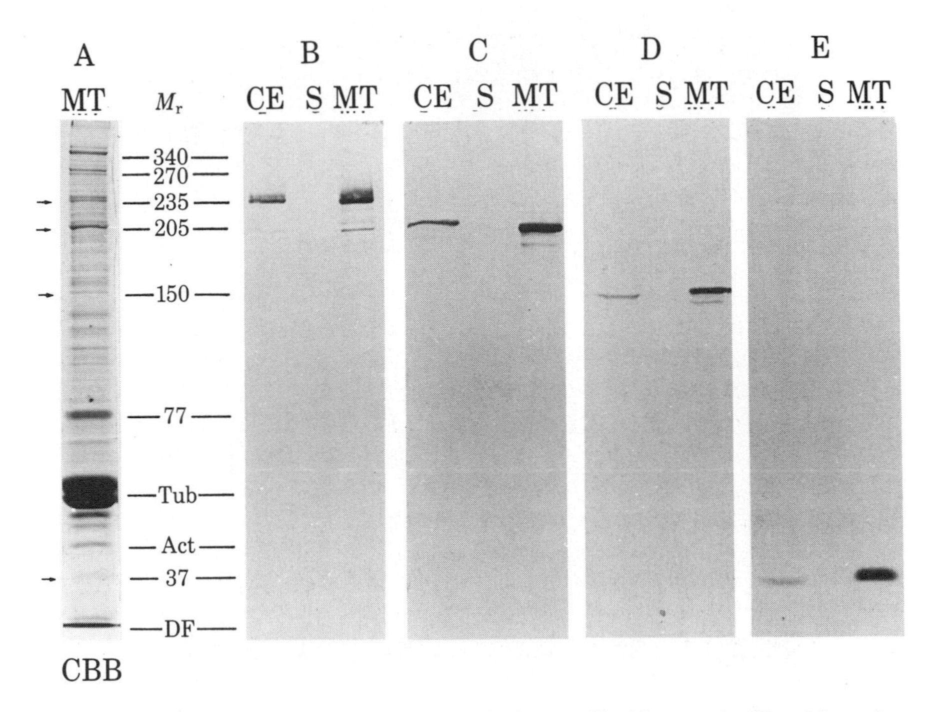

Figure 5.3. Immunoblot analysis of microtubules purified from unfertilized *L. variegatus* eggs. Samples from early stages of the microtubule purification procedure were resolved on 7% polyacrylamide gels and stained with Coomassie blue (A) or transferred electrophoretically to nitrocellulose (B–E). Taxol was added to the cytosolic extract (lanes CE) to stimulate microtubule polymerization. Centrifugation yielded a microtubule-depleted supernate (lanes S) and a microtubule pellet (lanes MT), which was resuspended in one-quarter the extract volume. The nitrocellulose sheets were stained with monoclonal antibodies to proteins of (B) 235 kDa, (C) 205 kDa, (D) 150 kDa, and (E) 37 kDa. Because equal volumes of the three fractions were loaded, it is evident that all four immunoreactive proteins co-purified quantitatively with microtubules. (From Vallee and Bloom, 1983.)

species. Some evidence of MAP proteolysis during microtubule purification was observed, indicated by the minor immunoreactive bands seen in immunoblots stained with antibodies to the 235 kDa, 205 kDa and 150 kDa proteins (Fig. 5.3). However, this was very limited, and indicated that we had successfully controlled this possible complicating factor in the analysis of the MAPs. There was no evidence that the various species derived from a common larger precursor. We did note that some, but not all, of the antibodies to the 77 kDa species reacted with a trace species of higher molecular weight in the

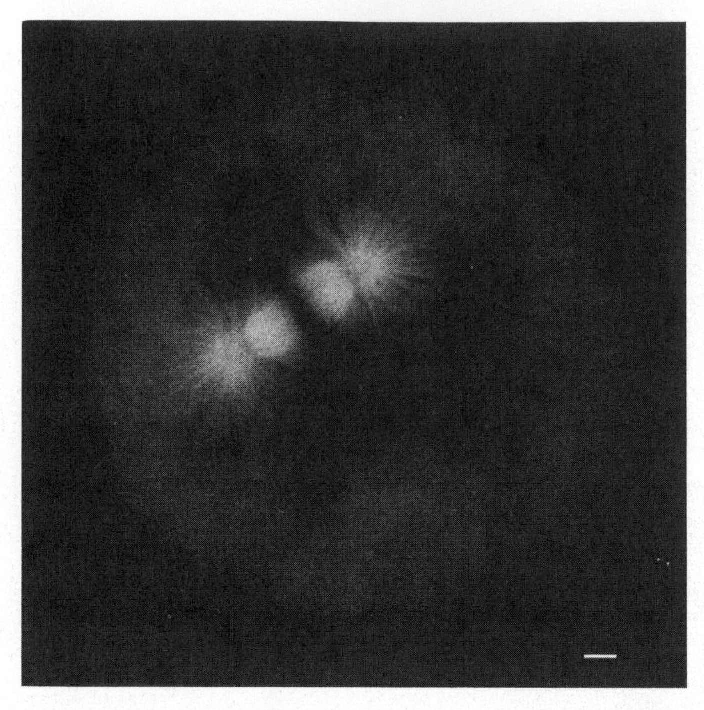

Figure 5.4. Immunofluorescence microscopy of a fertilized *L. variegatus* egg at anaphase using a monoclonal antibody to the 37 kDa protein. Bar = 10 μm. (Adapted from Vallee and Bloom, 1983.)

microtubule pellets, but not in the extracts (see Fig. 5.1C). This may reflect the presence of a minor cross-reactive species, oligomerization of the 77 kDa protein in SDS gels, or a precursor form of the protein.

Further evidence that the five immunoreactive proteins indeed represented MAPs was obtained by immunofluorescence microscopy. Figure 5.4 illustrates localization of the 37 kDa protein in the mitotic spindle of a fertilized egg at anaphase. Monoclonal antibodies to the 77 kDa (Fig. 5.5), 150, 205 and 235 kDa MAPs (Vallee and Bloom, 1983) also stain the spindle. Thus, by both biochemical and cytological criteria, the five immunoreactive proteins behave as expected for genuine spindle MAPs. A summary of the monoclonal antibodies we have raised against these proteins is presented in Table 5.1.

Figure 5.5. Immunofluorescence microscopy of mitotic, fertilized *S. purpuratus* eggs using five distinct monoclonal antibodies to the 77 kDa protein (panels 1–5) and a control monoclonal IgM (panel 6). Bar = 20 μm. (From Bloom *et al.*, 1985b.)

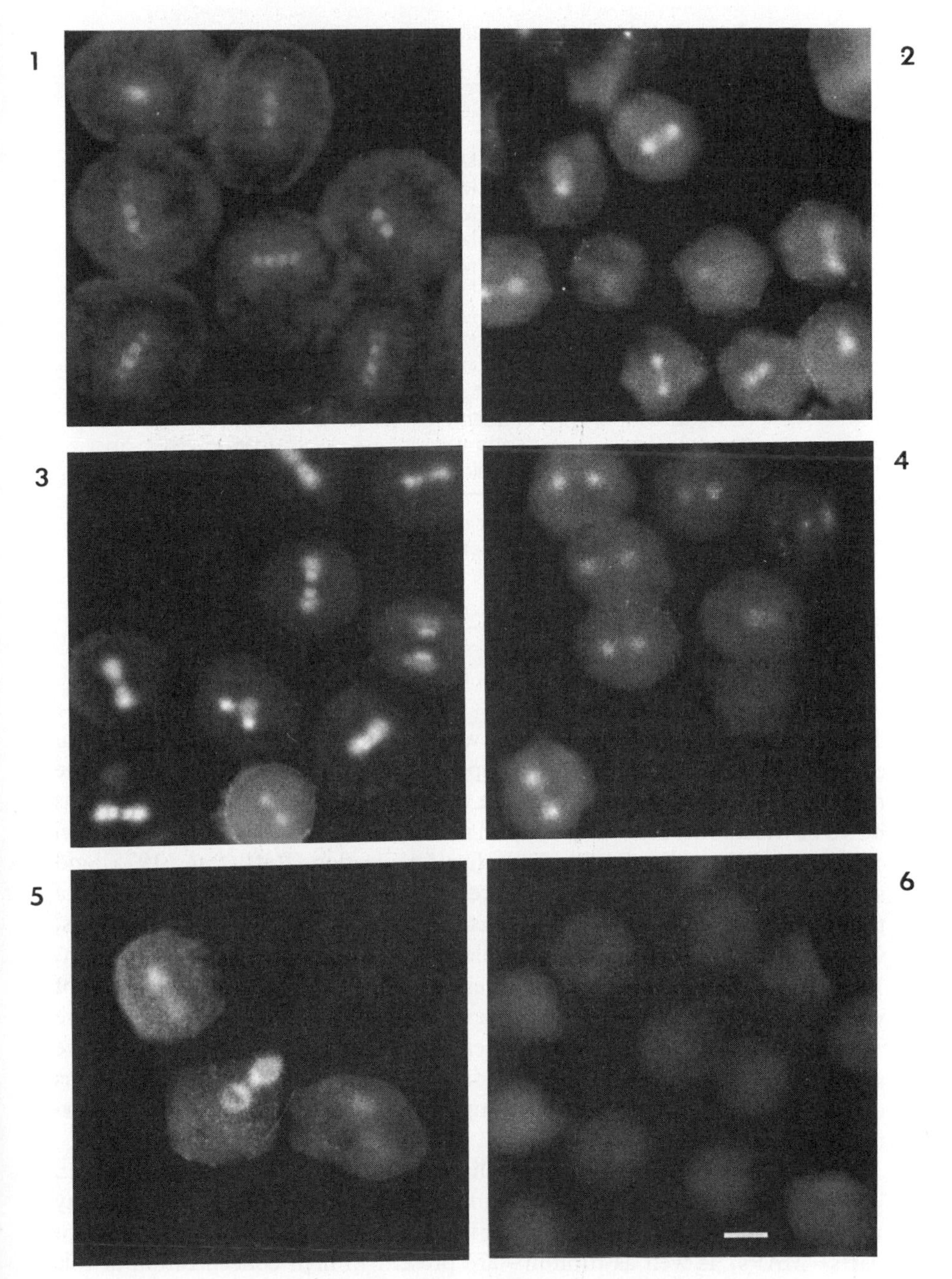

Figure 5.5.

Table 5.1. Monoclonal antibodies to sea urchin egg proteins

Antibody name	Reacts with	Isotype
L.v. HMW-1	235K MAP	IgG1
L.v. HMW-2	205K MAP	IgG1
L.v. 150-1	150K MAP	IgG1
L.v. 37-1	37K MAP	IgG1
S.p. 77-1	77K MAP	IgM
S.p. 77-2	77K MAP	IgG1
S.p. 77-3	77K MAP	IgM
S.p. 77-4	77K MAP	IgG1
S.p. 77-5	77K MAP	IgM
L.v. tub-1	Beta-tubulin > alpha	IgM
L.v. tub-2	Tubulin#	IgM
L.v. tub-3	Alpha-tubulin*	IgM
L.v. tub-4	Tubulin*#	IgM

*React with sea urchin tubulins, but not mammalian tubulins.
#Specificity for alpha- or beta-tubulin not determined.

VIII. The Unfertilized Egg is a Stockpile of Mitotic Precursors

The basis for undertaking this work was the supposition that the sea-urchin egg might contain microtubule components destined to play a role in the multiple mitotic divisions that ensue upon fertilization. Our reasoning was based on the known content of tubulin in the unfertilized egg, and the low level of new protein synthesis observed after fertilization (Raff *et al.*, 1975) and seems to have been borne out by the results we have obtained to date. Most of the biochemical work we have done has been performed with unfertilized eggs, which themselves contain no assembled microtubules (Bestor and Schatten, 1981). Nonetheless, we obtain an excellent yield of a variety of proteins, all of which behave as MAPs *in vitro*, and several of which we have found to behave as components of the spindle in the cell. In preliminary experiments we have also purified microtubules from first-division embryos. There was no obvious difference in microtubule protein composition from that obtained with unfertilized eggs.

The finding of MAPs in the unfertilized egg brings into yet sharper focus an intriguing problem in understanding microtubule behaviour in the egg: why are microtubules assembled only after fertilization if their requisite component parts—MAPs, as well as tubulin—are present in high concentration at this stage? MAPs are known to promote microtubule polymerization in other systems (Weingarten *et al.*, 1975; Sloboda *et al.*, 1976; Murphy *et al.*, 1977), and they presumably serve this role in the sea-urchin egg as well. Perhaps, as suggested by others (Bryan *et al.*, 1975; Naruse and Sakai, 1981), additional inhibitory factors must be invoked to explain totally the behaviour of egg tubulin *in vitro* and *in vivo*. Whether such factors might, themselves, be present in purified egg microtubules is an interesting possibility.

IX. The Mitotic Motor

It seems reasonable to expect that the proteins we have identified in the purified egg microtubule preparations will ultimately prove to be involved in organizing the spindle microtubules and, perhaps, in regulating their assembly as well. Do these preparations also contain proteins involved in spindle mechanochemistry?

The purified egg microtubules do contain ATPase activity (Scholey *et al.*, 1984; Collins and Vallee, 1986a). Some of this is due to the presence of a dynein-like enzyme present at high concentrations in egg cytosol, known both as "cytoplasmic dynein" and "egg dynein" (Weisenberg and Taylor, 1968; Pratt *et al.*, 1980). Only a small fraction of the total egg dynein co-sediments with microtubules (Scholey *et al.*, 1984), and most of the ATPase activity in the microtubule preparations is actually due to an unrelated enzyme (Collins and Vallee, 1986a; and see below).

It has been a matter of debate for many years whether egg dynein has a cytoplasmic function, as in mitosis, or instead represents a storage form of dynein to be used in the formation of cilia on the surface of the blastula-stage embryo. While this question has been difficult to resolve by direct analysis of egg dynein, recent work from other systems has shed light on this question. Dynein has now been found in cytosolic extracts of brain and other tissues (Paschal *et al.*, 1987; Vallee *et al.*, 1988; Collins and Vallee, 1988). This soluble form of the enzyme is believed to be responsible for retrograde axonal transport (Paschal and Vallee, 1987). It is presumed to be responsible for the centripetal movement of membranous organelles along microtubules toward the nucleus and could possibly be involved in anaphase A chromosome movement as well.

Analysis of the enzymological properties of this newly described form of dynein has revealed a number of important differences from egg dynein (Shpetner *et al.*, 1988; Collins and Vallee, 1988) which, however, is similar to ciliary and flagellar dynein. Along with evidence that antibodies to egg dynein react with blastula-stage cilia but not with cytoplasmic microtubules (Asai, 1986), it seems that at least the major form of dynein in egg cytoplasm, is a ciliary precursor.

Another ATPase capable of associating with egg microtubules is kinesin. This enzyme was originally identified in neuronal tissue, but has also been found in the sea-urchin egg (Scholey *et al.*, 1985). It binds to microtubules in a nucleotide-sensitive manner. It is present in only limited amounts in microtubules prepared by our normal methods, but can be induced to co-purify with microtubules by inclusion of the ATP analogue AMPPNP (Scholey *et al.*, 1985). Kinesin, like the other MAPs described in this article, is found in abundance in the unfertilized egg, and has been localized to the mitotic spindle of developing embryos by immunocytochemical means (Scholey *et al.*, 1985). However, a specific role in mitotic motility has not yet been established.

Recent work has demonstrated the existence of yet another distinct 10S ATPase, which is particularly abundant in sea-urchin eggs and which was first identified in egg microtubules (Collins and Vallee, 1986a). This enzyme seems to account for most of the ATPase activity in purified egg microtubules. Like kinesin and the brain cytoplasmic form of dynein, it is strongly activated by microtubules. Its activity, in fact, is normally undetectable in cytosolic extracts of sea-urchin eggs, but reaches a level comparable to that of egg dynein when assayed in the presence of microtubules. The enzyme co-sedimented with microtubules, but, unlike dynein and kinesin could not readily be dissociated by ATP. This could mean that the enzyme interacts with microtubules via distinct ATP-dependent and ATP-independent binding sites (model II shown in Fig. 5.6), as is the case for ciliary and flagellar dynein. This would be consistent with the expected properties of an enzyme involved in microtubule–microtubule sliding, which has been postulated to occur during anaphase B.

Because the enzyme remains attached to microtubules in the presence of ATP, it has been difficult to purify. However, recent work has led to the identification of a 100 kDa polypeptide in mammalian brain microtubule preparations that has several of the properties of the sea-urchin enzyme (Shpetner and Vallee, 1988, 1989). Work is in progress to purify the brain enzyme and to establish whether it is the same protein as the 100 kDa MAP found in our sea-urchin egg microtubule preparations (Fig. 5.1, next most prominent band above 77 kDa species; Vallee and Bloom, 1983).

The precise role of the several mechanochemical factors in the egg remains to be determined. However, what seems clear is that the purified egg microtubules have been, and should continue to be, a very useful system for further investigation into the biochemistry of mitosis. It is to be hoped that this will finally lead to an understanding of the molecular mechanisms underlying this most fundamental event in the life of the cell.

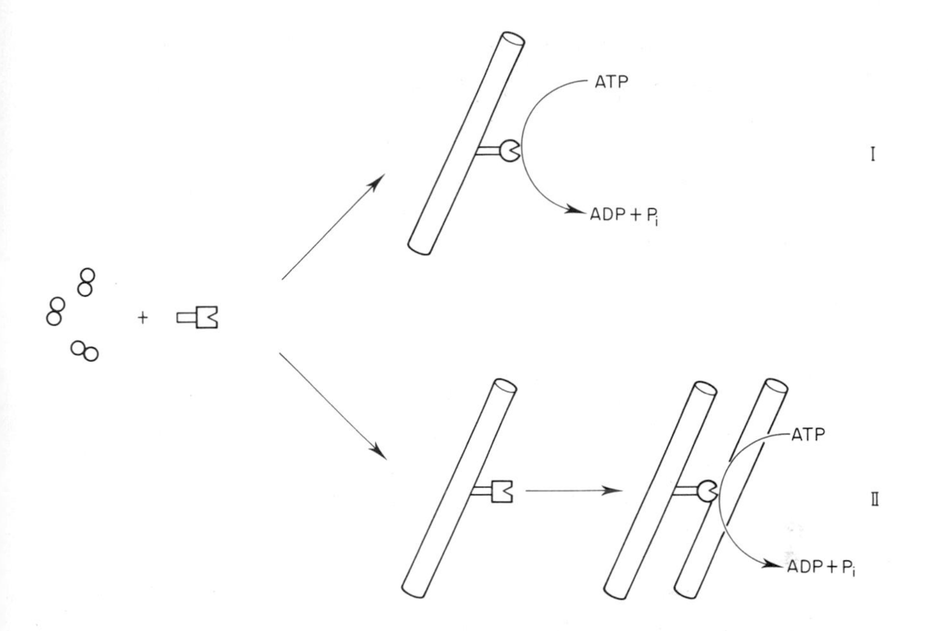

Figure 5.6. Alternative schemes to account for the biochemical behaviour of the 10 S microtubule-activated ATPase found in sea-urchin egg microtubules (Collins and Vallee, 1986a). Tubulin dimers are shown at left in an unassembled state along with inactive enzyme. Activation is envisaged to occur in either of two ways, each accounting for the lack of effect of ATP on microtubule binding. In scheme I, the enzyme binds to a microtubule via a non-catalytic site, and activation occurs by an allosteric mechanism. In scheme II, binding occurs via two distinct sites. Binding at a non-catalytic site would account for the ATP-independent binding of the enzyme to the microtubule. The observed microtubule activation of the ATPase activity would reflect a force-producing interaction with a second microtubule. (From Collins and Vallee, 1986b.)

Acknowledgements

This work was supported by NIH Grant GM32977 and March of Dimes Grant 5-388 to Richard B. Vallee, and by the Mimi Daron Greenberg Fund of the Worcester Foundation for Experimental Biology.

REFERENCES

Asai, D. J. (1986). *Develop. Biol.* **118**, 416–424.
Auclair, W. and Siegel, B. W. (1966). *Science* **154**, 913–915.

Bell, C. W., Frank, E. and Gibbons, I. R. (1979). *J. Supramol. Struct.* **11**, 311–317.
Bestor, T. H. and Schatten, G. (1981). *Develop. Biol.* **88**, 80–91.
Binder, L. I., Frankfurter, A. and Rebhun, L. I. (1985). *J. Cell Biol.* **101**, 1371–1378.
Bloom, G. S., Schoenfeld, T. A. and Vallee, R. B. (1984a). *J. Cell Biol.* **98**, 320–330.
Bloom, G. S., Luca, F. C. and Vallee, R. B. (1984b). *J. Cell Biol.* **98**, 331–340.
Bloom, G. S., Luca, F. C. and Vallee, R. B. (1985a). *Proc. Natl. Acad. Sci. USA* **82**, 5404–5408.
Bloom, G. S., Luca, F. C., Collins, C. A. and Vallee, R. B. (1985b). *Cell Motility* **5**, 431–446.
Brinkley, B. R. and Cartwright, J., Jr. (1971). *J. Cell Biol.* **50**, 416–431.
Bryan, J., Nagle, B. W. and Doenges, K. H. (1975). *Proc. Natl. Acad. Sci. USA* **72**, 3570–3574.
Bulinski, J. C. and Borisy, G. G. (1980). *J. Cell Biol.* **87**, 792–801.
Caceres, A., Binder, L. I., Payne, M. R., Bender, P., Rebhun, L. and Steward, O. (1984). *J. Neurosci.* **4**, 394–410.
Collins, C. A. and Vallee, R. B. (1986a). *Proc. Natl. Acad. Sci. USA* **83**, 4799–4803.
Collins, C. A. and Vallee, R. B. (1986b). *J. Cell Sci. Suppl.* **5**, 197–204.
Collins, C. A. and Vallee, R. B. (1988). *Cell Motil. Cytoskel.*, **11**, 195a (abstr.).
De Camilli, P., Miller, P. E., Navone, F., Theurkauf, W. E. and Vallee, R. B. (1984). *Neurosci.* **11**, 819–846.
Detrich, W. H., III and Wilson, L. (1983). *Biochemistry* **22**, 2453–2462.
Huber, G. and Matus, A. (1984). *J. Neurosci.* **4**, 151–160.
Kane, R. E. (1962). *J. Cell Biol.* **12**, 47–55.
Kane, R. E. (1975). *J. Cell Biol.* **66**, 305–315.
Keller, T. C. S., III and Rebhun, L. I. (1980). *J. Cell Biol.* **93**, 788–796.
Kuriyama, R. (1977). *J. Biochem.* **81**, 1115–1125.
McIntosh, J. R. (1974). *J. Cell Biol.* **61**, 166–187.
Miller, P., Walter, U., Theurkauf, W. E., Vallee, R. B. and De Camilli, P. (1982). *Proc. Natl. Acad. Sci. USA* **79**, 5562–5566.
Murphy, D. B., Vallee, R. B. and Borisy, G. G. (1977). *Biochemistry* **16**, 2598–2605.
Naruse, H. and Sakai, H. (1981). *J. Biochem.* **90**, 581–587.
Parness, J. and Horwitz, S. B. (1981). *J. Cell Biol.* **91**, 479–487.
Parysek, L. M., Asnes, C. A. and Olmsted, J. B. (1984a). *J. Cell Biol.* **99**, 1309–1315.
Parysek, L. M., Wolosewick, J. J. and Olmsted, J. B. (1984b). *J. Cell Biol.* **99**, 2287–2296.
Paschal, B. M., Shpetner, H. S. and Vallee, R. B. (1987). *J. Cell Biol.* **105**, 1273–1282.
Paschal, B. M. and Vallee, R. B. (1987). *Nature* **330**, 181–183.
Pfister, K. K., Fay, R. B. and Witman, G. B. (1982). *Cell Motility* **2**, 525–547.
Piperno, G. and Luck, D. J. (1979). *J. Biol. Chem.* **254**, 3084–3090.
Piperno, G., Huang, B., Ramanis, Z. and Luck, D. J. (1981). *J. Cell Biol.* **88**, 73–79.
Pratt, M. M., Otter, T. and Salmon, E. D. (1980). *J. Cell Biol.* **86**, 738–745.
Raff, R. A. and Kaumeyer, J. K. (1973). *Develop. Biol.* **32**, 309–320.
Raff, R. A., Brandis, J. W., Green, L. H., Kaumeyer, J. F. and Raff, E. C. (1975). *Ann. NY Acad. Sci.* **253**, 304–317.
Salmon, E. D. (1982). *Methods Cell Biol.* **25**, 69–105.
Salmon, E. D. and Segall, R. R. (1980). *J. Cell Biol.* **86**, 355–365.
Schiff, P. B., Fant, J. and Horwitz, S. B. (1979). *Nature* **277**, 665–667.
Scholey, J. M., Neighbors, B., McIntosh, J. R. and Salmon, E. D. (1984). *J. Biol. Chem.* **259**, 6516–6525.

Scholey, J. M., Porter, M. E., Grissom, P. M. and McIntosh, J. R. (1985). *Nature* **318**, 483–486.
Shpetner, H. S. and Vallee, R. B. (1988). *Cell Motil. Cytoskel.*, **11**, 199a (abstr.).
Shpetner, H. S. and Vallee, R. B. (1989). *J. Cell Biol.*, **107**, 673a (abstr.).
Shpetner, H. S., Paschal, B. M. and Vallee, R. B. (1988). *J. Cell Biol.* **107**, 1001–1009.
Silver, R. B., Cole, R. D. and Cande, W. Z. (1980). *Cell* **19**, 505–516.
Sloboda, R. D., Dentler, W. L. and Rosenbaum, J. L. (1976). *Biochemistry* **15**, 4497–4505.
Suprenant, K. A. and Rebhun, L. I. (1983). *J. Biol. Chem.* **258**, 4518–4525.
Vallee, R. B. (1982). *J. Cell Biol.* **92**, 435–442.
Vallee, R. B. and Bloom, G. S. (1983). *Proc. Natl. Acad. Sci. USA* **80**, 6259–6263.
Vallee, R. B., Paschal, B. M., Shpetner, H. S. and Wall, J. S. (1988). *Nature* **332**, 561–563.
Wani, M. C., Taylor, H. L., Wall, M. E., Coggon, P. and McPhail, A. T. (1971). *J. Amer. Chem. Soc.* **93**, 2325–2327.
Weingarten, M. D., Lockwood, A. H., Hwo, S. and Kirschner, M. W. (1975). *Proc. Natl. Acad. Sci. USA* **72**, 1858–1862.
Weisenberg, R. and Taylor, E. W. (1968). *Exp. Cell Res.* **53**, 372–384.
Wilson, H. J. (1969). *J. Cell Biol.* **40**, 854–859.
Witt, P. L., Ris, H. and Borisy, G. G. (1981). *Chromosoma* **83**, 523–540.

CHAPTER 6

Calmodulin Regulation of Spindle Function

MICHAEL J. WELSH and STUART C. SWEET

Department of Anatomy & Cell Biology and Cellular and Molecular Biology Program, The University of Michigan Medical School, Ann Arbor, Michigan, USA

I. Introduction

Mitosis is one of the most fundamental and significant activities of eukaryotic cells. In the process of mitosis the genetic material is separated, in a highly organized manner, into two daughter cells. Much is now known about the structure of the mitotic apparatus but less is known about the dynamic events that comprise the process of mitosis and almost nothing is known for certain about how the events of mitosis are regulated.

Mitosis is a complex process and the mitotic apparatus is a complex structure; many proteins of the mitotic apparatus probably still remain to be discovered. Even for those components currently known or suspected to be present in the mitotic apparatus, much is not understood about their role in mitosis. Thus, a discussion of the regulation of mitosis is, at this time, primarily speculative. None the less, it is useful occasionally to review a developing area of research to evaluate the results so far obtained and to speculate on those areas of study that might fruitfully be pursued to further knowledge in the area.

Other chapters in this volume describe current knowledge about spindle structure and dynamics, the existence and roles of microtubule-associated proteins, membranes and calcium in the mitotic apparatus, as well as other aspects of mitosis. A comprehensive discussion of the regulation of spindle function would require an integration of all these aspects of mitosis. Rather than attempt to duplicate contents of these other chapters, this chapter will be limited to a discussion of evidence for the presence of calmodulin in the

MITOSIS: Molecules and Mechanisms
ISBN 0-12-363420-2

spindle, calmodulin-mediated events that may operate during mitosis to regulate the process, and a brief mention of other possible constituents of the mitotic apparatus. Space limitations in some instances may not allow exhaustive discussion or citation of all studies relevant to a particular topic. If we should fail to cite some relevant publications, we hope our colleagues will forgive our, at times necessary, brevity.

II. Calmodulin—A Ubiquitous Calcium-dependent Regulatory Protein

It is not our intention to give a complete review of the calmodulin literature. A number of more or less extensive reviews have been written in the past several years that describe the state of our knowledge concerning calmodulin's structure and function (Cheung, 1980; Means and Dedman, 1980; Klee and Vanaman, 1982; Manalan and Klee, 1984). However, information about calmodulin regulation of other biochemical activities is probably relevant to understanding at least some of what calmodulin may be doing during mitosis.

Calmodulin, which is found in both plants and animals, is a highly conserved protein (Goodman *et al.*, 1979); the amino-acid sequences of calmodulins isolated from a number of organisms have been determined to be almost identical (see Goodman *et al.*, 1979; Klee and Vanaman, 1982). The three-dimensional structure of the protein has been determined to a resolution of 3 Å (Babu *et al.*, 1985) and numerous structural studies have helped to define the functional properties of the protein (for reviews, see Krebs, 1981; Klee and Vanaman, 1982). Calmodulin binds calcium (Teo and Wang, 1973) and, when doing so, changes conformation to expose a hydrophobic region (LaPorte *et al.*, 1980) that is believed to be the domain by which calmodulin binds to the other proteins it affects (reviewed by Klee and Vanaman, 1982). The dissociation constants of calmodulin for calcium ions (about 10^{-6}–10^{-5} M) are compatible with the concept of calmodulin being a physiologically appropriate intracellular receptor for and mediator of calcium action. Although calmodulin has been observed to bind to some proteins in a calcium-independent manner (Cohen *et al.*, 1978; Hertzberg and Gilula, 1981; Greenlee *et al.*, 1982; Welsh *et al.*, 1982), current dogma generally views calmodulin interaction with, and regulation of, other proteins to be dependent upon calcium first being bound by calmodulin, followed by changes in conformation of the protein that result in exposure of the hydrophobic binding site to which the other proteins bind.

As described below, calmodulin has been identified as being present in the mitotic apparatus of virtually all eukaryotic cells so far examined. As yet, no specific function for calmodulin in the spindle has been demonstrated unequivocally. It has been hypothesized, on the basis of its localization in the spindle by immunofluorescence methods, that calmodulin may regulate microtubule assembly/disassembly in the mitotic apparatus (Welsh *et al.*, 1978,

1979; Brinkley *et al.*, 1978). In fact, the localization of calmodulin in the spindle led directly to studies that demonstrated effects of calmodulin on microtubule assembly and disassembly *in vitro* (Marcum *et al.*, 1978; Brinkley *et al.*, 1978).

Certainly, other possible functions for calmodulin in the spindle can be envisioned. Indeed, some regulatory actions of calmodulin demonstrated to occur in other biochemical systems may be of relevance when considering the possible role or roles played by calmodulin in the process of mitosis. For example, calmodulin has been shown to regulate cellular protein kinases, phosphatases and calcium pumps, as well as to bind to a number of proteins, including tubulin and microtubule-associated proteins (MAPs) (reviewed by Klee and Vanaman, 1982; Manalan and Klee, 1984). All of these observations suggest that calmodulin may play multiple roles in regulation of spindle structure and function. There follows evidence for calmodulin's presence in the mitotic spindle and a more detailed discussion of its potential regulatory functions during mitosis.

III. Calmodulin Localization in the Mitotic Apparatus

A. Immunolocalization at the light-microscopic level

Originally, calmodulin was suggested by Welsh and co-workers (1978) to perform a regulatory function in mitosis and was implicated as a regulator of microtubule assembly/disassembly on the basis of immunofluorescence localization studies that showed calmodulin to be concentrated in the mitotic apparatus of all vertebrate cell types examined, including cells of mammalian, avian and amphibian origin (Welsh *et al.*, 1977, 1978). These observations were subsequently confirmed by Anderson *et al.* (1978) and, since that time, have been extended to higher plants including onion and pea meristem cells and *Haemanthus* endosperm as well (Wick *et al.*, 1985; Vantard *et al.*, 1985).

Typical immunolocalization patterns of calmodulin in mitotic PtK_1 cells are demonstrated in Fig. 6.1. The pattern of calmodulin localization in the mitotic apparatus resembles, but is not identical to, that of tubulin. Throughout metaphase and anaphase, calmodulin is found in the half-spindles, the area between the chromosomes and the spindle poles. Calmodulin appears to be relatively more concentrated toward the poles of the spindle. In many cells, calmodulin can be seen to form a halo around what is most probably the centriole pair at each pole, suggesting that calmodulin is present in the electron dense, amorphous pericentriolar material seen surrounding the centrioles at the electron-microscopic level (McDonald; Vandré and Borisy, this volume). As long as kinetochore microtubules exist in the spindle, calmodulin is

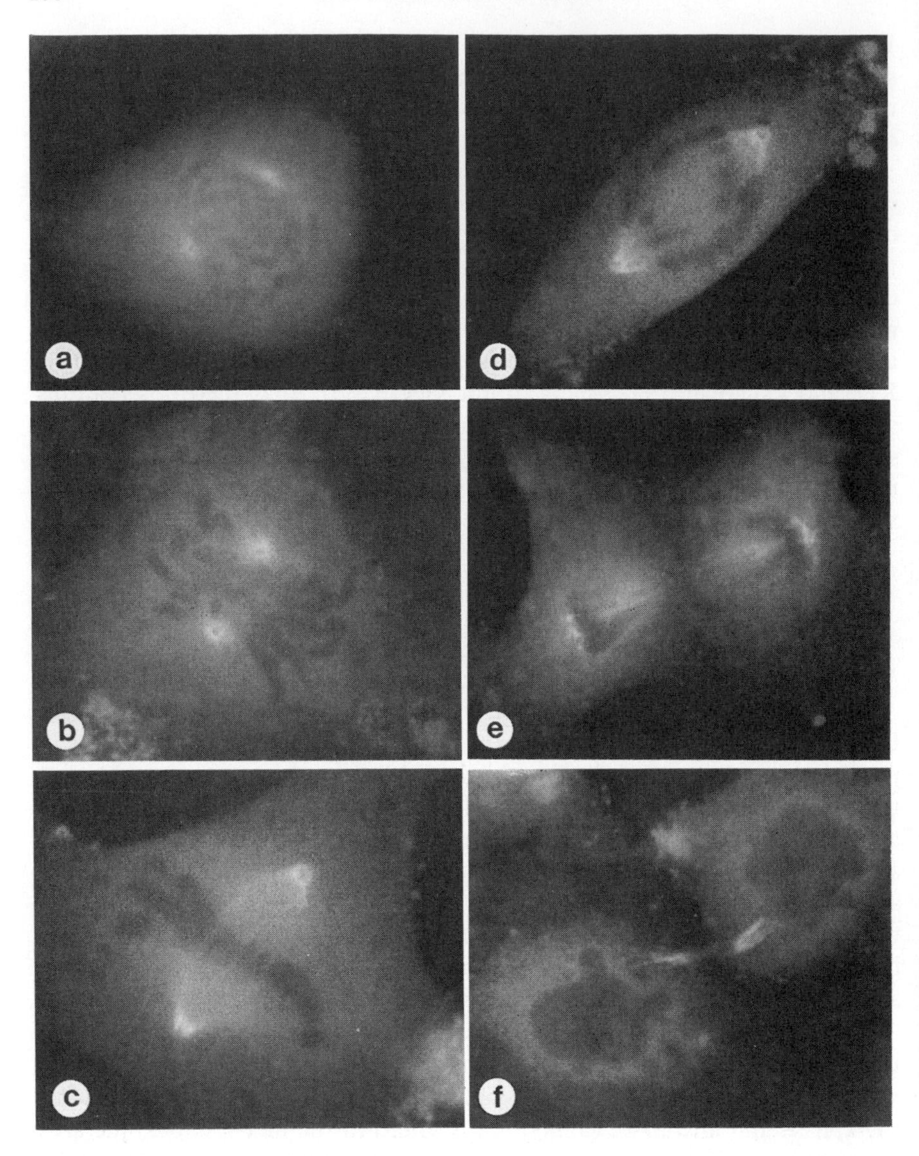

Figure 6.1. Immunofluorescence localization of calmodulin in mitotic PtK_1 cells. Cells were cultured on glass coverslips and prepared for immunofluorescence localization of calmodulin as described by Welsh (1983). Calmodulin can be localized to concentrations associated with the forming spindle poles in early prophase, as shown in (a) and (b). During metaphase (c), calmodulin appears to be concentrated around the spindle poles and in a pattern confined to the half-spindles and the pericentriolar region: no concentration of calmodulin is seen in the interzone region between the separating chromosomes. In late anaphase (e),

concentrated in the half-spindles. No concentration of calmodulin can be observed in the interzone region, between the separating sets of chromosomes, until very late in anaphase or during telophase. The area of interzone localization of calmodulin late in mitosis appears to become reduced in extent as cytokinesis proceeds, until it becomes localized to a small area at each end of the midbody.

Subsequent to the initial immunolocalization of calmodulin in the mitotic spindle (Welsh *et al.*, 1977, 1978), studies of the possible association of calmodulin with microtubules of the mitotic apparatus showed that calmodulin localization in the mitotic apparatus was coincident with a particular subpopulation of spindle microtubules (Welsh *et al.*, 1979). Indeed, at least two populations of microtubules can be said to exist in the mitotic apparatus (Brinkley and Cartwright, 1975). In one, that of the kinetochore microtubules, the microtubules extend from the spindle poles to the kinetochores on each chromosome at which they attach. The other population is that of the non-kinetochore microtubules, which comprise all remaining microtubules (McDonald, this volume). These microtubules extend from the spindle poles into the cell cytoplasm where they end with no identifiable attachment site. Many of the non-kinetochore microtubules extend toward the chromosomes but pass beyond them to end in the cytoplasm of the equatorial or interzone region of the cell.

Calmodulin concentration in the spindle appears to be dependent on the presence of kinetochore, but not non-kinetochore, microtubules (Welsh *et al.*, 1979). This conclusion was based upon studies of cells that had been incubated at 0–4°C to eliminate non-kinetochore microtubules, which are cold-labile. Calmodulin immunolocalization in cold-treated cells was identical to that in cells that were not cold-treated, suggesting coincidence of calmodulin with kinetochore microtubules. However, agents such as colcemid, when used at concentrations that eliminated all spindle microtubules, resulted in loss of calmodulin localization in the spindle (Welsh *et al.*, 1979). Interestingly, treatment of cells with nitrous oxide under pressure, which causes disorganization of the spindle but does not cause disassembly of the spindle microtubules, resulted in a diffuse, but none the less concentrated, localization of calmodulin that was similar to that of the disorganized spindle microtubules (Welsh *et al.*, 1979). These results concerning calmodulin localization as determined by

---

Figure 6.1 continued

calmodulin diminishes in the half-spindles as these structures become reduced in size. At this time, calmodulin also begins to become concentrated in the interzone region. Finally, in telophase (f), calmodulin can be seen to be concentrated at the distal ends of the midbody. In all figures some background fluorescence can be observed. We believe this is a true reflection of the presence of calmodulin remaining in the cell cytoplasm during mitosis. (Figures from Dedman *et al.*, 1982.)

immunofluorescence led to the hypothesis that calmodulin functions to regulate microtubule disassembly during mitosis (Brinkley *et al.*, 1978; Welsh *et al.*, 1978, 1979).

Similar preferential association of calmodulin with kinetochore microtubules in plant cells, which possess no centrioles and which have diffuse microtubule organizing centres during mitosis, has also been observed (Vantard *et al.*, 1985). These workers observed that when *Haemanthus* endosperm cells were incubated at 3°C or treated with an appropriately low concentration of colchicine (treatments that selectively eliminated non-kinetochore microtubules), calmodulin localization was similar to what was seen in untreated cells. Vantard *et al.* (1985) concluded that their results further support the hypothesis that calmodulin may function in regulating microtubule disassembly during mitosis.

One conclusion made by Vantard *et al.* (1985) deserves comment. They concluded, on the basis of their observations of plant cells and their interpretation of studies on animal cells reported by other workers, that calmodulin in plant cells localizes differently from that in animal cells. Specifically, they state that plant-cell calmodulin is always coincident with the kinetochore microtubules but that in animal cells calmodulin is not. We disagree with this interpretation. Our observations show that calmodulin in animal cells appears to be found coincident with kinetochore microtubules as long as this class of microtubule exists within the cell. The main difference in calmodulin localization during mitosis between plant cells and animal cells is that in animal cells calmodulin also becomes associated with the interzone microtubules in telophase (Welsh *et al.*, 1978; Dedman *et al.*, 1982; Zavortink *et al.*, 1983) while in plant cells, according to Vantard *et al.* (1985), calmodulin does not concentrate in the interzone area. Indeed, *Haemanthus* endosperm cells may be unusual in this lack of calmodulin association with interzone microtubules late in mitosis. Wick *et al.* (1985) have reported that calmodulin is concentrated in the interzone area, the phragmoplast region, in pea (*Pisum sativum*). As yet, the reason is not known for the discrepancy between the observations of Wick *et al.* (1985) and those of Vantard *et al.* (1985) concerning interzone calmodulin localization in telophase of mitosis in plant cells.

B. Localization of fluorescent analogues of calmodulin in live mitotic cells

It has been possible to synthesize biochemically active analogues of calmodulin and these have been used to observe calmodulin localization in live cells. Hamaguchi and Iwasa (1980) microinjected calmodulin coupled with *N*-(7-dimethylamino-4-methylcoumarinyl)-maleimide into living eggs of the sand-dollar (*Clypeaster japonicus*). They observed the fluorescent analogue to concentrate in the region of the spindle of the cell. Although Wang and Taylor (1979) found that, when echinoderm blastomeres were injected with proteins, such as BSA, the protein was often excluded from the yolk-rich cytoplasm and thus appeared to concentrate in the spindle region because of its greater accessible volume, Wadsworth and Sloboda (1983) did not observe this effect. Thus, the distribution of calmodulin observed by Hamaguchi and Iwasa (1980) may have reflected accurately calmodulin concentration in the mitotic spindle.

We prepared a biochemically active conjugate of calmodulin and tetramethylrhodamine isothiocyanate (CaM-TRITC) and microinjected it into cultured *Potorous tridactylis* PtK_1 (rat kangaroo) and BSC1 (African green monkey kidney) cells as well as 3T3 and HeLa cells (Welsh *et al.*, 1982; Zavortink *et al.*, 1983). The same pattern of distribution of microinjected calmodulin–rhodamine was seen in all types of cells. As seen in Fig. 6.2, microinjected calmodulin–rhodamine is seen to concentrate in the mitotic apparatus. This occurs within 5 seconds of microinjection of the fluorescent analogue. If interphase cells are microinjected, cells proceeding through mitosis at least 25 hours later can be observed to incorporate the fluorescent calmodulin into the mitotic apparatus (Fig. 6.2).

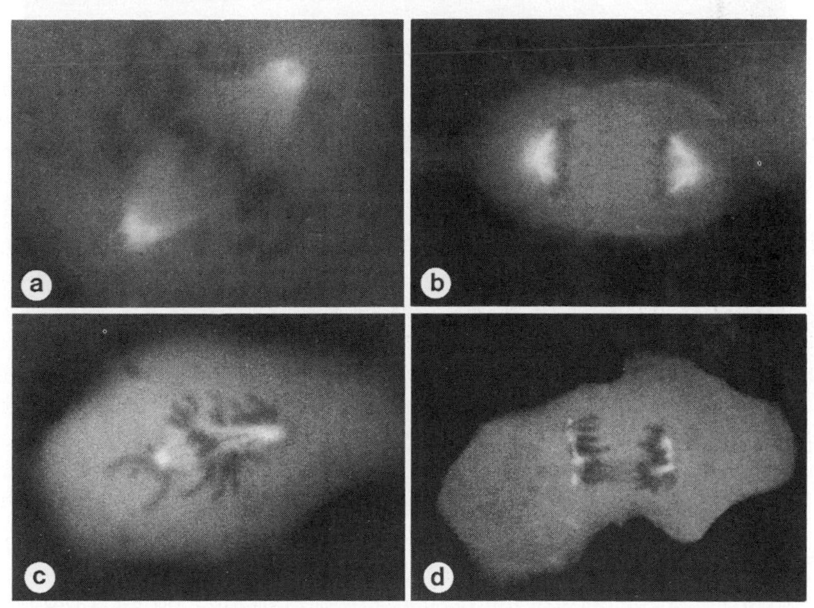

Figure 6.2. Cells microinjected with calmodulin–rhodamine isothiocyanate (CaM-TRITC). CaM-TRITC was prepared and cells were microinjected as described by Zavortink *et al.* (1983). (a) CaM-TRITC in a metaphase PtK_1 cell 10 min after microinjection. CaM-TRITC can be observed to concentrate in the spindle in as little as 5 s after microinjection. (b) A BSC1 cell in anaphase that was microinjected during interphase 25 h previously. (c) and (d) The same PtK_1 cell at 5 min and 15 min after microinjection of CaM-TRITC. As a cell proceeds through mitosis, microinjected CaM-TRITC remains associated with the spindle. (Parts (b), (c) and (d) from Zavortink *et al.*, 1983.)

Microinjected live cells can be observed and their images can be recorded using an ultralow-light-level video camera. Observing cells through mitosis shows that the calmodulin localizes in a similar, but not identical, manner to what has been observed using indirect immunofluorescence with fixed cells (Fig. 6.3). In live PtK_1 cells, calmodulin–rhodamine concentrates in the forming spindle very early in prophase, before the nuclear envelope breaks down. At this stage of mitosis, although a spindle as such has not yet formed, the centriole pairs have begun to separate, and even at this stage calmodulin can be observed to concentrate in the forming spindle poles. As prophase proceeds and the spindle develops, calmodulin continues to be concentrated in it. By prometaphase and continuing through metaphase, calmodulin concentrates in the spindle and can be seen to form a halo around the presumptive location of the centrioles. Calmodulin appears to be concentrated along bundles of

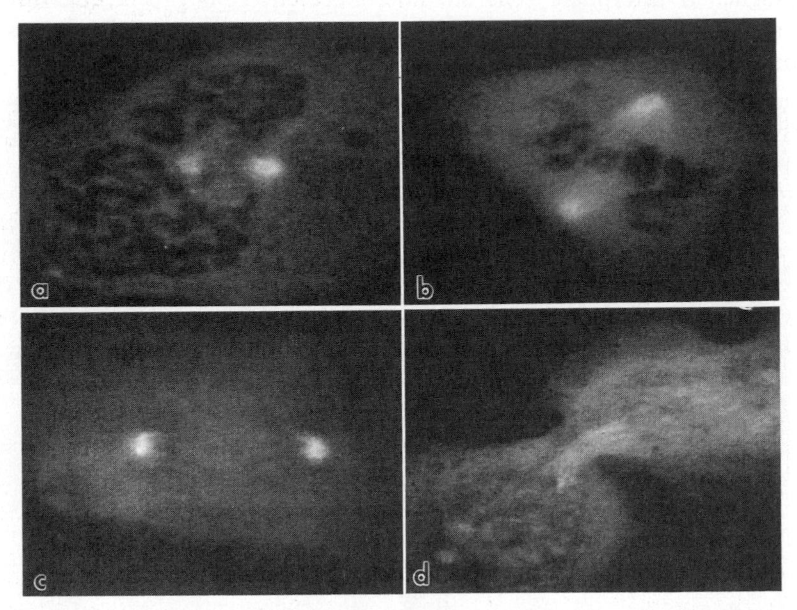

Figure 6.3. Localization of calmodulin–rhodamine isothiocyanate (CaM-TRITC) in live PtK_1 cells as seen by low-light-level video microscopy. (a) Microinjected CaM-TRITC concentrates at the nascent spindle poles during early prophase. During metaphase, as shown in (b), CaM-TRITC concentrates in the pericentriolar region and in a pattern similar to that of spindle microtubules. (c) During anaphase, CaM-TRITC concentrates in the half-spindles. (d) In telophase, CaM-TRITC is observed throughout the midbody. This is different from the pattern of calmodulin localization as revealed by the immunofluorescence localization technique (see Fig. 6.1(f)) or by the ferritin bridge technique (see Fig. 6.6). We believe that the immunolocalization pattern is the result of the inability of antibodies to penetrate the midbody and that calmodulin is, in fact, present throughout the midbody.

microtubules in the spindle, with a possible concentration gradient that increases toward the poles. During anaphase, calmodulin concentrates in the half-spindles, in a pattern that is similar to the pattern of half-spindle microtubules, as seen by indirect immunofluorescence in the same cell. At late anaphase, calmodulin can be seen to become concentrated in the interzone area. This localization in the interzone is not diffuse but appears to be in a fibrous or linear distribution that is arrayed parallel to the spindle axis. Calmodulin concentrated in the half-spindles may diminish at this time but it usually remains associated with the centrosomes, even after the kinetochore microtubules have disassembled. Thus, calmodulin appears to be present in the half-spindles as long as or longer than kinetochore microtubules are present. In some cells, this localization of calmodulin is apparent after the nuclear envelope re-forms in the daughter cells. At telophase, microinjected calmodulin concentrates in the midbody, but, in contrast to what is seen by immunofluorescence localization in fixed cells, it is found throughout the midbody rather than only at the ends of this structure (Fig. 6.3(d)).

A particular comment should be made concerning calmodulin distribution at the spindle pole in live cells. As seen in Fig. 6.3(b), the area around the pole binds a significant concentration of calmodulin. This localization is not identical to kinetochore microtubule distribution and suggests that while calmodulin may distribute largely in a pattern equivalent to that of kinetochore microtubules, additional binding sites or regions of concentration exist within the spindle. It may be possible that the calmodulin that is coincident with kinetochore microtubule distribution may be associated with one type of structure or function while the specifically pericentriolar calmodulin may be binding to a different component of the spindle and may serve a different function in this location.

Interestingly, when cells were fixed and permeabilized and then incubated with calmodulin–rhodamine, no binding of the fluorescent analogue could be detected in the spindle (Pardue *et al.*, 1981). Possibly, the cell's own endogenous calmodulin occupies all available binding sites for calmodulin in the spindle; thus, the additional exogenous fluorescent calmodulin could not bind to the spindle of fixed cells. We have observed that cells microinjected with calmodulin–rhodamine retain the calmodulin analogue in the spindle even if the cells are fixed and permeabilized (see Fig. 6.4). We have also observed that calmodulin–rhodamine will concentrate in the spindles of permeabilized cells (see Fig. 6.8) (Sweet *et al.*, 1988a). These observations support the idea that although calmodulin analogues may exchange with spindle-binding sites in live and permeabilized cells, fixed cells may have no available calmodulin-binding sites remaining in the mitotic apparatus.

Just as was observed with fixed cells, microinjected fluorescent analogues of calmodulin can be demonstrated in live cells to be preferentially associated

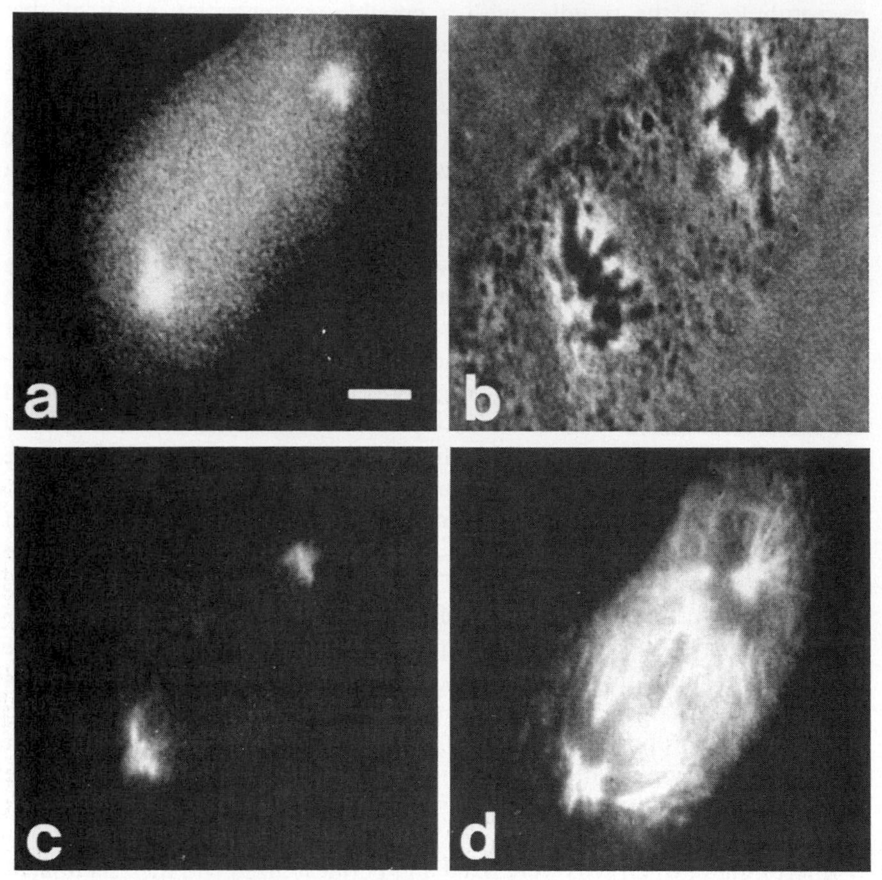

Figure 6.4. Localization of calmodulin–rhodamine isothiocyanate (CaM-TRITC) in living anaphase PtK_1 cell compared with immunolocalization of tubulin in the same cell. (a) The distribution of CaM-TRITC in the live cell. (b) The phase-contrast image of the cell. (c) The CaM-TRITC remaining after fixation and processing for immunolocalization of tubulin. (d) The distribution of tubulin in the fixed cell as seen by immunofluorescence. (From Sweet and Welsh, 1988.)

with a stable population of microtubules by using drugs or conditions that result in elimination of non-kinetochore microtubules. When live cells are incubated at 0–4°C for 2 hours, then microinjected with calmodulin–rhodamine while the culture is maintained at 4–6°C, calmodulin can be seen to concentrate in the spindle. If these cells are then fixed in the cold and prepared for immunolocalization of tubulin using a fluorescein-tagged second antibody, cold-stable microtubules can be seen to be coincident with the calmodulin. During anaphase, as shown in Fig. 6.5, microtubules in treated cells can be

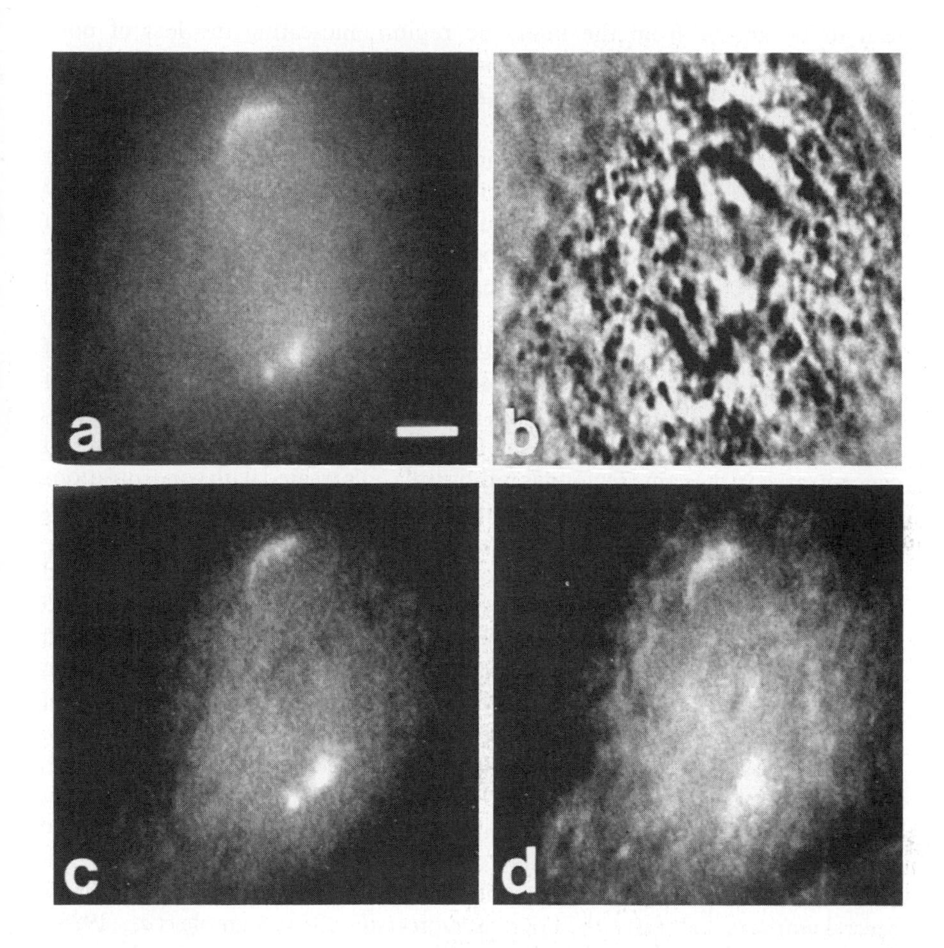

Figure 6.5. Localization of calmodulin–rhodamine isothiocyanate (CaM-TRITC) in living anaphase PtK_1 cell that was microinjected with CaM-TRITC while the cell was maintained at 4–6°C. (a) The distribution of CaM-TRITC in the live cell. (b) The phase-contrast image of the cell. (c) The CaM-TRITC remaining after fixation and processing for immunolocalization of tubulin. (d) The distribution of tubulin in the fixed cell as seen by immunofluorescence. Comparing (d) with Fig. 6.4(d), it can be seen that while untreated cells contain numerous interzonal microtubules (Fig. 6.4(d)), treatment with cold (d) results in cells that apparently lack the interzone microtubules. Both control and treated cells appear to have microtubules in their half-spindles. CaM-TRITC is concentrated, in both control and treated cells, in a distribution that is virtually identical to the distribution of half-spindle tubulin. Presumably, this half-spindle tubulin represents kinetochore microtubules. (From Sweet and Welsh, 1988.)

seen to be absent from the interzone region, indicating the loss of non-kinetochore microtubules. Cold-stable microtubules, which presumably are the kinetochore microtubules, remain in the half-spindles, as does the calmodulin.

More recently, the microtubule inhibitor nocodazole has also been employed to confirm that calmodulin localization in the mitotic apparatus is specifically linked to the presence of relatively drug-resistant microtubules. When cells in culture are treated with low concentrations (30 ng/ml) of nocodazole, localization of fluorescent calmodulin is virtually identical to localization in untreated cells. In the same cell, the remaining microtubules appear to be kinetochore microtubules (Sweet and Welsh, 1988). With higher concentrations of nocodazole (3 μg/ml), the spindle becomes disorganized, with remnants of kinetochore microtubules remaining, as seen by immunofluorescence localization of tubulin (Sweet and Welsh, 1988). In such cells, calmodulin is still concentrated in the same regions as is tubulin. Cells treated with concentrations of nocodazole high enough to eliminate all mitotic microtubules have no apparent concentrations of calmodulin.

Microinjection of fluorescent calmodulin into living cells largely confirmed results obtained by immunofluorescence localization of calmodulin in fixed cells but did not greatly advance our knowledge of calmodulin function in mitosis. One possible exception to this is the finding that calmodulin in living cells is found throughout the midbody of telophase cells. Microinjection of fluorescent analogues of calmodulin into live cells, either by direct microneedle injection or by red-cell fusion-mediated injection (Zavortink *et al.*, 1983), did, however, demonstrate the feasibility of employing a microinjected fluorescent analogue for studies of the intracellular mobility of the protein using the fluorescence redistribution after photobleaching (FRAP) technique.

FRAP studies of tubulin in the mitotic apparatus have been conducted by several workers (Saxton *et al.*, 1984; Salmon *et al.*, 1984; Stemple *et al.*, 1988). Results of these studies, as described in greater detail by Salmon in Chapter 4, indicate that microtubules in the spindle are highly dynamic structures in which tubulin rapidly exchanges between the structural elements of the spindle and the pool of subunits in the surrounding cytoplasm. Preliminary FRAP studies of calmodulin in the spindle of live PtK_1 cells has suggested that calmodulin may be more stable than tubulin (McIntosh *et al.*, 1985). The observation that calmodulin may be less mobile than tubulin in the mitotic apparatus was not expected; if calmodulin is bound to microtubules in the spindle, one would expect calmodulin to exhibit mobility equal to or greater than that of tubulin.

More recent studies conducted in collaboration with the McIntosh laboratory, with what we believe to be a better FRAP instrument system, have yielded results that indicate that calmodulin stability in the metaphase spindle of PtK_1 cells at 37°C is similar to that of tubulin (Stemple *et al.*, 1988). Interestingly, calmodulin mobility remains about the same in cells at 27°C or 37°C. In contrast, tubulin mobility in the spindle of cells at 27°C is lower (i.e. the tubulin is more stable) than in cells at 37°C. These results would be compatible with a model for the spindle in which calmodulin could be directly associated with the microtubules, with calmodulin associating with and dissociating from the microtubule at similar rates at either 27°C or 37°C while the microtubule becomes more stable or turns over less rapidly at 27°C than at 37°C. However, these results do not rule out the possibility that calmodulin may associate with a spindle component other than microtubules. If this does happen, one candidate for an alternative structure with which calmodulin might be associated is the smooth endoplasmic reticulum system, which seems to be associated with the mitotic apparatus (see Chapter 7). A further discussion of the possible significance of calmodulin association with smooth endoplasmic reticulum is given later.

Comparisons of tubulin and calmodulin dynamics as measured by FRAP are complicated by the fact that calmodulin is associated specifically with kinetochore microtubules. Ideally, calmodulin dynamics would be compared with the dynamics of the kinetochore microtubules alone. Unfortunately, reported measurements of tubulin FRAP rates have not yet discriminated between kinetochore and non-kinetochore microtubule dynamics. Thus, these measurements may not yield accurate information about the turnover rates of the more stable kinetochore microtubules. Indeed, Salmon *et al.* (1984) made the explicit distinction that they assumed that the values they observed for tubulin mobilities in spindles of sea-urchin eggs (*Lytechnius variegatus*) was attributable to non-kinetochore microtubules. They did not draw any conclusions about, nor did they discuss stability of, kinetochore microtubules, although they implied that the kinetochore microtubules would be more stable. As yet, no experiments have been conducted or published concerning efforts to examine explicitly the dynamics of kinetochore microtubules. It will be important in the future to perform FRAP studies on both calmodulin and tubulin in live cells at different stages of mitosis and under conditions that allow only kinetochore microtubules to remain in the mitotic apparatus (e.g. low nocodazole or 0–4°C).

C. Calmodulin localization at the electron-microscopic level

Because the resolution of the light-microscope is not sufficient to allow conclusions to be drawn about which organelle or structure calmodulin may be associated with in the mitotic apparatus, calmodulin has been localized in the mitotic apparatus at the electron-microscopic level. DeMey *et al.* (1980), using the peroxidase–antiperoxidase (PAP) method of Sternberger *et al.* (1979), on cultured PtK_2 cells, found calmodulin to be associated with some but not all microtubules of the spindle. Because they observed PAP reaction product only with intact microtubules in the half-spindle regions (Fig. 6.6), these researchers concluded that calmodulin is associated with microtubules that are participating in lateral interaction with neighbouring microtubules. Thus, DeMey *et al.* (1980) proposed that calmodulin might be involved in the process of maintaining lateral interaction of microtubules and that calmodulin or this lateral interaction may be involved in generation of poleward force during mitosis. This hypothesis certainly has its attraction in that, as discussed earlier, calmodulin is seen at the light-microscopic level to be coincident with the cold- and drug-stable, kinetochore microtubule population rather than with the more labile non-kinetochore microtubules. In addition, that microinjected

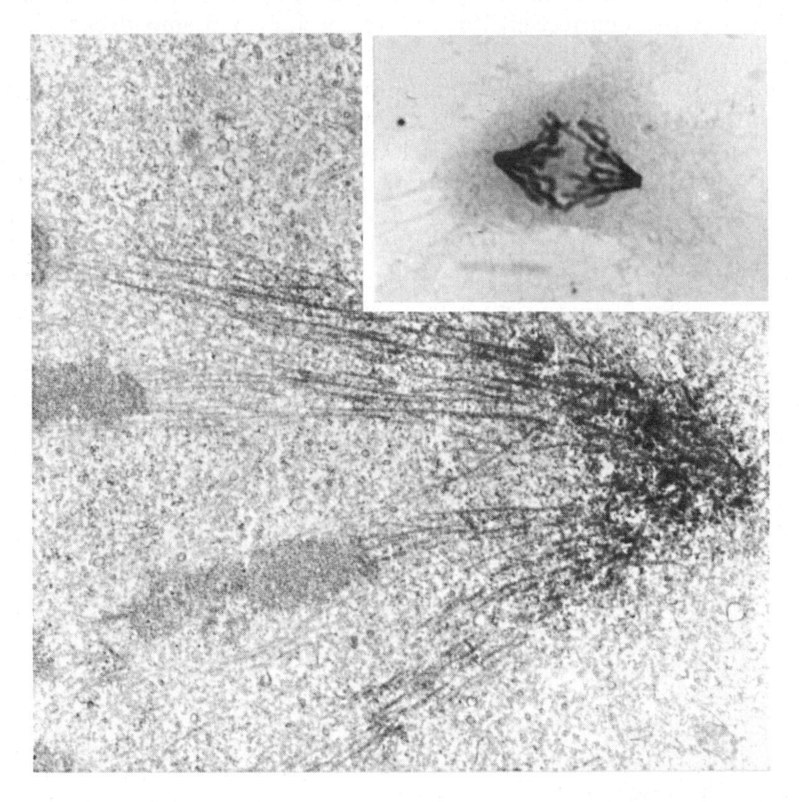

Figure 6.6. Pre-embedding immunoperoxidase (PAP-method) labelling of calmodulin in a mitotic (mid-anaphase) PtK_2 cell. The inset shows a cell in a similar stage of mitosis. The reaction product shows a gradient towards the spindle pole and is mainly associated with kinetochore microtubules. (Courtesy of Dr. Jan DeMey: from DeMey *et al.*, 1980.)

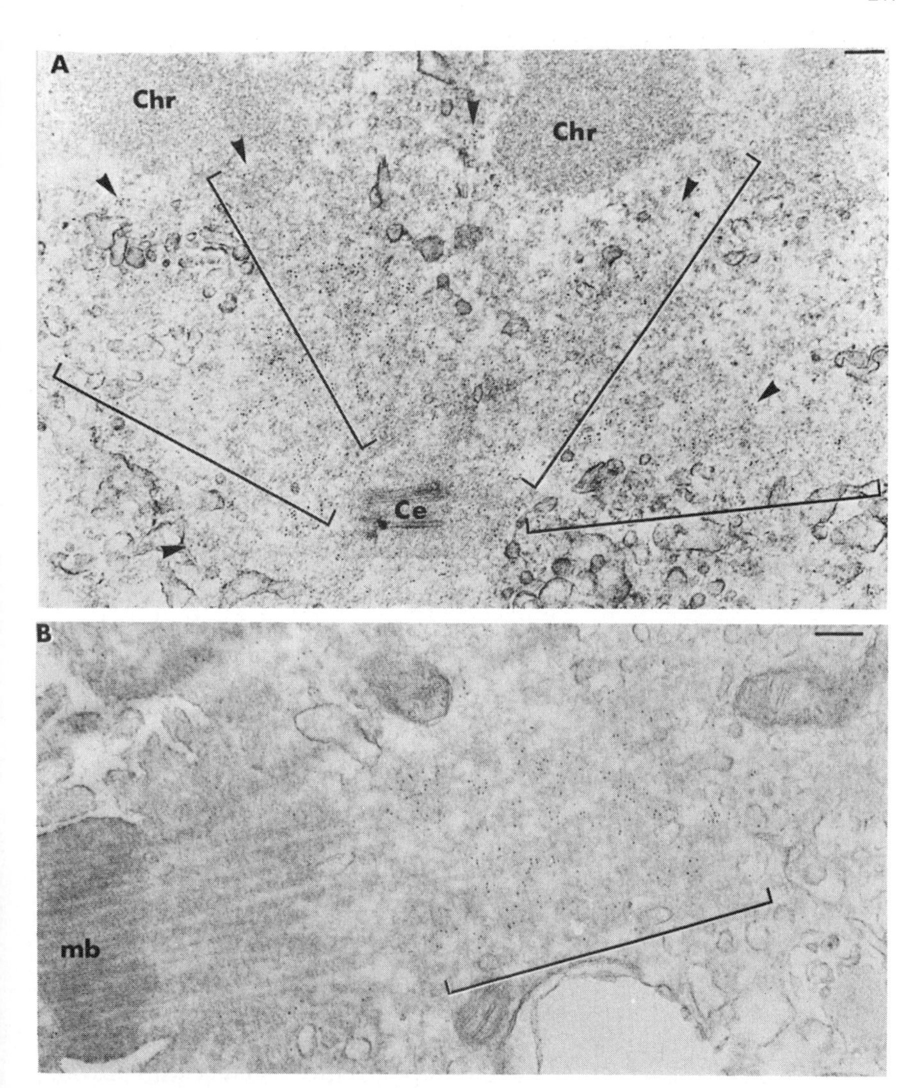

Figure 6.7. Ultrastructural immunocytochemical localization of calmodulin in the pericentriolar region and intercellular bridge (midbody) in the mitotic cells. Swiss 3T3 cells were fixed and processed for immunolocalization of calmodulin using the ferritin bridge technique (Willingham *et al.*, 1983). (A) In addition to concentrations of calmodulin surrounding the centriole (Ce), concentrations of calmodulin (arrowheads) appear to radiate out (brackets) from the centriolar region towards the chromosomes (Chr.). (B) Calmodulin can be seen to be localized in the intercellular bridge region with a large gap between the edge of the calmodulin concentration (brackets) and the edge of the midbody (mb). (Bars = 0.2 μm.) (Courtesy of Dr. Mark Willingham: from Willingham *et al.*, 1983.)

calmodulin–rhodamine localizes throughout the midbody in live cells, rather than just at the ends of the midbody bundle of microtubules (as is seen by immunofluorescence localization), tends to support the concept of calmodulin being present in areas of stable microtubules and/or lateral interaction of microtubules.

Willingham *et al.* (1983) have also localized calmodulin in mitotic cells (Fig. 6.7). The method used for fixation by these workers did not well preserve the microtubules of the mitotic spindle and clear conclusions about subcellular localization of the calmodulin could not be made with certainty. In general, however, their results confirmed the earlier results about calmodulin localization as seen by immunofluorescence methods.

Vantard *et al.* (1985) have published one figure of calmodulin localization, using colloidal gold as a marker, at the electron-microscopic level in an anaphase *Haemanthus* endosperm cell. This figure may be the best figure as yet published depicting calmodulin localization at the electron-microscopic level in any cell type. They show that calmodulin is localized very close to, if not on, the microtubules in the half-spindle. The localization of calmodulin appears to be as much with an irregular electron-dense material associated with the microtubules as with the microtubule itself. Most but not all microtubules are labelled; those microtubules that are labelled do not show an even distribution of gold particles. Labelling appears to be more irregular along the length of the microtubules. The authors suggest that the labelling is periodic, but do not show evidence of a rigorous analysis of this possibility.

IV. Biochemical Evidence for Calmodulin Regulation of the Mitotic Apparatus

A. Calmodulin effects on microtubule assembly/disassembly *in vitro*—direct interactions with microtubule proteins and activation of protein kinases and phosphatases

The discovery of calmodulin in the mitotic apparatus, together with previous observations that microtubule assembly *in vitro* depended upon the absence of calcium from the assembly buffer, suggested that the effect of calcium on microtubule assembly might be mediated by calmodulin. In order to investigate this possibility, Marcum *et al.* (1978) prepared bovine brain microtubules by the warm-polymerization–cold-depolymerization cycling method (Borisy *et al.*, 1975) and used this preparation of cold-labile microtubule protein to study polymerization kinetics, as indicated by measurement of turbidity of the protein solution, under different conditions. Addition of either calcium-free calmodulin or 11 μM free calcium had relatively little effect on the polymerization kinetics of this cold-labile microtubule protein. However, addition of calmodulin and calcium together completely inhibited formation of microtubules. A similar effect was seen if the calcium-binding protein troponin C was substituted for calmodulin. If polymerized microtubules were treated with 11 μM calcium, a slow and incomplete depolymerization of microtubules resulted. However, when calcium (11 μM) plus calmodulin was added, rapid depolymerization of the microtubules ensued. This effect was duplicated by the calcium-binding protein troponin C plus calcium. Maximum effect, which resulted in depolymerization of virtually all the microtubules, was accomplished at a molar ratio of calcium binding protein to tubulin dimer of 8 and 2, respectively, for calmodulin and troponin C. That troponin C was effective at a lower stoichiometry than calmodulin was unanticipated and no explanation for this result has ever been proposed. Although no mechanism was proposed for the observed effects of calmodulin on microtubule assembly/disassembly, the results of Marcum *et al.* (1978) supported the concept of calmodulin acting in the spindle to regulate microtubule assembly/disassembly.

The major shortcoming of the Marcum *et al.* (1978) study, in relation to understanding the role of calmodulin in spindle function, was that cold-labile brain microtubules were studied. Microtubules in the spindle are not neuronal microtubules and are not all cold-labile (Brinkley and Cartwright, 1975). In mitotic cells, calmodulin concentrates with the cold-stable kinetochore microtubules (Welsh *et al.*, 1979). Thus, results using cold-labile brain microtubules could be misleading. Indeed, in brain preparations, just as in the mitotic apparatus, both cold-labile and cold-stable microtubules exist (Webb and Wilson, 1980). Significantly, Webb and Wilson (1980) found the cold-stable brain microtubules to depolymerize in response to calcium. Their observations suggested that cold stability was conferred on the brain microtubules by a low-molecular-weight factor whose effect was calcium-dependent (Webb and Wilson, 1980). The exact size of the low-molecular-weight factor was not determined, but information given by Webb and Wilson (1980) indicates that the factor would pass through a dialysis membrane, thus implying that the factor was too small to be calmodulin.

The most extensive examination of cold-stable microtubules has come from a series of studies conducted by the laboratory of Margolis. These workers have studied cold-stable microtubules isolated from the brains of rats (Job *et al.*, 1981, 1982, 1983; Margolis and Rauch, 1981), sheep (Pirollet *et al.*, 1983) or cow (Pabion *et al.*, 1984). Job *et al.* (1981) first demonstrated that the cold-stable microtubules from rat brain were much more sensitive than the cold-labile microtubules to the depolymerizing effect of calcium/calmodulin. In this study, Job *et al.* (1981) showed that calmodulin in substoichiometric amounts, in the presence of calcium, would cause depolymerization of the cold-stable microtubules. In comparison, Marcum *et al.* (1978) reported that a calmodulin-to-tubulin molar ratio of about 8 was required for maximum effect to be observed.

A study from the same laboratory (Margolis and Rauch, 1981), demonstrated the presence of a phosphoprotein that was present in its non-phosphorylated form in the cold-stable subpopulation of microtubules but was phosphorylated in the total microtubule population from rat brain. Margolis and Rauch (1981) proposed that this phosphoprotein of 64 kDa and its state of phosphorylation is the determinant for cold stability or lability of the microtubules. This hypothesis was based only on the coincidence of the protein being phosphorylated in total microtubule fractions but being dephosphorylated in the cold-stable population of microtubules. The protein was not purified nor was it added back to microtubules or tubulin to investigate specifically its possible action.

Several subsequent studies endeavoured to define what were named STOPs (stable tubule-only polypeptides) (Job *et al.*, 1982, 1983; Pirollet *et al.*, 1983; Pabion *et al.*, 1984). Job *et al.* (1982) identified several proteins from cold-stable microtubules isolated from crude supernatants of gently homogenized rat brains. These were described as having molecular masses of 56 kDa, three proteins between 70 kDa and 82 kDa, supposedly different from Tau, another microtubule-associated protein, and a protein of about 135 kDa. It seems probable that these proteins could be mistaken for Tau if the only criterion is of molecular weight estimation from one-dimensional polyacrylamide gels. In the presence of calcium, these proteins bound to a calmodulin affinity column and when they were removed from the microtubule preparation by calmodulin affinity chromatography the microtubules were rendered cold-labile. It was proposed that these STOPs functioned to stabilize microtubules by binding to the microtubule and forming a stable cap or block to disassembly of the microtubule. Subsequently, Job *et al.* (1983) showed that the STOPs could be added back to cold-labile microtubules and would render the microtubules cold-stable. The proteins of 56 kDa and 72 kDa could be phosphorylated in a calcium/calmodulin-dependent manner and, when phosphorylated, the microtubules were cold-labile. If the brain tissue used for isolation of the cold-stable microtubules was vigorously homogenized, other proteins of the cold-stable microtubule population could be phosphorylated in a calmodulin-independent manner (Job *et al.*, 1983). That only gentle homogenization would support the calmodulin-dependent phosphorylations alone suggests that these phosphorylation events may be more physiological than the calmodulin-independent phosphorylations, which require vigorous homogenization.

Pirollet *et al.* (1983) extended these observations to sheep brain. Some differences were observed between results obtained with sheep and rat brain cold-stable microtubules, but these were negligible. In this publication, they also offered evidence and arguments for believing that the microtubule-associated protein Tau has no effect on microtubule stability and that the 70–82 kDa STOPs are not, in fact, Tau. Pabion *et al.* (1984) further extended these observations to bovine brain cold-stable microtubules and presented evidence for the STOPs being able to translocate or slide along the microtubules even while maintaining the cold-stable state of the microtubules.

While it is difficult to compare results from other laboratories, some studies may be relevant to a better understanding of the significance of the results from Margolis' laboratory. The laboratory of Miyamoto (Fukunaga *et al.*, 1982; Yamamoto *et al.*, 1983) has studied a calcium/calmodulin-dependent kinase of 640 kDa from rat brain. The laboratory of DeLorenzo (Goldenring *et al.*, 1983) has also isolated from rat brain a calcium/calmodulin-dependent protein kinase complex of about 600 kDa, using purification steps that included affinity chromatography on a calmodulin column. When Goldenring *et al.* (1983) subjected the enzyme to gel electrophoresis, the complex was resolved into two subunits of approximately 52 kDa and 63 kDa. Each of the subunits was shown to be able to bind calmodulin. The kinase was able to autophosphorylate both of its subunits and also phosphorylated both α and β tubulin subunits, as well as the high-molecular-weight MAP 2.

Although the apparent sizes of the enzyme complex isolated by Miyamoto's group and that isolated in DeLorenzo's laboratory are similar, Goldenring *et al.* (1983) offered evidence (substrate specificities, subunit size and composition, and K_m for ATP) for the two enzymes being different. It seems possible to us that the proteins described by these two laboratories may be the same protein and that the differences may be due to the fact that the work was conducted in different laboratories. The kinase described by workers in DeLorenzo's laboratory virtually certainly is the same as the calcium/calmodulin-dependent protein kinase II, which has been extensively studied by the laboratories of Greengard (Kennedy *et al.*, 1983; McGuinness *et al.*, 1983; Ouimet *et al.*, 1984) or Cuatrecasas (LeVine *et al.*, 1985; Sahyoun *et al.*, 1985). The calmodulin-dependent kinase II from rat brain has been shown to consist of subunits of about 50 kDa and 60 kDa, which assemble into a complex of about 600 kDa molecular weight (Goldenring *et al.*, 1983; Larson *et al.*, 1985). The subunits autophosphorylate and phosphorylate tubulin (Yamauchi and Fujisawa, 1984), as well as MAP 2 and other proteins (Yamamoto *et al.*, 1983; McGuiness *et al.*, 1983; Vallano *et al.*, 1985).

The 63 kDa subunit of the kinase described by Goldenring *et al.* (1983), or the 60 kDa subunit by workers in Greengard's (Kennedy *et al.*, 1983; Ouimet *et al.*, 1984) or Cuatrecasas' laboratory (Sahyoun *et al.*, 1985) may be the same as the 64 kDa phosphoprotein described by Margolis and Rauch (1981). Additionally, careful examination of figure 7 from Margolis and Rauch (1981) reveals a band of ^{32}P-labelling, under experimental conditions in which the

64 kDa protein is also phosphorylated, at a molecular weight at about the level of the α-tubulin subunit (approximate molecular mass 50 kDa) in the gels. This apparent phosphorylation of an additional protein of approximately 50 kDa molecular mass was not discussed by Margolis and Rauch (1981), and may have been judged by these workers to be part of the α-tubulin band. However, in light of the more recent report concerning the calmodulin-dependent kinase II, it may be useful to reinterpret the results from Margolis' laboratory. Thus, the calcium/calmodulin-dependent phosphorylations observed by Margolis' laboratory may be evidence for the presence, in their preparations of cold-stable microtubules, of the subunits of the calcium/calmodulin-dependent kinase (kinase II) described by others (Kennedy *et al.*, 1983; Goldenring *et al.*, 1983) (i.e. the 52 kDa and 63 kDa components of the calmodulin-dependent kinase of Goldenring *et al.* may be the same as the 64 kDa protein and the band at the level of α-tubulin in the Margolis and Rauch (1981) preparation).

Of distinct relevance to the comparison of results of Margolis and Rauch (1981) with studies of the calmodulin-dependent kinase II by other laboratories is the observation by LeVine *et al.* (1985) that the autophosphorylation of the subunits of the kinase II decreased their affinity for calmodulin and thus might serve as a negative feedback mechanism on the enzyme. If the calmodulin-dependent kinase II is to be most sensitively responsive to calmodulin and functional in cold-stable microtubules, it would be expected that it should be present in its unphosphorylated state. Thus, it would be expected that these proteins (assuming that these phosphoproteins are, in fact, the subunits of the calmodulin-dependent kinase II) would not be phosphorylated in cold-stable microtubule fractions as, indeed, was shown by Margolis and Rauch (1981).

The studies from Miyamoto's (Fukunaga *et al.*, 1982; Yamamoto *et al.*, 1983), DeLorenzo's (Goldenring *et al.*, 1983), Greengard's (Kennedy *et al.*, 1983; Ouimet *et al.*, 1984) or Cuatrecasas' laboratories (Sahyoun *et al.*, 1985) are difficult to compare with the results of Margolis and Rauch (1981) because of the differences in the way their respective preparations were obtained. Although all groups were, at one level, studying rat brain cytosol, some were interested in the calcium/calmodulin-dependent kinase isolated from rat brain cytosol, while Margolis and Rauch (1981) were investigating rat brain cold-stable microtubules.

Subsequent reports from DeLorenzo's laboratory (Vallano *et al.*, 1985; Larson *et al.*, 1985) appear to be more comparable with the study of phosphorylation of cold-stable microtubules by calmodulin-dependent and calmodulin-independent kinases as described by Job *et al.* (1983). Vallano *et al.* (1985) prepared microtubules from rat brain and assayed for the presence of the calcium/calmodulin-dependent kinase activity described by Goldenring *et al.* (1983). The cold-stable microtubule fraction was measured to be enriched

16-fold (Vallano *et al.*, 1985) or 21-fold (Larson *et al.*, 1985) over the crude cytosol for the calmodulin-dependent kinase activity. Under appropriate conditions, the kinase activity appeared to be associated with a complex composed of tubulin rings and MAP 2 (Vallano *et al.*, 1985). The kinase activity was also reported to be significantly inactivated by incubation of the cytosol at 37°C (Vallano *et al.*, 1985), and thus it may have been fortuitous that the studies reported by Margolis' laboratory never concerned preparations incubated at more than 30°C, to promote microtubule polymerization.

The studies of cold-stabile brain microtubules described above may or may not be relevant to an understanding of calmodulin regulation of microtubule assembly/disassembly in mitotic cells. The presence of calmodulin-dependent kinase as described by Vallano *et al.* (1985) has not been reported to exist in the mitotic apparatus. Indeed, Greengard generously supplied us with antibody raised against the brain calmodulin-dependent kinase II. Using this antibody we were unable to localize the kinase II in the mitotic spindles of mitotic PtK_1 or HeLa cells by indirect immunofluorescence methods (unpublished results). Microtubule-associated proteins of brain may differ antigenically from those that are present in the mitotic apparatus (Bloom and Vallee, this volume), although functionally the proteins may act similarly. Thus, although the kinase II was not seen to be specifically localized in the spindles of cultured cells, it is probably too early to rule out completely a role for a calmodulin-dependent kinase II-like activity in regulation of the spindle. Without greater knowledge of spindle microtubule-associated proteins, it is difficult to know how to compare results of studies of brain microtubules with models for spindle microtubule behaviour. Obviously, the best way to understand calmodulin's role in mitosis is to study the mitotic apparatus and the microtubules of this structure.

We are aware of only one study that has concerned a direct examination of spindle microtubules and the effects of calmodulin and calcium on spindle microtubule assembly/disassembly *in vitro*. Keller *et al.* (1982), from the laboratory of Lionel Rebhun, prepared microtubules from the spindles of the sea urchin *Strongylocentrotus purpuratus*. Unfortunately, considering what is now known about the characteristics of brain cold-stable microtubules and the fact that calmodulin in the spindle seems to be associated with cold-stable microtubules, the results of Keller *et al.* (1982) may not contribute as much as hoped to an understanding of calmodulin regulation of the spindle. The tubulin studied by Keller *et al.* (1982) was obtained by the temperature-dependent assembly method. Using brain as a source of microtubule protein, this method results in the complete loss of the cold-stable microtubules (Webb and Wilson, 1980). Thus, the microtubule population studied by Keller *et al.* (1982) was specifically selected to not be the cold-stable population.

Secondly, the studies of Keller *et al.* (1982) involved polymerizing the microtubules at a temperature of 37°C, a temperature at which the calmodulin-dependent kinase from brain is said to be inactivated (Vallano *et al.*, 1985). Interestingly, at temperatures at which the sea-urchin embryos naturally exist (15–18°C), spindle microtubules are sensitive to micromolar concentrations of calcium (Keller *et al.*, 1982). It is not possible to draw valid conclusions about the effect of calmodulin on sea-urchin spindle microtubules, because no experiments were reported about calmodulin effects at the lower temperature of 15–18°C. At 37°C, the only temperature at which the effect of calmodulin was examined, a slight effect was reported. We do not understand why Keller *et al.* (1982) did not examine the effect of calmodulin at the lower, physiological, temperature for sea urchins when an effect of physiological levels of calcium was only observed in the lower temperature range. Thus, because Keller *et al.* (1982) examined cold-labile microtubules and did not study calmodulin effects at a physiological temperature for the organism from which the microtubules were isolated, these results have not really answered the relevant questions about calmodulin regulation of spindle microtubules.

A definitive answer cannot yet be given concerning the possibility of spindle microtubule assembly/disassembly being regulated by a calcium/calmodulin-dependent protein kinase. The results of studies of rat brain microtubules described above indicate that calmodulin can perform the role of an activator of protein kinase in the brain. Phosphorylation of microtubule-associated proteins by the calmodulin-dependent kinase results in decreased microtubule stability. A similar effect on microtubule assembly/disassembly kinetics has been observed to occur after cyclic AMP-dependent phosphorylation of MAP 2 (Jameson *et al.*, 1980). As yet, insufficient data exist to justify applying this model to the spindle.

One other possible role for phosphorylation of microtubule-associated proteins is worth mentioning. It has been shown that MAP 2 is extensively phosphorylated (Jameson *et al.*, 1980; Vallee, 1980) by both cAMP-dependent and cAMP-independent (possibly calmodulin-dependent kinase?) kinases (Theurkauf and Vallee, 1983). Indeed, Vallee *et al.* (1981) have reported that a cAMP-dependent kinase, with properties virtually identical with the cytosolic cAMP-dependent protein kinase, is associated with MAP 2. The kinase was also shown to be inhibited by the cAMP-dependent protein kinase inhibitor described by Walsh *et al.* (1971). Vallee *et al.* (1981) suggested that phosphorylation of MAP 2 may control not only the ability of MAP 2 to associate with, and thus stabilize, microtubules, but also may regulate the ability of MAP 2 to interact with other cellular components. MAP 2 is phosphorylated in as many as 22 sites that span both the portion of the molecule that binds to tubulin as well as the projecting portion, which may be available for interacting with other cell structures (Theurkauf and Vallee, 1983). Apparently some of these phosphorylation sites are those under the control of the calmodulin-dependent kinase (Yamamoto *et al.*, 1983; Schulman *et al.*, 1985) while others are clearly subject to the action of the cAMP-dependent kinase (Jameson *et al.*, 1980; Theurkauf and Vallee, 1983). With so many sites available and at least two different kinases, and presumably phosphatases, also being responsible for regulating the phosphorylation state of the MAP 2, it is easy to envision that the actions of MAP 2 could be highly varied. Furthermore, if the phosphorylation state of sites on the projecting portion of the protein do affect the ability of the MAP to interact with other cellular components, it would seem possible that MAP 2, or a similar protein, might function to cross-link spindle microtubules. If this should occur in the kinetochore microtubule bundles in particular, calmodulin-dependent kinase could be hypothesized as being a potential regulator of microtubule interaction. It should be remembered, that DeMey *et al.* (1980) have indicated that calmodulin's localization in the spindle has led them to the conclusion that calmodulin may be involved in a mechanism of establishing lateral interactions between microtubules in the half-spindles or kinetochore microtubules.

Indirect evidence that the cAMP-dependent kinase may actually be present on microtubules in cells comes from the work of Tash *et al.* (1980). These workers prepared an antibody to the cAMP-dependent protein kinase inhibitor and, by indirect immunofluorescence, localized the inhibitor in cultured cells. In cells of rat, mouse and rat kangaroo, the inhibitor was localized to both interphase and mitotic microtubules. However, in contrast to the pattern seen with calmodulin, there appeared to be no preferential localization to any subpopulation of microtubules.

In addition to calmodulin regulation of a protein kinase in the spindle, there has been a suggestion that calmodulin might regulate a protein phosphatase in the spindle of mitotic cells. Brady *et al.* (1984) reported the presence of a 60 kDa calmodulin-binding protein in spindles isolated from CHO cells. This protein co-migrated with brain calcineurin, a calcium/calmodulin-dependent phosphoprotein phosphatase, on one-dimensional polyacrylamide gels and an antibody to calcineurin localized to spindles of mitotic cells.

Other studies by several groups have indicated that calmodulin has an affinity for Tau protein (Sobue *et al.*, 1981), for both MAP 2 and Tau of brain microtubules (Lee and Wolff, 1984a,b; Erneux *et al.*, 1984) or for tubulin dimer itself (Kumagai *et al.*, 1982). Thus, at least in brain, calmodulin may both mediate phosphorylation of MAP 2 and Tau as well as bind directly to these MAPs. How these interactions may relate to spindle microtubule structure and function is not clear. The affinities of calmodulin for these microtubule components are low compared to those measured for calmodulin and the enzymes calmodulin is known to regulate. For example, calmodulin affinity is reported to be about 4×10^{-6} M for tubulin (Kumagai *et al.*, 1982) and about 7×10^{-6} M for MAP 2 (Lee and Wolff, 1984b). In comparison, the affinity of calmodulin for enzymes, including phosphodiesterase, myosin light-chain kinase and calcineurin, is in the range of 10^{-10} M (reviewed by Manalan and Klee, 1984).

Kakiuchi and Sobue (1981) have proposed a simple "flip-flop" mechanism in which calmodulin, when binding calcium, competes with tubulin for Tau binding, thus destabilizing the microtubules. This model does not take into account the possible binding of calmodulin to MAP 2, nor does it address in any way possible phosphorylation reactions that may further regulate microtubule protein interactions. In our opinion, the mechanism for calmodulin regulation of microtubule assembly and disassembly must be considerably more complex than suggested in the "flip-flop" mechanism.

B. Calcium in the spindle and calmodulin regulation of calcium pumps

In all calmodulin-regulated systems studied thus far, it has been shown that the action of calmodulin is dependent upon the prior binding of calcium by calmodulin (see reviews by Means and Dedman, 1980; Klee and Vanaman, 1982). Thus, if calmodulin mediates the action of calcium in the spindle as in other systems, the regulation of calcium is probably the key regulatory event. A system of vesicles associated with the spindle is believed to be a source of calcium ion during mitosis and to be responsible for regulating the concentration of free calcium in the mitotic apparatus during mitosis (Harris, 1975, 1978, 1982; Hepler and Wolniak, 1984). A comprehensive discussion of this membrane vesicle system is given by Hepler in Chapter 7 and will not be discussed extensively here. A few brief comments, however, will be made concerning calcium in the spindle and possible calmodulin involvement in the mechanisms of regulation of calcium ions in the mitotic apparatus.

As mentioned immediately above, a system for regulation of calcium in the spindle has been proposed to exist. A vesicular system associated with the mitotic apparatus has been shown at the electron-microscopic level, using an antimonate precipitation method, to contain calcium (Hepler, 1980; Wick and Hepler, 1980). The antimonate precipitation method for calcium localization has shown calcium to be relatively concentrated in a smooth endoplasmic reticulum associated with the spindle in plant cells (Hepler, 1980; Wick and Hepler, 1980). The method does not seem to allow localization of calcium in the cytoplasm of the plant cells, however. Slocum and Roux (1982) have applied a similar method to mammalian cells but were also unable to detect precipitates in the cytoplasm. Indeed, Slocum and Roux (1982) have good evidence that the method can be used to give at least semi-quantitative measurements of calcium, but only within the range of 2×10^{-5} to 1×10^{-3} M. Thus, the failure of the antimonate precipitation method for localizing calcium in mitotic cell cytoplasm, or in the membrane vesicle system of mitotic animal cells, appears to be the result of calcium levels being below the limit of sensitivity of this procedure. The method does show, as mentioned above, that the vesicle reticulum associated with the spindle of plant cells is high in calcium. Presumably, the mitotic reticulum in animal cells is also a repository for calcium, but at a lower concentration than in plants. In analogy to the mechanism of calcium regulation in muscle, this vesicular system, like the sarcoplasmic reticulum, could serve as a source of calcium ion and as a calcium-sequestering system during mitosis.

Several fluorescent probes have also been used for localizing calcium in live cells. Chlorotetracycline has been used to locate calcium associated with the mitotic apparatus in living plant cells (Wolniak *et al.*, 1980). Considering the lipid solubility of chlorotetracycline, this calcium is probably in the vicinity of membranes or is membrane-associated. More recently, the fluorescent calcium indicator quin-2 has also been employed in live plant (Keith *et al.*, 1985b) and animal cells (Keith *et al.*, 1985a). The results obtained with *Haemanthus* endosperm cells are in agreement with the concept that concentrations of calcium are present in the mitotic apparatus; specifically, that increased concentrations of calcium seem to be distributed similarly to kinetochore microtubules. More recent experiments using the fluorescent calcium indicator fura-2 suggest that the calcium concentration in PtK_2 may rise during the transition from metaphase to anaphase, in the region between the poles and the chromosomes (Ratan *et al.*, 1986). These methods have, of course, been possible only with observation at the light-microscopic level. Thus, it is not yet possible to confirm directly the exact location, at the ultrastructural level, of the calcium being indicated by the fluorescent compounds.

Several lines of evidence support the model of the spindle in which calcium plays a significant role. Mitotic spindles have been isolated that retain an ability to sequester calcium in an ATP-dependent manner (Silver *et al.*, 1980). Treatment of isolated spindles with calcium causes rapid disassembly of microtubules (Salmon and Segall, 1980). Similarly, when live, intact mitotic cells are microinjected with calcium, spindle microtubules rapidly disassemble, suggesting that calcium can, in fact, cause disassembly of microtubules *in vivo* (Kiehart, 1981). The observed *in vivo* disassembly caused by microinjected calcium is rapidly reversed in the living cell, with microtubule reassembly occurring within minutes after the microinjection of calcium. This is interpreted as an indication of sequestering of the injected calcium (Kiehart, (1981). Microinjection of caffeine, a potent inhibitor of the skeletal muscle sarcoplasmic reticulum calcium pump ATPase (Weber, 1968; Weber and Herz, 1968), also results in rapid disassembly of mitotic microtubules (Kiehart, 1981). Thus, calcium appears to be released from a source in or near the mitotic apparatus by an agent that can inhibit calcium uptake by sarcoplasmic reticulum. Overall, these results support the concept that calcium affects microtubules in the mitotic spindle and they support the analogy drawn between the vesicular calcium-regulatory system of muscle and that of the mitotic apparatus (Harris, 1975, 1978, 1982, 1983; Petzelt, 1979; Silver, 1986; Silver *et al.*, 1980; Inoué, 1981; Kiehart, 1981).

Results indicate that calmodulin may participate in the regulation of calcium levels within skeletal and cardiac muscle as well as in erythrocytes and other cell types. Calmodulin was first shown to regulate the calcium pump of erythrocytes (reviewed by Vincenzi and Larsen, 1980). Calmodulin also stimulates the uptake of calcium by cardiac microsome preparations (Lopaschuk *et al.*, 1980). Calcium uptake by cardiac sarcoplasmic reticulum is associated with the stimulation of calcium-dependent ATPase activity and phosphorylation of a 22 kDa molecular weight membrane protein called phospholamban (Tada *et al.*, 1979; Tuana *et al.*, 1981). The rate of calcium uptake by cardiac sarcoplasmic reticulum is increased when phospholamban is specifically phosphorylated by a sarcoplasmic reticulum-associated calmodulin-dependent protein kinase (Le Peuch *et al.*, 1979; Davis *et al.*, 1983). The calmodulin-dependent calcium pump of heart sarcoplasmic reticulum has a K_m for calcium of about 0.5 μM (Caroni and Carafoli, 1981; Davis *et al.*, 1983). Calmodulin has also been shown to stimulate the phosphorylation of three specific proteins in skeletal muscle sarcoplasmic reticulum (Chiesi and Carafoli, 1983). Presumably, these phosphorylations are related to calcium regulation within the muscle cytoplasm.

Calmodulin has also been shown to stimulate calcium transport in rat brain synaptosomal membrane vesicle preparations (Ross and Cardenas, 1983). This uptake system demonstrated a K_m for calcium of 0.4 μM. A similar, high-affinity (K_m for calcium = 0.1 μM) calmodulin-stimulated calcium pump has been observed in baso-lateral plasma membranes of kidney cortex (Gmaj *et al.*, 1983). More recently, calcium/calmodulin-dependent phosphorylation of a 20 kDa protein of light microsomal fractions of liver has been correlated with calcium uptake by the microsome fractions (Famulski and Carafoli, 1984), suggesting a parallel with the more extensively studied muscle calcium pump systems.

As yet, only one study of the possibility of calmodulin regulation of mitotic apparatus calcium pump activity has been described (Nagle and Egrie, 1981). This calcium-ATPase was shown to not be directly affected by calmodulin. However, Nagle and Egrie (1981) were not able to rule out the possibility that the activity could be affected indirectly by calmodulin, possibly acting through a calmodulin-dependent kinase, as is seen in cardiac sarcoplasmic reticulum (Le Peuch *et al.*, 1979; Lopaschuk *et al.*, 1980; Plank *et al.*, 1983; Davis *et al.*, 1983), skeletal muscle sarcoplasmic reticulum (Chiesi and Carafoli, 1983) or liver microsomes (Famulski and Carafoli, 1984). Also, the activity studied by Nagle and Egrie (1981) exhibited a much lower affinity for calcium (1.0 mM) than the calcium pumps examined in the tissues described above. One might expect a mitotic calcium pump to demonstrate a K_m for calcium that is within the range of calcium concentrations that occur in the cell. This would be on the order of micromolar values, as has been shown for other calcium pumps. It is not entirely comprehensible to us how a calcium pump with a millimolar affinity for calcium would function in the mitotic apparatus of animal cells, where calcium concentrations are probably in the micromolar range and definitely considerably below the millimolar range. The particular calcium-ATPase activity examined by Nagle and Egrie (1981) may not represent the functional, physiological calcium pump of the mitotic apparatus. Thus, the question has not yet been answered of whether calmodulin regulates a calcium pump that may be associated with the vesicle system seen associated with the mitotic apparatus.

C. Other regulators and components of the spindle

A number of other proteins have been indicated, by a variety of methods, to be present in the mitotic apparatus (McIntosh, 1984; Vallee and Bloom, this volume). In addition to those listed by McIntosh (1984), other proteins may also be present in the spindle, including protein kinase inhibitor (Tash *et al.*, 1980), calcineurin (Brady *et al.*, 1984) and ankyrin (Bennett and Davis, 1981; Davis and Bennett, 1984). Many of these have been suggested to be present in the spindle only on the basis of evidence from immunolocalization studies. Because antibodies may cross-react with epitopes on proteins other than the one against which the antibody was presumably raised, immunolocalization results should be viewed as only tentative evidence for the existence of a particular protein in the spindle. As yet, relatively few of the proteins localized to the spindle have been rigorously demonstrated to have a necessary function in the mitotic apparatus. Thus, it is not yet obvious how these proteins should be integrated into a model of how the spindle is regulated.

V. Direct Evidence for an Influence of Calmodulin on the Mitotic Apparatus

In spite of what we consider irrefutable evidence that calmodulin is a significant component of the mitotic apparatus, there are remarkably few reports that provide direct evidence about the nature and effect of the interaction of calmodulin with its spindle binding sites.

A. Calcium dependence of the interaction of calmodulin with the mitotic apparatus

The vast majority of the interactions of calmodulin with its target proteins are dependent on the concurrent presence a sufficient concentration of calcium to induce the calcium-bound calmodulin conformation. We are aware of only a few instances in which calcium-independent interactions of calmodulin with target proteins have been demonstrated (Cohen *et al.*, 1978; Hertzberg and Gilula, 1981; Greenlee *et al.*, 1982; Welsh *et al.*, 1982).

The interaction of calmodulin with the mitotic apparatus has generally been assumed to be calcium-dependent, although there has been speculation that calmodulin could interact with microtubules independently of calcium (Deery *et al.*, 1984; Rebhun *et al.*, 1980). However, we have recently demonstrated that the interaction of calmodulin with its binding sites in the mitotic apparatus is not dependent on the presence of calcium (Sweet *et al.*, 1988). Using permeabilized PtK_1 cells, we found that calmodulin–rhodamine localized in the mitotic apparatus in a manner identical to that found with microinjection or immunofluorescence. This incorporation occurred when the permeabilization buffers included either 10 mM EGTA (Fig. 6.8(a)–(d)) or 10 μM calcium (Fig. 6.8(e)–(f)). We thus concluded the interaction of calmodulin with its binding proteins in the mitotic apparatus is calcium-independent.

B. Calcium/calmodulin-induced depolymerization of microtubules *in vivo*

There are two reports that indicate that calmodulin may have a depolymerizing effect on microtubules *in vivo*. In the first report, interphase microtubules were found to be depolymerized following microinjection of "calcium-saturated" calmodulin (Keith *et al.*, 1983). In the second report, the prometaphase transit time was found to be increased following injection of calcium or "calcium-saturated" calmodulin as compared to calcium-free controls (Keith, 1987). In this last paper, the author suggests that the increased transit time is due to a transient shortening of the mitotic apparatus, presumably mediated by calmodulin. We have found these studies difficult to interpret, and our attempts

---

Figure 6.8. Incorporation of calmodulin–rhodamine isothiocyanate (CaM-TRITC) into permeabilized cells in the presence and absence of calcium. Cells were lysed by a method modified from Candé *et al.* (1981) as follows. Prior to permeabilization, the culture was rinsed twice with 2 ml of a solution containing 85 mM PIPES pH 6.94 (buffer A) and either 10 μM calcium (A/C) or 10 mM EGTA (A/E). The cells were then lysed for 90 s with 2 ml of A/C or A/E containing 0.08% Brij 58. The lysis solution was removed and replaced with 20 μl of CaM-TRITC that had been dialysed against the lysis solution. The solution was spread over the dish by placing an 18 mm diameter coverslip over the drop. After 90 s this solution was rinsed off with 2 ml of A/C or A/E and the cells were either observed immediately or fixed with 3% formaldehyde in A/C or A/E. (a, b, c, d) Permeabilization in the presence of 10 mM EGTA. (a, b) Metaphase cell: (a) CaM-TRITC fluorescence; (b) phase contrast. (c, d) Anaphase cell: (c) CaM-TRITC fluorescence; (d) phase contrast. (e, f) Permeabilization in the presence of 10 μM calcium. (e) CaM-TRITC fluorescence; (f) phase contrast. (Bars = 5 μm.) (From Sweet *et al.*, 1988.)

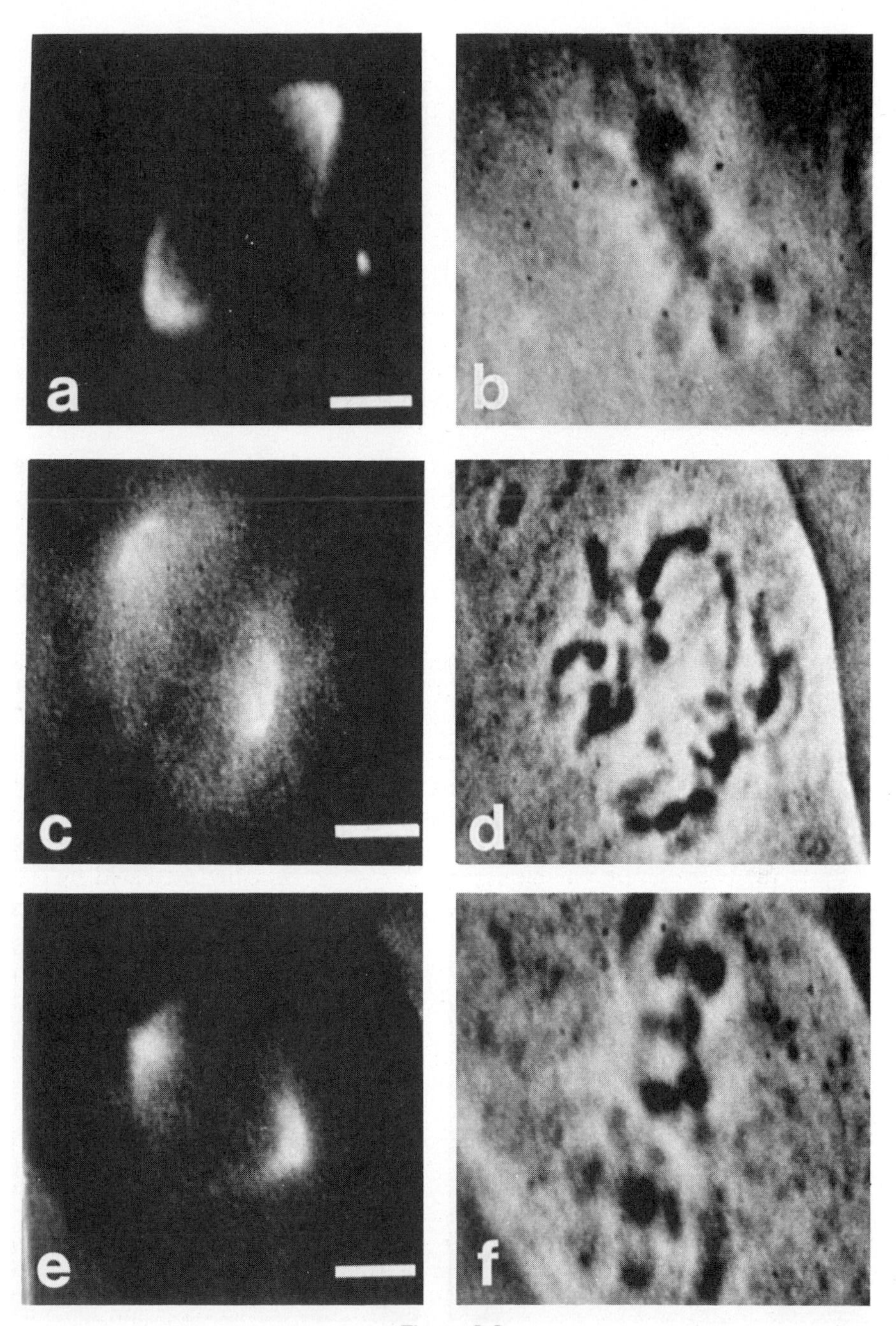

Figure 6.8.

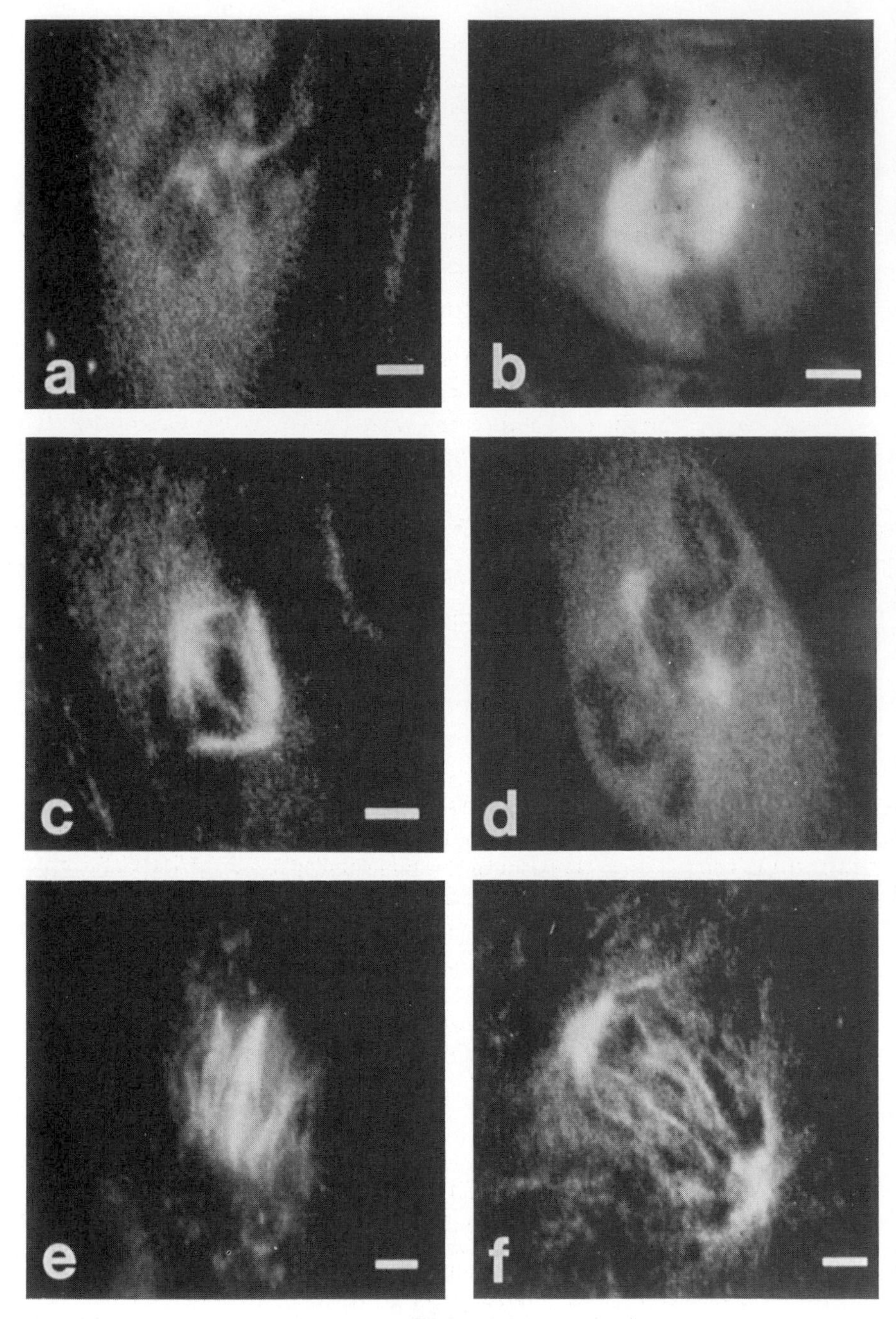

Figure 6.9.

to repeat parts of this work in our own laboratory have failed. None the less, the studies indicate that exogenous calmodulin may be able to influence microtubules *in vivo*.

C. Calmodulin stabilization of kinetochore microtubules

Recent work in our laboratory has led to a contrasting view of the effect of microinjected calmodulin on the mitotic apparatus. We have found that injection of calmodulin, CaM-TRITC or an apparently calcium-insensitive analogue of calmodulin will transiently protect kinetochore microtubules from depolymerizing during treatment with nocodazole (Sweet and Welsh, 1986; Sweet *et al.*, 1988). Using a dosage of nocodazole that would normally produce complete loss of normal microtubule organization in the mitotic apparatus (Fig. 6.9(a)), we found that cells microinjected with calmodulin before exposure to nocodazole specifically retained kinetochore microtubules (Fig. 6.9(c), (d)). When we examined populations of cells that had been injected with calmodulin, calmodulin analogues or controls for nocodazole-induced alterations in spindle structure, we concluded that injection of calmodulin provided a stabilizing influence on kinetochore microtubules (Fig. 6.10).

D. Calmodulin association with re-forming microtubules

Several years ago, DeMey *et al.* (1981) reported in abstract form that calmodulin was localized, by immunoperoxidase staining, to regions in which microtubules were re-forming following release of mitotic cells from nocodazole treatment. Recently, we have confirmed this report using mitotic PtK_1 cells that had been injected with both calmodulin–rhodamine and a fluorescein analogue of tubulin. Following treatment in nocodazole to remove all microtubules, injection of the fluorescent analogues into cells that had entered mitosis in the presence of nocodazole, and finally removal of the nocodazole, calmodulin–rhodamine and tubulin–fluorescein localized in mitotic cells in regions containing re-forming microtubules (Sweet *et al.*, 1987; 1989).

Figure 6.9. Protection of kinetochore microtubules from depolymerizing effect of nocodazole. (a) Metaphase cell, treated with nocodazole (1 μM, 15 min, 25°C); tubulin immunofluorescence. (b) Metaphase control cell; tubulin immunofluorescence. (c, d) Metaphase cell, injected with calmodulin–rhodamine isothiocyanate (CaM-TRITC) prior to treatment with nocodazole: (c) tubulin immunofluorescence; (d) CaM-TRITC fluorescence. (e) Anaphase cell, treated with nocodazole; tubulin immunofluorescence. (f) Anaphase control cell; tubulin immunofluorescence. (Bars = 5 μm.) (From Sweet *et al.*, 1988.)

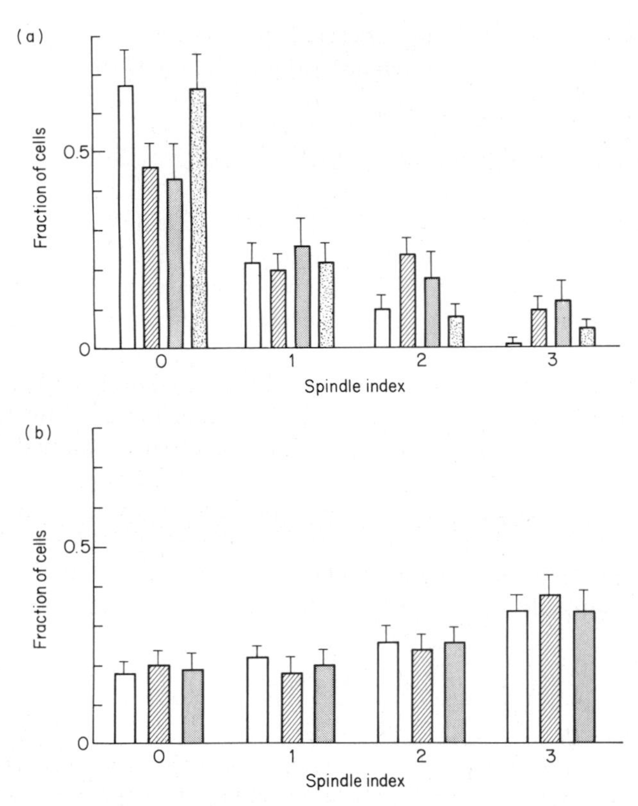

Figure 6.10. Spindle index distribution in populations of cells following various injections and nocodazole treatment (1 μM, 15 min, 25°C). (a) Treatments without stabilizing effect: □, no injections (n = 89); ▨ cells injected with free TRITC (n = 176); ▩, cells injected with BSA-TRITC (n = 66); ▦, cells injected with a performic acid oxidized analogue of CaM (n = 97). (b) Treatments with stabilizing effect: □, cells injected with native CaM (n = 239); ▨, cells injected with CaM-TRITC (n = 142); ▩, cells injected with heat-altered CaM-TRITC (n = 137). Results are presented as fraction of cells ±SD. (From Sweet *et al.*, 1988a.)

Figure 6.11. PtK_1 cell treated with nocodazole (3.0 μg/ml, 4 h, 37°C), injected with tubulin–dichlorotriazinyl fluorescein (tubulin-DTAF) and calmodulin–rhodamine isothiocyanate (CaM-TRITC), then allowed to recover from nocodazole treatment for 10 min at 37°C. (a) Tubulin-DTAF fluorescence image: area in rectangle labelled (c) is shown in panel (c) and area in rectangle labelled (d) is shown in panel (d). (Bar = 5 μm.) (b) CaM-TRITC fluorescence image. (c) TEM image from a serial section of the region indicated in (a), showing a kinetochore with growing microtubules. (Bar = 0.25 μm.) (d) TEM image from a serial section of the region indicated in (a) showing a centrosome. (Bar = 0.33 μm.) (From Sweet *et al.*, 1989).

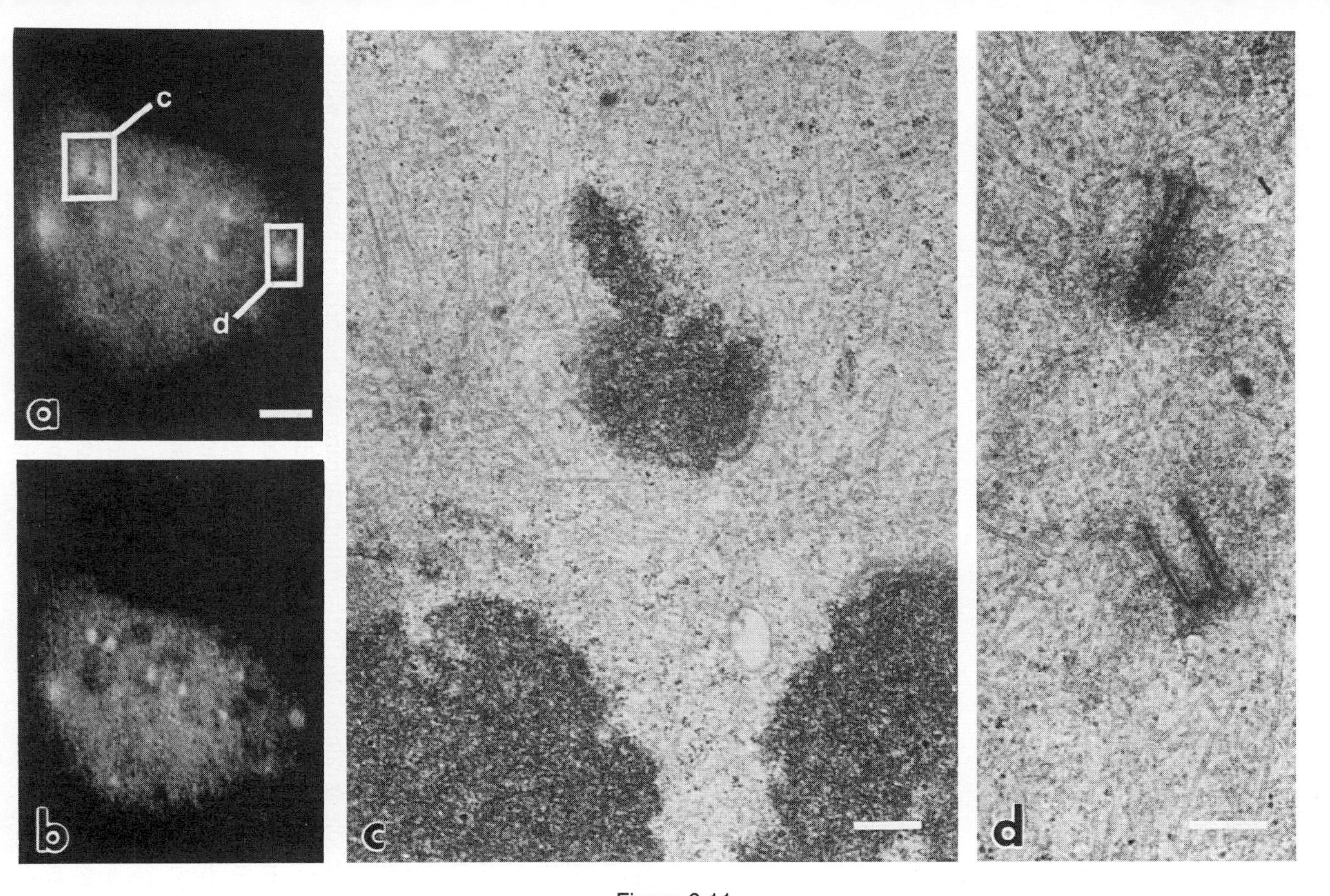

Figure 6.11.

Calmodulin was associated with microtubules in all regions of the cell where microtubules were found; not only with microtubules that had nucleated near centrosomes, but also with microtubules that had nucleated in the vicinity of the kinetochores (Fig. 6.11). Because calmodulin is not normally found to be highly concentrated near kinetochores, we concluded that the presence of calmodulin indicated a physical association of calmodulin with these microtubules. Moreover, the appearance of calmodulin-free microtubules later during the recovery process suggested that calmodulin-associated microtubules are inherently more stable. These results further indicate that an association of calmodulin with microtubules may contribute to their stability.

VI. Summary

Calmodulin has been demonstrated convincingly to be a component of the mitotic apparatus of all eukaryotic cells examined. Calmodulin appears to maintain a physical association with kinetochore microtubules and the spindle poles. Although evidence from studies of brain microtubules suggests that calmodulin may mediate calcium-dependent microtubule disassembly, it seems clear that the presence of calmodulin in the mitotic apparatus does not unequivocally indicate regions containing depolymerizing microtubules. Indeed, calmodulin appears, at least prior to anaphase, to be able to enhance the stability of kinetochore microtubules. The mechanism by which calmodulin influences microtubule assembly/disassembly kinetics may be directly mediated by a protein kinase and/or protein phosphatase. Calmodulin could also influence the mitotic apparatus indirectly by modulating other activities in the spindle. For example, calmodulin regulation of the calcium-ATPase of calcium pumps could alter the levels of calcium ion in the vicinity of the spindle. Finally, calmodulin could be interacting with elements of the spindle that are presently poorly understood or not recognized at all.

At this moment it is difficult or impossible to present a complete model for the role of calmodulin in the mitotic apparatus. The original hypothesis which suggested that calmodulin might mediate the anaphase disassembly of kinetochore microtubules seems to be contradicted by recent evidence that kinetochore microtubules lose subunits at the kinetochore end rather than the spindle pole (Gorbsky *et al.*, 1987; Mitchison *et al.*, 1986). None the less, realizing that our model is largely speculative, we hypothesize that calmodulin is an integral component of the overall kinetochore microtubule structure, contributing to its biochemical character. Perhaps the ability of calmodulin-associated microtubules to form at the onset of mitosis, while the remainder of the microtubules in the cell are disappearing, is due to a calmodulin-induced stability. The microtubule attachment site on the kinetochore may recognize only calmodulin-associated microtubules. Finally, the decreased stability of kinetochore microtubules during anaphase may be induced by calcium and mediated by kinetochore microtubule-associated calmodulin, perhaps via a phosphorylation event.

Clearly, much remains to be learned about what calmodulin is doing in the mitotic apparatus. Our current concepts about the regulation of spindle function are probably deficient in a number of ways. A further understanding of mitosis awaits the development of more complete *in vitro* model systems (Candé *et al.*, 1981; Candé, this volume), as well as more specific probes and approaches for examining the spindle *in vivo*.

REFERENCES

Anderson, B., Osborn, M. and Weber, K. (1978). *Cytobiologie* **17**, 354–364.
Babu, Y. S., Sack, J. S., Greenhough, T. J., Bugg, C. E., Means, A. R. and Cook, W. J. (1985). *Nature* **315**, 37–40.
Bennett, V. and Davis, J. (1981). *Proc. Natl. Acad. Sci. USA* **78**, 7550–7554.
Borisy, G. G., Marcum, J. M., Olmsted, J. B., Murphy, D. B. and Johnson, K. A. (1975). *Ann. NY Acad. Sci.* **253**, 107–132.
Brady, R. C., Dedman, J. R. and Cabral, F. R. (1984). *J. Cell Biol.* **99**, 442a. (Abstr.).
Brinkley, B. R. and Cartwright, J. (1975). *Ann. NY Acad. Sci.* **253**, 428–439.
Brinkley, B. R., Marcum, J. M., Welsh, M. J., Dedman, J. R. and Means, A. R. (1978). In *Cell Reproduction: Daniel Mazia Dedicatory Volume* (ed. E. R. Dirksen, D. Prescott and C. F. Fox), pp. 299–314. Academic Press, New York.
Candé, W. Z., McDonald, K. and Meusen, R. L. (1981). *J. Cell Biol.* **88**, 618–629.
Caroni, P. and Carafoli, E. (1981). *J. Biol. Chem.* **256**, 3263–3270.
Chiesi, M. and Carafoli, E. (1983). *Biochemistry* **22**, 985–993.
Cheung, W. Y. (1980). *Science* **207**, 198–207.
Cohen, P., Burchell, A., Foulkes, J. G., Cohen, P. T. W., Vanaman, T. C. and Nairn, A. (1978). *FEBS Lett.* **92**, 287–293.
Davis, B. A., Schwartz, A., Samaha, F. J. and Kranias, E. G. (1983). *J. Biol. Chem.* **258**, 13 587–13 591.
Davis, J. Q. and Bennett, V. (1984). *J. Biol. Chem.* **259**, 13 550–13 559.
Dedman, J. R., Welsh, M. J., Kaetzel, M. A., Pardue, R. L. and Brinkley, B. R. (1982). In *Calcium and Cell Function*, Vol. III (ed. W. Cheung), pp. 455–472. Academic Press, New York.
Deery, W. J., Means, A. R. and Brinkley, B. R. (1984). *J. Cell Biol.* **98**, 904–910.
DeMey, J., Moeremans, M., Geuens, G., Nuydens, R., van Belle, H. and DeBrabander, M. (1980). In *Microtubules and Microtubule Inhibitors 1980* (ed. M. DeBrabander and J. DeMey), pp. 227–240. Elsevier/North Holland, Amsterdam.
DeMey, J., DeBrabander, M., Geuens, G., Nuydens, R. and Moermans, M. (1981). *J. Cell Biol.* **91**, 332a. (Abstr.).
Erneux, C., Passareiro, H. and Nunez, J. (1984). *FEBS Letts.* **172**, 315–320.
Famulski, K. S. and Carafoli, E. (1984). *Eur. J. Biochem.* **141**, 15–20.
Fukunaga, K., Yamamoto, H., Matsui, K., Higashi, K. and Miyamoto, E. (1982). *J. Neurochem.* **39**, 1607–1617.
Gmaj, P., Zurini, M., Murer, H. and Carafoli, E. (1983). *Eur. J. Biochem.* **136**, 71–76.

Goldenring, J. R., Gonzalez, B., McGuire, J. S., Jr. and DeLorenzo, R. J. (1983). *J. Biol. Chem.* **258**, 12632–12640.
Goodman, M., Pechere, J. F., Haiech, J. and Demaille, J. G. (1979). *J. Mol. Evol.* **13**, 331–352.
Gorbsky, G. J., Sammak, P. J. and Borisy, G. G. (1987). *J. Cell Biol.* **104**, 9–18.
Greenlee, D. V., Andreasen, T. J. and Storm, D. R. (1982). *Biochemistry* **21**, 2759–2764.
Hamaguchi, Y. and Iwasa, F. (1980). *Biomed. Res.* **1**, 502–509.
Harris, P. (1975). *Exp. Cell Res.* **94**, 409–425.
Harris, P. (1978). In *Cell Cycle Regulation* (ed. Jeter *et al.*), pp. 75–104. Academic Press, New York.
Harris, P. (1982). *Cell Differ.* **11**, 357–358.
Harris, P. (1983). *Develop. Biol.* **96**, 277–284.
Hepler, P. K. (1980). *J. Cell Biol.* **86**, 490–499.
Hepler, P. K. and Wolniak, S. M. (1984). *Int. Rev. Cytol.* **90**, 169–238.
Hertzberg, E. L. and Gilula, N. B. (1981). *Cold Spring Harbor Symp. Quant. Biol.* **46**, 639–645.
Inoué, S. (1981). *J. Cell Biol.* **91**, 131s–147s.
Jameson, L., Frey, T., Zeeberg, B., Dalldorf, F. and Caplow, M. (1980). *Biochemistry* **19**, 2472–2479.
Job, D., Fischer, E. H. and Margolis, R. L. (1981). *Proc. Natl. Acad. Sci. USA* **78**, 4679–4682.
Job, D., Rauch, C. T., Fischer, E. H. and Margolis, R. L. (1982). *Biochemistry* **21**, 509–515.
Job, D., Rauch, C. T., Fischer, E. H. and Margolis, R. L. (1983). *Proc. Natl. Acad. Sci. USA* **80**, 3894–3898.
Kakiuchi, S. and Sobue, K. (1981). *FEBS Letts.* **132**, 141–143.
Keith, C. H. (1987). *Cell Motil. Cytoskel.* **7**, 1–9.
Keith, C. H., DiPaola, M., Maxfield, F. R. and Shelanski, M. L. (1983). *J. Cell Biol.* **97**, 1918–1924.
Keith, C. H., Maxfield, F. R. and Shelanski, M. L. (1985a) *Proc. Natl. Acad. Sci. USA* **82**, 800–804.
Keith, C. H., Ratan, R., Maxfield, F. R., Bajer, A. and Shelanski, M. L. (1985b). *Nature* **316**, 848–850.
Keller, T. C. S., Jemiolo, D. K., Burgess, W. H. and Rebhun, L. I. (1982). *J. Cell Biol.* **93**, 797–803.
Kennedy, M. B., McGuinness, T. and Greengard, P. (1983). *J. Neurosci.* **3**, 818–831.
Kiehart, D. P. (1981). *J. Cell Biol.* **88**, 604–617.
Klee, C. B. and Vanaman, T. C. (1982). *Adv. Protein Res.* **35**, 213–321.
Krebs, J. (1981). *Cell Calcium* **2**, 295–311.
Kumagai, H., Nishida, E. and Sakai, H. (1982). *J. Biochem.* **91**, 1329–1336.
LaPorte, D. C., Wierman, B. M. and Storm, D. R. (1980). *Biochemistry* **19**, 3814–3819.
Larson, R. E., Goldenring, J. R., Vallano, M. L. and DeLorenzo, R. J. (1985). *J. Neurochem.* **44**, 1566–1574.
Lee, Y. C. and Wolff, J. (1984a). *J. Biol. Chem.* **259**, 1226–1230.
Lee, Y. C. and Wolff, J. (1984b). *J. Biol. Chem.* **259**, 8041–8044.
Le Peuch, C. J., Haiech, J. and Demaille, J. G. (1979). *Biochemistry* **18**, 5150–5157.
LeVine, H., Sahyoun, N. E. and Cuatrecasas, P. (1985). *Proc. Natl. Acad. Sci. USA* **82**, 287–291.

Lopaschuk, G., Richter, B. and Katz, S. (1980). *Biochemistry* **19**, 5603–5607.
Manalan, A. S. and Klee, C. B. (1984). *Adv. Cyclic Nucleotide Protein Phosphoryl Res.* **18**, 227–278.
Marcum, J. M., Dedman, J. R., Brinkley, B. R. and Means, A. R. (1978). *Proc. Natl. Acad. Sci. USA* **75**, 3771–3775.
Margolis, R. L. and Rauch, C. T. (1981). *Biochemistry* **20**, 4451–4458.
McGuinness, T. L., Lai, Y., Greengard, P., Woodgett, J. R. and Cohen, P. (1983). *Fedn. Eur. Biochem. Soc.* **163**, 329–334.
McIntosh, J. R. (1984). *TIBS* **9**, 195–198.
McIntosh, J. R., Saxton, W. M., Stemple, D. L., Leslie, R. J. and Welsh, M. J. (1985). *Ann. NY Acad. Sci.* **466**, 566–579.
Means, A. R. and Dedman, J. R. (1980). *Nature* **285**, 73–77.
Mitchison, T., Evans, L., Schultze, E. and Kirschner, M. (1986). *Cell* **45**, 525–527.
Nagle, B. W. and Egrie, J. C. (1981). In *Mitosis/Cytokinesis* (ed. A. M. Zimmerman and A. Forer), pp. 337–361. Academic Press, New York.
Ouimet, C. C., McGuinness, T. L. and Greengard, P. (1984). *Proc. Natl. Acad. Sci. USA* **81**, 5604–5608.
Pabion, M., Job, D. and Margolis, R. L. (1984). *Biochemistry* **23**, 6642–6648.
Pardue, R. L., Kaetzel, M. A., Hahn, S. H., Brinkley, B. R. and Dedman, J. R. (1981). *Cell* **23**, 533–542.
Petzelt, C. (1979). *Int. Rev. Cytol.* **60**, 53–92.
Pirollet, F., Job, D., Fischer, E. H. and Margolis, R. L. (1983). *Proc. Natl. Acad. Sci. USA* **80**, 1560–1564.
Plank, B., Pifl, C., Hellmann, Gj., Wyskovsky, W., Hoffmann, R. and Suko, J. (1983). *Eur. J. Biochem.* **136**, 215–221.
Ratan, R. R., Shelanski, M. L. and Maxfield, F. R. (1986). *Proc. Natl. Acad. Sci. USA* **83**, 5136–5140.
Rebhun, L. I., Jemiolo, D., Keller, T., Burgess, W. and Kretsinger, T. (1980). In *Microtubules and Microtubule Inhibitors* (ed. M. DeBrabander and J. DeMey), pp. 243–252. Elsevier/North Holland, Amsterdam.
Ross, D. H. and Cardenas, H. L. (1983). *J. Neurochem.* **41**, 161–171.
Sahyoun, N., LeVine, H., III, Bronson, D., Siegel-Greenstein, F. and Cuatrecasas, P. (1985). *J. Biol. Chem.* **260**, 1230–1237.
Salmon, E. D. and Segall, R. R. (1980). *J. Cell Biol.* **86**, 355–365.
Salmon, E. D., McKeel, M. and Hays, T. (1984). *J. Cell Biol.* **99**, 1066–1075.
Saxton, W. M., Stemple, D. L., Leslie, R. J., Salmon, E. D., Zavortink, M. and McIntosh, J. R. (1984). *J. Cell Biol.* **99**, 2175–2186.
Schulman, H., Kuret, J., Jefferson, A. B., Nose, P. S. and Spitzer, K. H. (1985). *Biochemistry* **24**, 5232–5327.
Silver, R. B. (1986). *Proc. Natl. Acad. Sci. USA* **83**, 4302–4306.
Silver, R. B., Cole, R. D. and Candé, W. Z. (1980). *Cell* **19**, 505–516.
Slocum, R. D. and Roux, S. J. (1982). *J. Histochem. Cytochem.* **30**, 617–629.
Sobue, K., Fujita, M., Muramoto, Y. and Kakiuchi, S. (1981). *FEBS Letts.* **132**, 137–140.
Stemple, D. L., Sweet, S. C., Welsh, M. J. and McIntosh, J. R. (1988). *Cell Motil. Cytoskel.* **9**, 231–242.
Sternberger, L. A., Hardy, P. H., Cuculis, J. J. and Meyer, H. G. (1979). *J. Histochem. Cytochem.* **18**, 315–333.
Sweet, S. C. and Welsh, M. J. (1986). *J. Cell Biol.* **103**, 140a. (Abstr.).
Sweet, S. C. and Welsh, M. J. (1988). *Eur. J. Cell Biol.*, **47**, 88–93.

Sweet, S. C., Rogers, C. M. and Welsh, M. J. (1987). *J. Cell Biol.* **105**, 281a. (Abstr.).
Sweet, S. C., Rogers, C. M. and Welsh, M. J. (1988). *J. Cell Biol.*, **107**, 2243–2251.
Sweet, S. C., Rogers, C. M. and Welsh, M. J. (1989). *Cell Motil. Cytoskel.* **12**, 113–122.
Tada, M., Ohmori, F., Yamada, M. and Abe, H. (1979). *J. Biol. Chem.* **254**, 319–326.
Tash, J. S., Welsh, M. J. and Means, A. R. (1980). *Cell* **21**, 57–65.
Teo, T. S. and Wang, J. H. (1973). *J. Biol. Chem.* **248**, 5950–5955.
Theurkauf, W. E. and Vallee, R. B. (1983). *J. Biol. Chem.* **258**, 7883–7886.
Tuana, B. S., Dzurba, A., Panagia, V. and Dhalla, N. S. (1981). *Biochem. Biophys. Res. Commun.* **100**, 1245–1250.
Vallano, M. L., Goldenring, J. R., Buckholz, T. M., Larson, R. E. and DeLorenzo, R. J. (1985). *Proc. Natl. Acad. Sci. USA* **82**, 3202–3206.
Vallee, R. (1980). *Proc. Natl. Acad. Sci. USA* **77**, 3206–3210.
Vallee, R. B., BiBartolomeis, M. J. and Theurkauf, W. E. (1981). *J. Cell Biol.* **90**, 568–576.
Vantard, M., Lambert, A. M., DeMey, J., Picquot, P., Van Eldik, L. J. (1985). *J. Cell Biol.* **101**, 488–499.
Vincenzi, F. F. and Larsen, F. L. (1980). *Fedn. Proc.* **39**, 2427–2431.
Wadsworth, P. and Sloboda, R. D. (1983). *J. Cell Biol.* **97**, 1249–1254.
Walsh, D. A., Ashby, C. D., Gonzalez, C., Calkins, D., Fischer, E. H. and Krebs, E. G. (1971). *J. Biol. Chem.* **246**, 1977–1985.
Wang, Y.-L. and Taylor, D. L. (1979). *J. Cell Biol.* **82**, 672–679.
Webb, B. C. and Wilson, L. (1980). *Biochemistry* **19**, 1933–2001.
Weber, A. (1968). *J. Gen. Physiol.* **52**, 760–772.
Weber, A. and Herz, R. (1968). *J. Gen. Physiol.* **52**, 750–759.
Welsh, M. J. (1983). *Methods Enzymol.* **102**, 110–121.
Welsh, M. J., Dedman, J. R., Brinkley, B. R. and Means, A. R. (1977). *J. Cell Biol.* **75**, 262a. (Abstr.).
Welsh, M. J., Dedman, J. R., Brinkley, B. R. and Means, A. R. (1978). *Proc. Natl. Acad. Sci. USA* **74**, 1867–1871.
Welsh, M. J., Dedman, J. R., Brinkley, B. R. and Means, A. R. (1979). *J. Cell Biol.* **81**, 624–634.
Welsh, M. J., Aster, J., Ireland, M., Alcala, J. and Maisel, H. (1982). *Science* **216**, 642–644.
Wick, S. M. and Hepler, P. K. (1980). *J. Cell Biol.* **86**, 500–513.
Wick, S. M., Muto, S. and Duniec, J. (1985). *Protoplasma* **126**, 198–206.
Willingham, M. C., Wehland, J., Klee, C. B., Richert, N. D., Rutherford, A. V. and Pastan, I. H. (1983). *J. Histochem. Cytochem.* **31**, 445–461.
Wolniak, S. M., Hepler, P. K. and Jackson, W. T. (1980). *J. Cell Biol.* **87**, 23–32.
Yamamoto, H., Fukunaga, K., Tanaka, E. and Miyamoto, E. (1983). *J. Neurochem.* **41**, 1119–1125.
Yamauchi, T. and Fujisawa, H. (1984). *Arch. Biochem. Biophys.* **234**, 89–96.
Zavortink, M., Welsh, M. J. and McIntosh, J. R. (1983). *Exp. Cell Res.* **149**, 375–385.

CHAPTER 7

Membranes in the Mitotic Apparatus

PETER K. HEPLER

Department of Botany, University of Massachusetts, Amherst, Massachusetts, USA

I. Introduction

Membranes comprise a significant part of the mitotic apparatus (MA). They occur throughout plants and animals and have been noted in almost all dividing cell types that have been examined. Together with the spindle fibres, they undergo dramatic transformations and redistributions during mitosis and in addition they may form specific structural associations with the mitotic microtubules. In brief, membranes are an integral part of the MA; it becomes important now to understand their contribution to the process of mitosis.

A commonly held view is that membranes control the calcium ion concentration ([Ca^{2+}]) within the MA and thereby regulate the formation and function of the spindle fibres. In addition, membranes may act as a structural scaffolding to which spindle fibres attach and generate shearing forces. It is thus possible to imagine that membranes contribute in a fundamental way to the structure and operation of the MA. Deciphering their role may help us understand basic aspects of the mitotic mechanism.

There have been several reviews on membranes in the MA (Harris, 1978; Hepler *et al.*, 1981; Paweletz, 1981; Heath, 1980; Kubai, 1978); this topic has recently been covered at length (Hepler and Wolniak, 1984). For this reason, the present chapter will be more limited in scope and will focus instead on the highlights of the problem and on current developments, especially as they apply to higher eukaryotic organisms in which dispersal of the nuclear envelope (NE) occurs.

MITOSIS: Molecules and Mechanisms
ISBN 0-12-363420-2

II. Endoplasmic Reticulum (ER)

A. Morphology

The endoplasmic reticulum (ER) is usually the most conspicuous membrane system in the MA. Although there is great variability in its particular distribution among dividing cells of different organisms, there is nevertheless considerable evidence revealing aggregation of ER in three locations: (1) the MA pole, (2) the MA interior, and (3) surrounding the MA (Hepler and Wolniak, 1984) (Figs. 7.1, 7.2, 7.3). The ER appears to be well disposed to create compartments or domains, both for separating the MA from the rest of the cell and for creating subcompartments within the MA itself such as the spindle pole or along specific membrane intrusions (Hepler, 1980).

In detailed morphology the ER appears in three general forms: lamellar, reticulate tubular, and vesiculate. Lamellar ER often resides at the spindle periphery and in the pole. In some organisms, for example *Drosophila* spermatocytes, many layers of lamellar ER, called the parafusorial lamellae, virtually encase the MA (Church and Lin, 1982). One often notes discontinuities in the ER lamellae and these may provide a limited pathway across the membrane system (Fig. 7.3). Lamellar ER may also possess fenestrations or pores. These are much less regular in appearance than nuclear pores but nevertheless may be avenues of communication between the MA and its surrounding cytoplasm (Hepler, 1980; Hepler and Wolniak, 1984).

ER that resides within the MA interior, but can also be present at the pole, often appears as a tubular reticulum, lacking attached ribosomes. Cells from organisms as diverse as barley leaf (Hepler, 1980) and cultured rat kangaroo (McDonald, 1984) (Fig. 7.4) possess an extensive tubular network of ER that intimately intermingles with the spindle microtubules. In some cell types, such as spermatocytes of the heteropteran insect *Dysdercus*, the membrane system forms an "ordered framework" throughout the mitotic apparatus (Motzko and Ruthmann, 1984), and in others, such as oocytes from the strepsipteran parasite *Xenos*, the spindle is virtually filled with tubular and vesiculate ER (Rieder and Nowogrodzki, 1983).

Vesiculate ER has been reported in the MAs of a wide number of cell types (Harris, 1978; Ryan, 1984) (Figs. 7.1, 7.2). Whether these membranes naturally occur in this form or whether they are transformed from lamellar or tubular ER during fixation is a matter of controversy. Serial-section analysis of some cells reveals that membranes, which may appear as vesiculate in a given section, are actually continuous to other elements and form a highly interconnected three-dimensional matrix (Hepler, 1980).

One of the striking features of the mitotic ER is the extent to which it resembles the sarcoplasmic reticulum of muscle. It seems pertinent to note, for

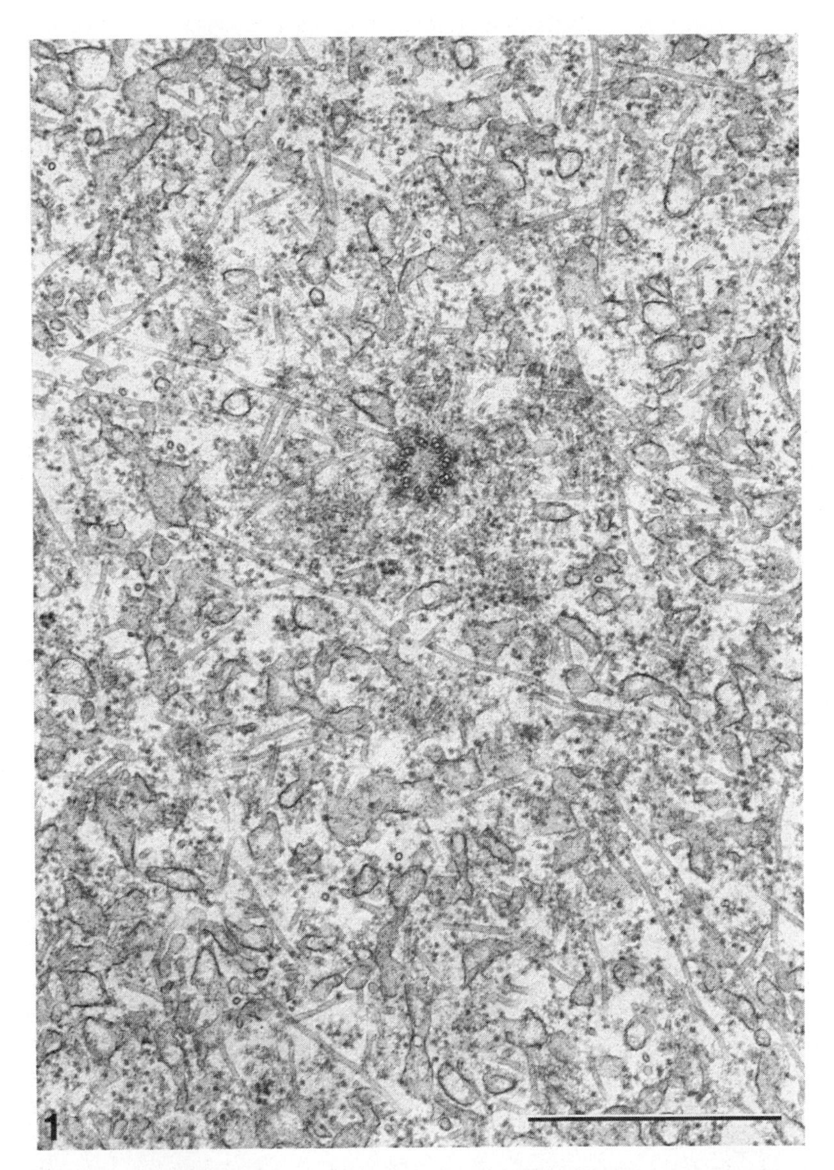

Figure 7.1. The spindle pole at metaphase in the sea urchin *Arbacia*. The centriole is surrounded by radiating microtubules. Elements of ER occur in great profusion among the microtubules. Bar = 1 μm. (From P. K. Hepler and E. D. Salmon, unpublished.)

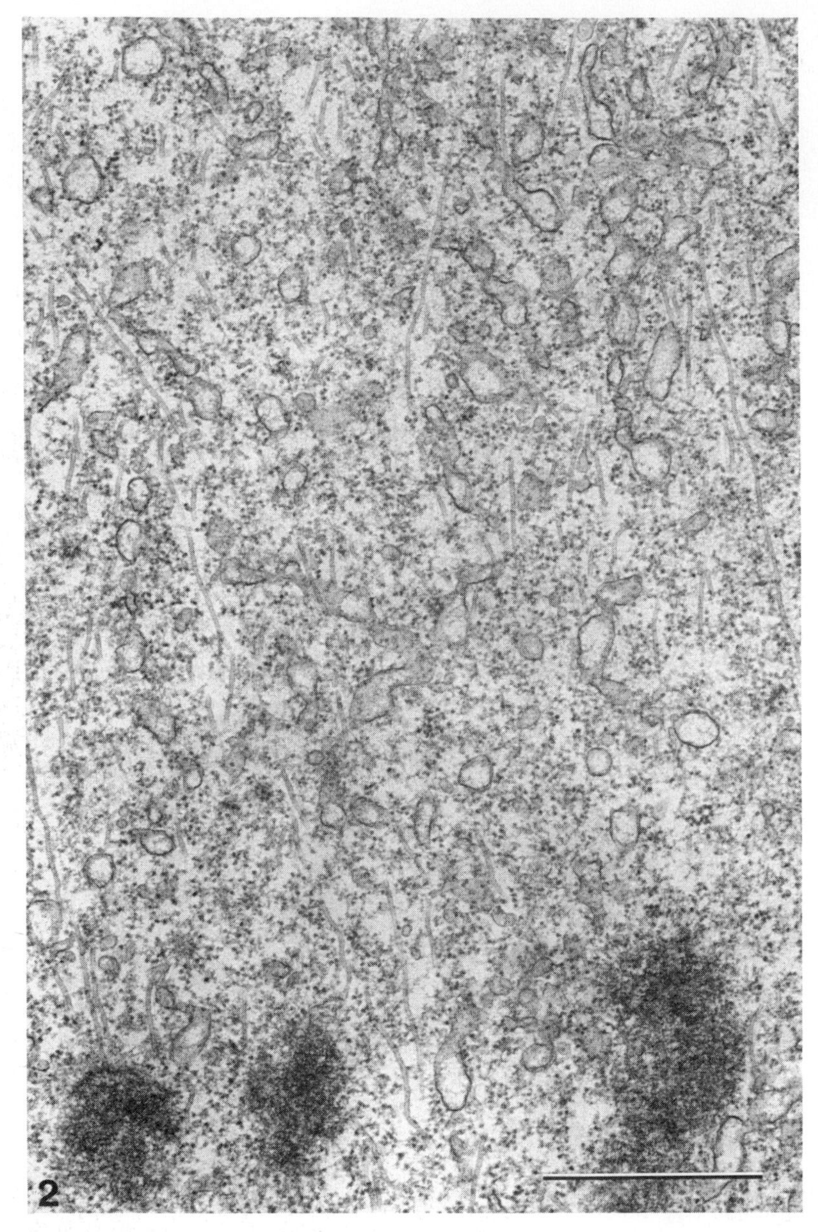

Figure 7.2. In a metaphase MA of *Arbacia* many membrane elements are observed among the spindle microtubules. Bar = 1 μm. (From P. K. Hepler and E. D. Salmon, unpublished.)

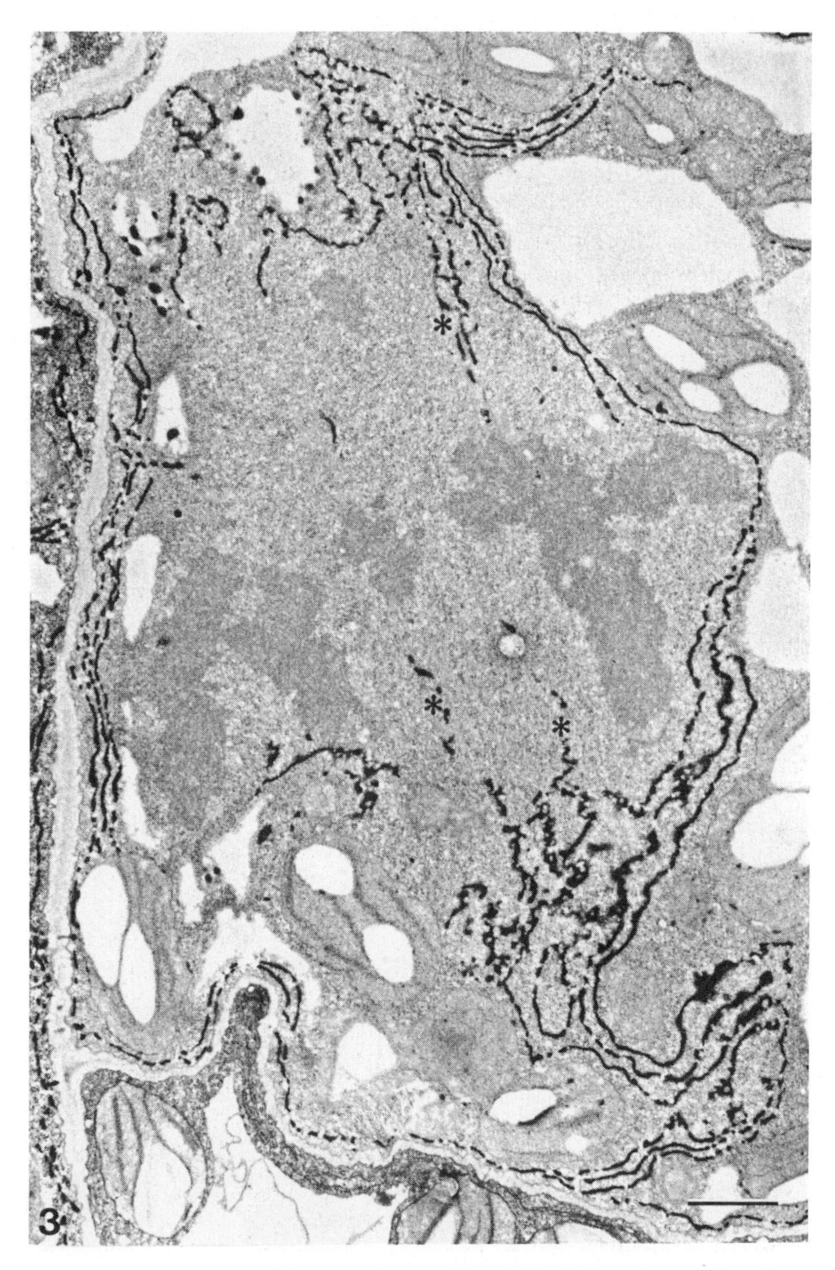

Figure 7.3. A barley (*Hordeum*) leaf mesophyll cell in metaphase, which has been post-fixed in OsFeCN, contains an extensive endomembrane system that surrounds the MA, clusters at the poles and extends into the MA along kinetochore microtubules (*). Bar = 1 μm. (From Hepler, 1980.)

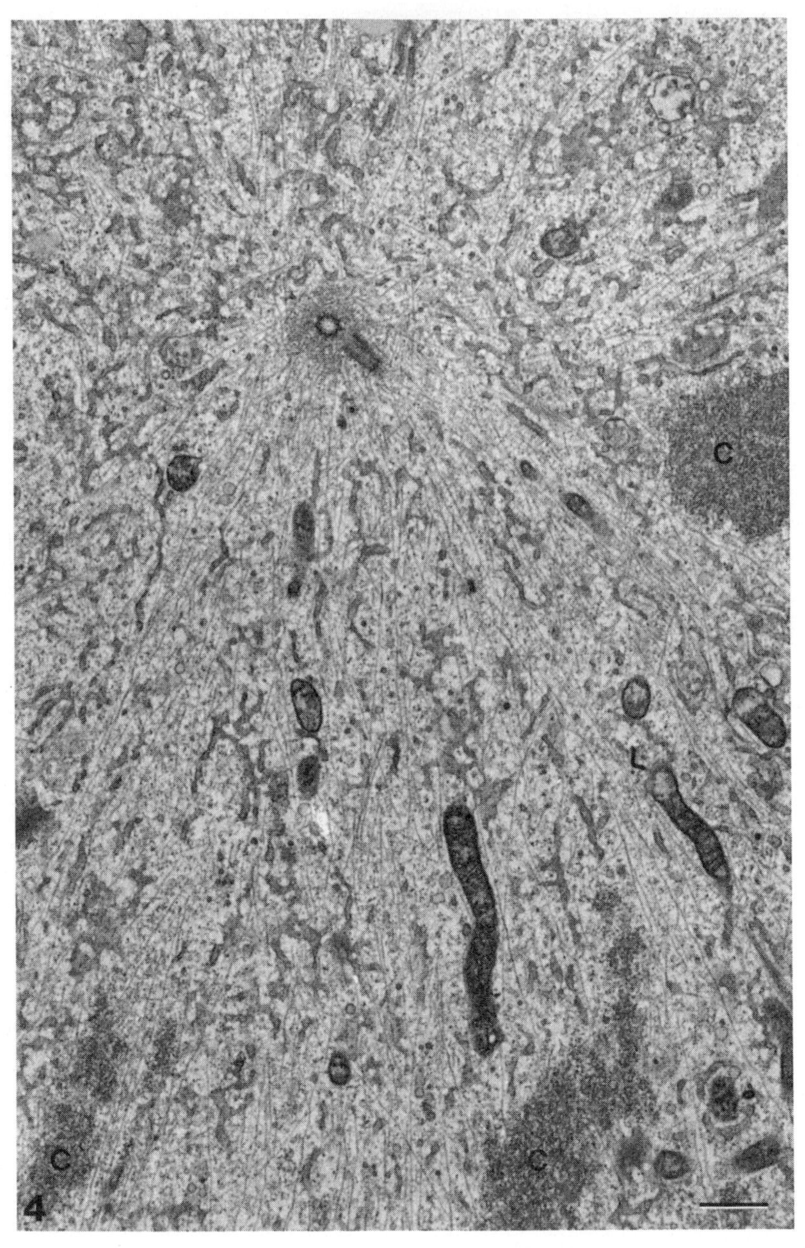

Figure 7.4. A prometaphase PtK cell, stained with OsFeCN, contains an extensive system of smooth, tubular ER that ramifies among the spindle microtubules. Chromosomes (C). Bar = 1 μm. (From K. L. McDonald, unpublished.)

early prometaphase is the growth or intrusion of ER elements into the MA interior (Figs. 7.3, 7.5). The amount of membrane varies from those cells that are virtually filled with smooth tubular ER (e.g. oocytes of *Xenos* (Rieder and Nowogrodzki, 1983)) to those that contain no detectable internal ER (spermatocytes of *Nephrotoma* (Fuge, 1977); see Section II.C). The bulk of cell types examined has some internal ER (Hepler and Wolniak, 1984). Using the osmium ferricyanide (OsFeCN) reagent to contrast ER elements, it has become apparent in some organisms (e.g. barley) that the tubular reticulum extends from the poles to the chromosomes specifically along the kinetochore microtubule bundles (Hepler, 1980) (Fig. 7.5). A similar ER–kinetochore relationship occurs in *Drosophila* spermatocytes (Church and Lin, 1982). Continuing studies are revealing additional examples of membrane–microtubule alignments (Hepler and Wolniak, 1984). Perhaps the most differentiated system so far observed occurs in Wolf-spider spermatocytes, in which each chromosome and its attached kinetochore microtubules are encased in a separate sheath of ER (Wise, 1984).

How ER moves into the MA and forms its specific microtubule associations is not known, but it may be related to the formation of the kinetochore itself. Recent studies provide new evidence for the idea that the kinetochore acquires microtubules through capture of ones growing from the centrosome (Mitchison and Kirschner, 1985; see Chapters 3 and 4). During prophase microtubules growing into the MA interior deform the NE (Hepler, 1976); they may even contribute to the breakdown of the NE. It seems possible therefore that these inwardly extending microtubules carry with them ER and fragments of the NE, and therefore at the moment of capture by the chromosome the kinetochore microtubules already possess a clearly associated membrane component. In addition spindle microtubules may actively transport membranes into the spindle interior (Paweletz and Fehst, 1984).

Metaphase is a time when little change in membrane or spindle structure appears to occur. However, when the chromosomes separate at the onset of anaphase and begin their movement to the poles, in many species there may be progressive accumulation of ER in the spindle pole region (Hepler and Wolniak, 1984) (Fig. 7.6). Microtubules depolymerize and the specific kinetochore membrane relationships disappear. At the end of anaphase, as the chromosomes approach the poles and begin to coalesce and de-condense, elements of lamellar ER become appressed to the distal surfaces of the chromosomes and transformed back to NE. The polar accumulations eventually disperse and the cell returns to an interphase condition.

C. MAs that lack internal membranes

Having made a general claim that the MAs of eukaryotic cells, in which the NE breaks down (i.e. open MA), have an internal membrane system, I must state

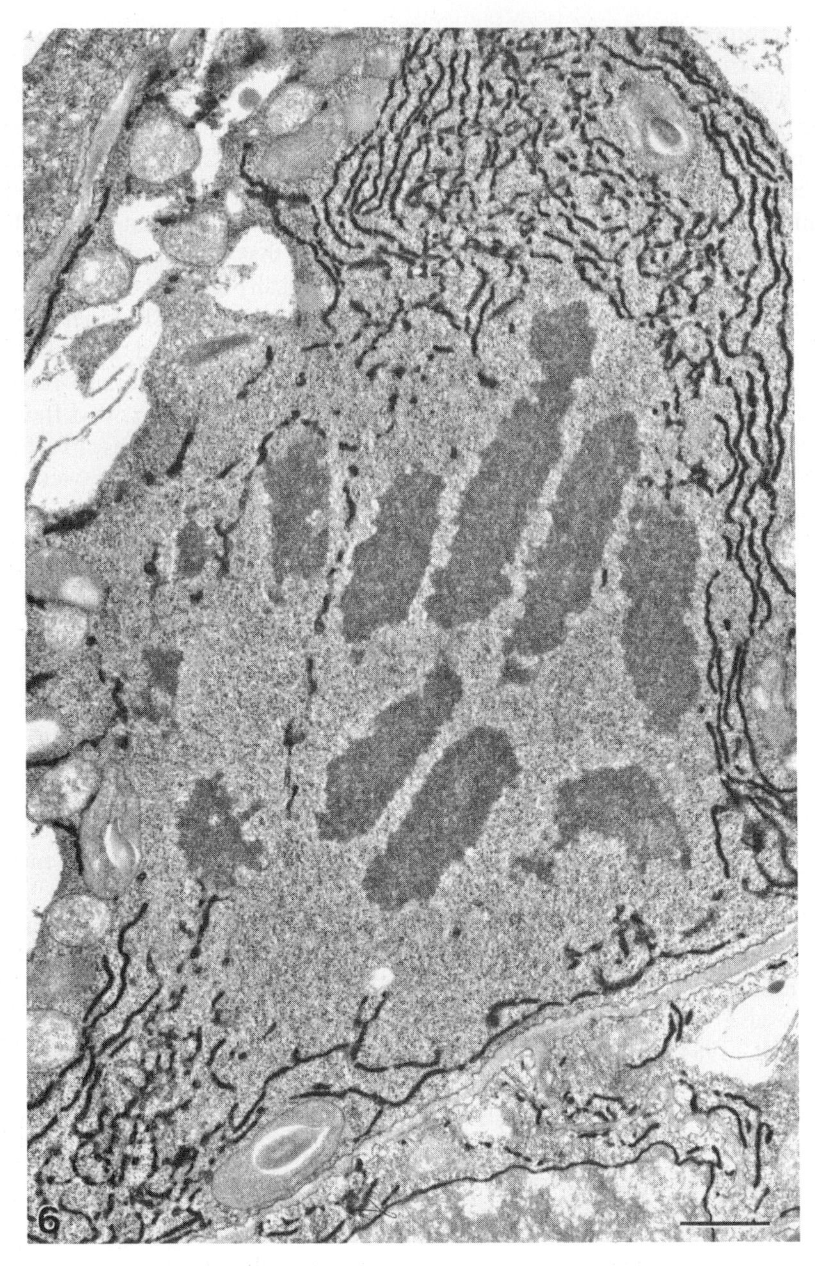

Figure 7.6. During anaphase in a barley (*Hordeum*) leaf cell, ER elements accumulate in the polar regions with relatively few in the interzone. Bar = 1 μm. (From Hepler, 1980.)

there is at least one exception and probably more. Extensive studies on cranefly (*Nephrotoma*) spermatocytes fail to reveal membranes deep into the MA interior (Fuge, 1977; personal communication). There is a sheath of ER surrounding the MA and there are a few elements of ER at the pole that may even extend for a short distance into the MA, but the overall impression is one in which the spindle is devoid of an ER system. These conditions do not hold for other insect spermatocytes, as *Drosophila* (Church and Lin, 1982), *Dysdercus* (Motzko and Ruthmann, 1984) and *Dissosteira* (Nicklas *et al.*, 1979) have extensively developed internal membranes.

One explanation for the lack of membranes within the MA of crane-fly spermatocytes may relate to its relatively small size ($\sim 10\,\mu m$ in diameter). The argument has been put forward that membrane intrusions may have arisen phylogenetically as a means of effectively delivering or removing Ca^{2+} in large MAs where the distance between the centre of the spindle and its perimeter may have been greater than an effective diffusion limit (Hepler and Wolniak, 1984). Cells with small MAs might be able to control their intraspindle $[Ca^{2+}]$ entirely through release and sequestration of the ion by the membranes at the spindle perimeter, thus eliminating the need for an internal membrane system.

III. Membranes Other Than ER

A. Golgi membranes

In addition to the ER elements, there are many other membranes in the MA: in particular, vesicles derived from the Golgi apparatus. It has been argued for some cells that the Golgi-derived vesicles comprise the bulk of the MA-associated membranes (Paweletz, 1981). The problem, however, is one of deciding a particular source for the membrane element in question. The OsFeCN reagent, by staining the ER, allows one to identify Golgi-derived vesicles because of their lack of stain. More recent attempts have used markers for certain enzyme activities such as acid phosphatase and thiamine pyrophosphatase that are known to reside in the Golgi membranes. Use of these reagents reveals that Golgi vesicles occur throughout the MA at all stages of mitosis in PtK cells (Schroeter *et al.*, 1985).

The Golgi cisternae themselves, however, are usually not found in the MA. Rather, in animal cells, they cluster around the centrioles in the region of the spindle pole. In several different cell types the Golgi cisternae undergo a profound transformation as the cell progresses through mitosis. By early prophase in cartilaginous cells grown *in vitro*, the Golgi dictyosomes begin to disperse from the centrosome (Moskalewski *et al.*, 1977). Subsequently, the cisternal membranes become fewer in number until none is evident. When the cell completes mitosis at telophase, the dictyosomal membrane stacks reappear and regroup in the juxtanuclear centrosomal region.

The interaction between dictyosomal membranes and microtubules may play an important role in stabilizing the organization of the Golgi apparatus. The dispersal of the dictyosome, for example, begins at the time when the cytoplasmic microtubules break down and the spindle microtubules appear. Also treatment of cells with anti-microtubule agents (colchicine and vinblastine) causes Golgi membranes to disperse in non-dividing cells. A microtubule cytoskeleton thus appears important for the integrity of the Golgi apparatus, although the latter may have no effect on the structure of the former (Thyberg and Moskalewski, 1985).

Not all cells show the dramatic dispersal and disappearance of the Golgi apparatus during mitosis that has been reported above. In several cell types, dictyosomes of normal size and structure persist throughout mitosis (Hepler and Wolniak, 1984). In plants also, dictyosomes retain their structure throughout division, although during cytokinesis they may aggregate in the vicinity of the growing cell plate, a structure to which they provide vesicles (Hepler, 1982). During the earlier phases of mitosis, one commonly notes dictyosomal vesicles throughout the MA. However, they appear to be randomly scattered and not to possess any particular structural association with components of the MA.

B. Plasma membrane

The plasma membrane (PM) exhibits a highly variable relationship to the MA, ranging from those examples such as oocytes, zygotes and pollen mother cells in which it is separated by large distances to those, such as cultured rat kangaroo cells or *Tradescantia* stamen hair cells in which it is very close. Although the PM is not a component of the MA, recent studies reveal that it may undergo profound structural changes in concert with the cell cycle. To the extent that the PM controls ion fluxes in the cell and even in the MA itself, it is important to understand the meaning of cell cycle-dependent changes in structure.

Morphometric analysis on dividing sea-urchin zygotes reveals that there are large and abrupt changes in the amount of PM preceding entry into the first mitotic cycle (Schroeder, 1978). Microvilli on the cell surface dramatically increase in length prior to mitosis. It is clear that PM must increase as cells divide and thus the idea has been put forward that these microvilli represent a storage component of membrane that will be used during cleavage as the requirement for new surface increases (Schroeder, 1978).

A role for the PM in the mitotic cycle of mammalian cells can be inferred from the well known fact that many cell types in culture become rounded in shape during mitosis (Porter *et al.*, 1973). Examination of the cell surface of Chinese hamster ovary (Porter *et al.*, 1973) and HeLa cells (Paweletz and Schroeter, 1974) with the scanning electron microscope reveals changes in the microvillar structure that correlate with events in the cell cycle. In Chinese hamster ovary cells during G-1 the microvilli are evident but decrease as the cell enters S (Porter *et al.*, 1973). At G-2 however, the microvilli markedly increase in number and the cells become rounded in anticipation of mitosis. Examination of PtK cells reveals additional detail about cell-surface changes that occur during mitosis itself (Sanger *et al.*, 1984). Because PtK cells remain flattened during mitosis, with the cell membrane closely appressed to the underlying structures, it is possible, in the scanning electron microscope, to clearly observe nuclei and chromosomes and to correlate their changes with that of the PM. The results show that as the cells progress from prophase to metaphase the PM is smooth and essentially devoid of microvilli. However, when the cells enter anaphase there is an abrupt increase in microvilli. Of special interest is that the microvilli appear on that portion of the PM that directly overlies the chromosome-to-pole region, with little or none over the interzone (Sanger *et al.*, 1984). Again the suggestion is made that these membrane outgrowths are stored PM that will be used during cytokinesis. Without refuting this notion, it seems reasonable to suggest that other functions might be related to microvilli formation. Given their spatial localization and temporal appearance with chromosome separation, the microvilli may be related to the signal(s), e.g. Ca^{2+} fluxes, that might participate in chromosome separation.

IV. Membrane Function

A. Calcium regulation

1. *General considerations*

It seems likely that spindle membranes regulate the Ca^{2+} concentration and thereby control one or more processes involved in the formation and function of the MA (Harris, 1978; Hepler and Wolniak, 1984). That Ca^{2+} controls mitosis has been considered for many years and especially so since the discovery that elevated levels of Ca^{2+} caused depolymerization of microtubules *in vitro* (Weisenberg, 1972). It has seemed plausible, for example, that the $[Ca^{2+}]$ would be low during prophase, to permit formation of the spindle microtubules, but that during the transition from metaphase to anaphase the $[Ca^{2+}]$ would increase, thereby facilitating the depolymerization of the spindle microtubules that must occur in order for the chromosomes to move to the poles. The remarkably specific topological arrangement of membranes with microtubules that occurs in the MAs of some organisms makes these arguments even more attractive, since it becomes apparent that the presumed calcium-regulatory system, the ER, is positioned precisely along the structure, the kinetochore microtubules, that may be Ca^{2+} regulated (Welsh and Sweet, this volume). Although a review of the possible Ca^{2+}-dependent processes during mitosis is not appropriate in this chapter, a few brief statements should alert the reader to the existence of mechanisms besides the regulation of microtubule polymerization that might be influenced by membrane-controlled changes in $[Ca^{2+}]$ (Hepler and Wolniak, 1983). For example, Ca^{2+} might participate in a microtubule-based shear-force-generating system, in which a mechanochemical ATPase provides the driving force. There is evidence for dynein activity in the MA (Pratt, 1984; Pratt *et al.*, 1980; Bloom and Vallee, this volume) and for its function, especially during anaphase B or the separation of the poles (Candé, 1982; this volume). By comparison with that from cilia or flagella, the dynein in the MA would be expected to be Ca^{2+}-regulated. Calcium might influence the structure and function of the MA through control of actin microfilaments. Microfilaments are present in the MA (Forer, 1985) and Ca^{2+} is known to play a role both in the maintenance of F-actin organization through Ca^{2+}-sensitive actin-binding proteins (Stossel *et al.*, 1985) and in its force-generating interaction with myosin. If such a shear force system were operating, Ca^{2+} might achieve its regulation through the phosphorylation of a myosin light chain. In support of this notion, fluorescently tagged antibodies to myosin light-chain kinase have been localized in the chromosome-to-pole region of 3T3, PtK_2, and Muntjac cells (Guerriero *et al.*, 1981).

Ca^{2+} fluctuations may also contribute to the control of the extensive membrane transformations that accompany the MA during mitosis. The breakdown of the NE, for example, involves the phosphorylation of the NE-associated lamins followed by an extensive depolymerization of these proteins (Gerace and Blobel, 1980; Ottaviano and Gerace, 1985). In view of the observation that nuclear envelope breakdown in stamen hair cells of *Tradescantia* is retarded or even inhibited by agents that restrict Ca^{2+} entry (Hepler, unpublished), it would not be unreasonable to suggest that the phosphorylation of the lamin is controlled by a Ca^{2+}–calmodulin-dependent kinase.

Finally, it seems prudent to consider the chromosomes themselves as potential targets for Ca^{2+}-dependent processes. Chromosomes constitute a large reservoir of bound Ca^{2+} in the cell (Chandra *et al.*, 1984) and it seems likely that the ion is needed for the process of condensation. The maturation and splitting of sister chromosomes also appears to involve Ca^{2+}, since restriction of ion entry into the cells selectively retards or prevents the transition from metaphase to anaphase (Hepler, 1985).

There are thus a variety of structures and processes within the MA that could be Ca^{2+}-regulated; microtubules continue to occupy our attention as prime targets for Ca^{2+} action but they are not the only ones.

2. Ca^{2+} localization in the MA

The MA, like other cellular compartments, contains a highly discontinuous distribution of Ca^{2+} (Keith *et al.*, 1985b; Wick and Hepler, 1980; Wolniak *et al.*, 1980). The free cytosolic levels are undoubtedly very low (0.1 μM), whereas certain structures and membrane-delimited compartments may have rather high (1.0 mM) concentrations. Examination of total Ca^{2+} in a dividing cell using secondary-ion mass spectroscopy reveals that the chromosomes are the most dominant reservoir of this ion within the MA, probably owing to interaction with the phosphate of DNA (Chandra *et al.*, 1984). Other portions of the MA do not stand out by this method as containing enriched amounts of Ca^{2+}.

While the secondary-ion mass spectroscopic procedure reveals total Ca^{2+} and tends to direct our attention to those regions or structures like the chromosomes that contain massive deposits of the ion, there are other methods that allow us to identify subcomponents of total cellular Ca^{2+}. Pyroantimonate, for example, has been used to identify Ca^{2+} deposits at the electron-microscopic level (Wick and Hepler, 1980). With this method one is able to observe that component of the ion which is exchangeably bound. The results show localization of Ca^{2+} in cisternal spaces of the NE, ER, mitochondria and Golgi membranes (Wick and Hepler, 1980). The PM, cell wall and vacuole, while not associated with the MA, show a strong reaction product; however, almost none is found on the chromosomes. The results thus obtained with secondary ion mass spectroscopy and with pyroantimonate are quite different and may be due to the inability of the latter to react with the tightly, non-exchangeably, bound deposits of Ca^{2+}. From a regulatory point of view, the tightly bound Ca^{2+} on the chromosomes observed by secondary-ion mass spectroscopy may simply not be available during normal cellular or mitotic activities and hence may not contribute to ion fluxes.

Examination of Ca^{2+} in living, dividing cells has been achieved with chlorotetracycline (CTC), a fluorescent chelate probe (Wolniak, *et al.*, 1980, 1981, 1983). When CTC binds Ca^{2+} in a non-polar, i.e. membrane, environment, it fluoresces and therefore has been designated as a probe for membrane-associated Ca^{2+} (Caswell, 1979), although recent studies indicate that it identifies the free unbound Ca^{2+} that is enclosed in a membrane vesicle (Dixon *et al.*, 1984). Endosperm cells of *Haemanthus* labelled with CTC show staining of the MA and the spindle poles (Wolniak *et al.*, 1980) (Fig. 7.7). There is also staining in the non-spindle cytoplasm, some of which is punctate in nature and probably is due to mitochondria and other membrane organelles that contain large stores of Ca^{2+}. Internal membranes stain with CTC, as evidenced by a distinct delineation of NE. Within the MA there is further differentiation of the staining that may relate to Ca^{2+} regulation. Specifically, in *Haemanthus* endosperm cells in metaphase there are regions of stain that localize over

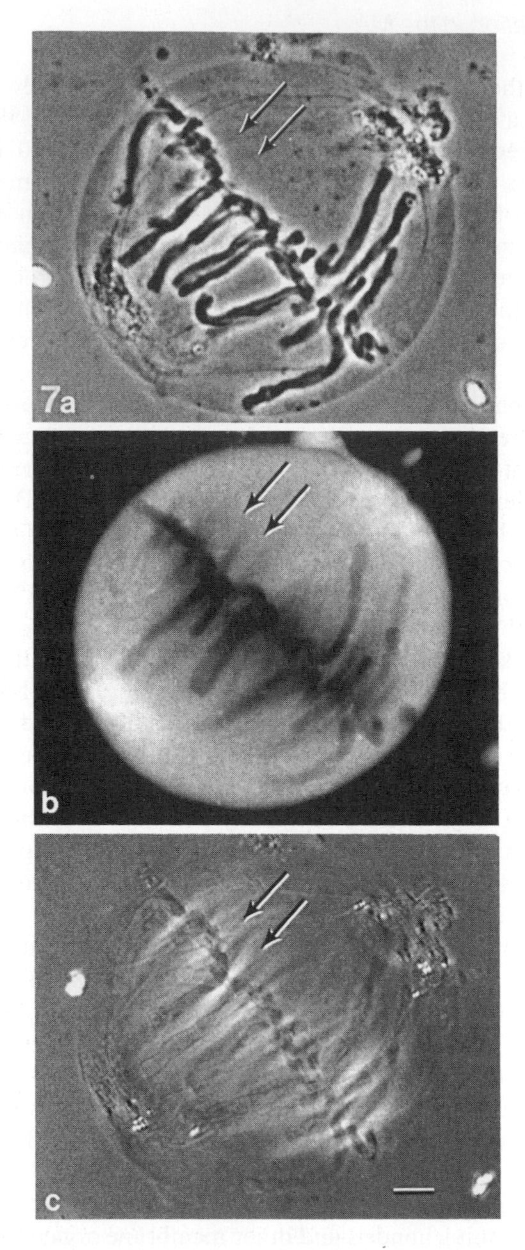

Figure 7.7. A living metaphase endosperm cell of *Haemanthus* observed by (a) phase contrast, (b) fluorescence and (c) polarized-light optics. The cell has been stained with CTC to reveal membrane-associated Ca^{2+} (b). The arrows indicate the position of prominent birefringent kinetochore bundles (c). These same regions are occupied by CTC fluorescence (b). Bar = 10 μm. (From Wolniak *et al.*, 1981.)

kinetochore spindle fibres (Wolniak *et al.*, 1981) (Fig. 7.7). The observation of these "cones" of fluorescence is made all the more interesting in the light of the studies at the electron-microscopic level showing that ER membranes occur throughout the MA (Jackson and Doyle, 1982). Thus, those elements associated with kinetochore bundles would appear to have more available Ca^{2+} than those among the non-kinetochore regions of the spindle.

One of the clearest correlations between membrane distribution and CTC staining has been obtained from studies of the Wolf-spider spermatocyte. Mention has already been made of the ER sheath that bounds each chromosome and its kinetochore fibre. Following CTC incubation these membranes show positive staining, indicating conclusively the close apposition of Ca^{2+} to the kinetochore spindle fibres (Wise and Wolniak, 1984).

While the particular distribution of CTC fluorescence within the MA observed in *Haemanthus* endosperm and Wolf-spider spermatocytes has not been observed widely, the MAs of other cells nevertheless show Ca^{2+}-dependent CTC fluorescence (Schatten *et al.*, 1982; Sisken *et al.*, 1981). In sea-urchin MAs that have been isolated without use of detergents, membrane elements are abundantly present and CTC produces a strong signal (Schatten *et al.*, 1982). Cultured HeLa cells at metaphase have relatively low CTC fluorescence from the central portion of the spindle, with the signal being much stronger at the edge of the MA, probably reflecting the build-up of membranes and organelles in that region (Sisken *et al.*, 1981). CTC has thus proved useful for looking at Ca^{2+} in the membrane domain. While this represents a sequestered or storage form of the ion, it is presumably that compartment from which the fluxes of free Ca^{2+} are derived.

Of all the Ca^{2+} fractions, that which is free in the cytoplasm is the most difficult to observe spatially because its concentration is extremely low. Recently, however, attempts have been made using the fluorescent Ca^{2+} indicators quin-2 (Keith *et al.*, 1985b) and fura-2 (Poenie *et al.*, 1986; Ratan and Shelanski, 1986; Ratan *et al.*, 1986; Tsien and Poenie, 1986) to localize free Ca^{2+} spatially. In *Haemanthus* endosperm cells loaded with quin-2, the images show an elevated fluorescence in the chromosome to pole region of the MA during anaphase (Keith *et al.*, 1985b). Cultured PtK cells stained with fura-2, on the other hand, yield a variety of images including pole labelling (Ratan *et al.*, 1986), a circumferential ring of fluorescence around the MA (Ratan *et al.*, 1986), and a uniform distribution of fluorescence (Poenie *et al.*, 1986; Tsien and Poenie, 1986). There is reason to be cautious about the interpretation of the observations in as much as the dyes, especially in their acetoxymethyl ester form, may load into various membrane compartments (Malgoroli *et al.*, 1987; Williams *et al.*, 1985) and no longer act as indicators of cytoplasmic Ca^{2+}. It is especially troubling in *Haemanthus* that the regions of high quin-2 fluorescence coincide with the region of the MA known to contain an extensive system of ER. Furthermore, since the quin-2 image looks remarkably similar to the CTC fluorescence images, one is additionally concerned about the possibility of the dye compartmentation. These initial studies on the spatial localization of free Ca^{2+} thus need to be extended using different probe molecules and different cell types.

3. *Calmodulin localization in the MA*

Closely related to Ca^{2+} distribution in the MA is the matter of the localization of the calcium-regulatory protein, calmodulin. Several approaches, using both calmodulin antibodies on fixed cells and fluorescently labelled calmodulin injected into living cells, consistently reveal its localization in the MA (Welsh and Sweet, this volume). In living sea-urchin (Hamaguchi and Iwasa, 1980) or PtK cells (Zavortink *et al.*, 1983) fluorescently tagged calmodulin first distributes evenly throughout the cell but quickly accumulates in the MA, where it most heavily stains the spindle pole region. Staining also extends in a fibrous pattern towards the chromosomes in a pattern similar to that of the kinetochore microtubule bundles during metaphase. As cells progress into anaphase, the localization of calmodulin, either in living cells or in those reacted with antibody (DeMey *et al.*, 1980; Means and Dedman, 1980; Vantard *et al.*, 1985; Willingham *et al.*, 1983; Wick *et al.*, 1985), is most evident in the chromosome-to-pole region. However, in late anaphase, staining may appear in the interzone associated with the distal portions of the midbody in animal cells (Means and Dedman, 1980; Willingham *et al.*, 1983) and with the phragmoplast in plant cells (Wick *et al.*, 1985). The phragmoplast association is not universal, since observations on *Haemanthus* endosperm cells in late anaphase/early telophase fail to reveal anticalmodulin staining despite the fact that an extensive array of microtubules is present (Vantard *et al.*, 1985).

Studies at the ultrastructural level provide more detail about the association of anticalmodulin to the spindle elements. In PtK cells a distinct staining of the chromosome-to-pole spindle fibres has been noted in which poleward ends of the microtubules are heavily labelled (DeMey *et al.*, 1980). Staining has also been observed in the MA of *Haemanthus* endosperm in which some microtubules are decorated with gold particles but others are not (Vantard *et al.*, 1985). The contention is made that the staining is restricted to the kinetochore microtubules (Welsh and Sweet, this volume), but additional work, e.g. analysis of serial sections, will be required to prove the point. In a variety of cultured animal cells, staining with ferritin-labelled anticalmodulin is observed in the chromosome-to-pole region, but unfortunately the microtubules are not preserved, making it impossible to determine what structures are being labelled (Willingham *et al.*, 1983).

In general, the staining at the electron-microscopic level agrees well with that at the light-microscopic level. These observations have led to the conclusion that microtubules are the target for calmodulin antibodies. Despite the apparent consistency in these results, there are certain observations and considerations that cause us to question some of the conclusions that have been drawn. In studies on living sea-urchin embryos that have been injected with labelled calmodulin, it is noted that the fluorescence distribution differs distinctly from the spindle fibre pattern observed using polarized-light microscopy (Hamaguchi and Iwasa, 1980). Calmodulin is heavily accumulated in the spindle pole, while spindle birefringence is most intense close to the chromosomes. The studies on *Haemanthus* endosperm also raise questions since in that study it is shown that ". . . CTC fluorescence clearly coincides with the immunocytochemical localization of calmodulin" (Vantard *et al.*, 1985). Since CTC stains membranes containing associated Ca^{2+} and not microtubules, it seems reasonable to suggest that some of the calmodulin might be localized on the spindle membrane system.

Observations from the electron-microscopic studies also indicate that the microtubules and calmodulin do not always co-localize (Willingham *et al.*, 1983) and cause us to consider additional targets for calmodulin staining. The detection of calmodulin localized on smooth ER lying parallel to microtubules in the half-spindle (Means and Dedman, 1980) provides direct evidence for membrane staining. The strict attention given to the role of calmodulin in regulating microtubules has prevented other interpretations of the results from emerging. Calmodulin, for example, might be regulating a membrane-associated CaATPase (Means and Dedman, 1980) and might indirectly control microtubule assembly by regulating the concentration of free Ca^{2+}. Future work on this problem should strive to keep in mind the complexity of the MA structure. Techniques need to be developed that allow one to localize antibodies to calmodulin at the electron-microscopic level in cells in which the microtubules, membranes and all associated structures have been preserved.

4. *Ca^{2+} transport in the MA*

Considerable evidence indicates that the endomembrane systems in a wide variety of cell types possess CaATPase activity and are able to remove Ca^{2+} with high affinity from the cytosol. Entirely consistent with these results is the observation that the endomembranes of the MA contain CaATPase (Petzelt, 1979). When Ca^{2+} is microinjected into the MA, the concentration of the ion is rapidly lowered to its resting level and microtubules that had momentarily depolymerized quickly grow back (Kiehart, 1981; Izant, 1983). Direct confirmation of the Ca^{2+}-sequestering capacity of mitotic membranes has been made on mechanically isolated MAs of sea urchins; Ca^{2+} uptake is enhanced 180% by ATP (Silver *et al.*, 1980). A CaATPase has been identified in membrane fractions from other cell types as well as sea urchins and attempts have been made to correlate its activity with the cell cycle (Petzelt, 1979; Petzelt and Wulfroth, 1984). In sea urchins a peak in enzyme activity is reported at fertilization and subsequently at successive mitoses. Studies on electrically permeabilized sea-urchin embryos further suggest that the peak activity may correlate closely with telophase, with the possibility of an earlier peak at prophase (Suprynowicz and Mazia, 1985). However, there are some problems with reproducibility in these studies, since prophase activity may be very low in one experiment but very high in another. Considerably more work is needed on this important point to establish the regularity in fluctuations in activity during the mitotic cycle. Also, it becomes important to identify these fluxes in activity in cells other than sea-urchin embryos.

More-detailed work on the molecular nature of the CaATPase has involved the production and use of antibodies (Hafner and Petzelt, 1987; Petzelt and Hafner, 1986; Silver, 1986). Silver (1986) has produced an antibody to the 105 kDa CaATPase of skeletal and smooth muscle, which on immunoblotting identifies only one protein from an extract of echinoderm MAs. The antibody inhibits calcium uptake in both sarcoplasmic reticulum and MA vesicular membranes. When injected into sand-dollar embryos it promotes rapid loss of spindle birefringence, suggesting an increase in [Ca^{2+}], and also inhibits mitosis. Petzelt and Hafner (1986) have instead prepared a series of monoclonal antibodies against HeLa cell membrane proteins. Of three that inhibited Ca^{2+} uptake in membrane vesicles, one antibody in particular attracted further attention because of its ability to react with the MA *in situ* (Petzelt and Hafner, 1986) and inhibit mitosis in sea urchins, while causing an increase in intracellular [Ca^{2+}] (Hafner and Petzelt, 1987). The antibody of Petzelt and Hafner (1986) differs from that examined by Silver (1986) by reacting only with a 46 kDa protein. The relationship between the 46 kDa and 105 kDa proteins remains unresolved, but it is interesting that both cause Ca^{2+} release and inhibit mitosis.

Localization of immunofluorescent antibodies in MAs of different cells reveals some novel patterns, not all of which are expected on the basis of our knowledge of membrane distribution. Earlier work by Petzelt (1984), using an antibody prepared against a CaATPase of Ehrlich ascite tumour cells, shows a punctate pattern of fluorescence in PtK cells with no staining of the nuclear envelope or the MA during mitosis. Rather, the staining is largely confined to bright vesicles, sometimes lying alongside the MA but also throughout the non-spindle cytoplasm. Since the enzyme activity from which the antibody is made is not inhibited by NaN_3 or Ruthenium red, the assertion is made that the staining cannot be due to mitochondria but is rather due to the endomembrane system (Petzelt, 1984). Nevertheless, there are large disparities between these localization patterns and what is known about the spindle-associated membrane system. The failure to stain the NE, which is attached to and derived from the ER, provides evidence that the ER is not being stained. Also, the tubular ER associated with the MA, of which PtK cells possess a striking example, has not reacted with the antibody. More recently Petzelt and Hafner (1986), examining the localization of the monoclonal antibody to the 46 kDa protein in sea urchins, again find a punctate pattern of fluorescence around the nucleus without any apparent accumulation in the MA during mitosis. In anaphase, punctate staining concentrates in the asters but does not follow the aster rays, indicating that the microtubules have not been stained. They further report that following mitosis the perinuclear staining disappears in the transition from telophase to interphase and reappears at prophase. Silver (1986) has examined the localization of the 105 kDa antibody in isolated MAs from echinoderm embryos and compared the staining pattern with that obtained by CTC. They both stain the asters but react only faintly with components in the MA, with the exception of the antibody, which intensely labels a local region immediately around the chromosomes. The argument is made that this local staining is due to ER associated with the chromosomes, but electron-microscopic examination of sea urchins shows membrane elements throughout the MA region and not just adjacent to the chromosomes (Hepler and Wolniak, 1984). The localization studies must be carried out at the electron-microscopic level to establish more definitively the identity of the inclusions that produce the positive reaction with the CaATPase antibodies.

5. *[Ca^{2+}] changes during mitosis*

The foregoing results provide evidence that membranes of the MA contain Ca^{2+} and that they possess the enzymology to regulate ion levels. What evidence supports the idea that the [Ca^{2+}] changes during mitosis? Changes in membrane-associated Ca^{2+} have been detected in dividing *Haemanthus* endosperm through quantitative measurement of CTC fluorescence (Wolniak *et al.*, 1983). The results show that the fluorescence in the MA declines 5–10 min prior to the onset of anaphase. That the fluorescence reduction of CTC is not due simply to a change in membrane distribution is shown from analysis of cells labelled with the membrane marker *N*-phenylnaphthylamine, in which the fluorescence levels remain constant throughout metaphase/anaphase (Wolniak *et al.*, 1983). By comparison with studies on muscle sarcoplasmic reticulum, in which a reduction in CTC fluorescence has been shown to correspond to a release of Ca^{2+} from the membrane compartment (Caswell and Brandt, 1981), it is attractive to postulate for the MA that the reduction in CTC fluorescence similarly indicates that Ca^{2+} has been released from the membrane system and that the free cytoplasmic [Ca^{2+}] has risen.

Additional evidence that marked ion changes from the spindle endomembrane system occur at the metaphase/anaphase transition is provided by the results with two permeant voltage-sensitive dyes, dipentyldioxacarbocyanine(+) and anilinonaphthalene sulphonate(−), that both show marked increases in fluorescence beginning precisely at the onset of anaphase (Wolniak *et al.*, 1983). Besides their temporal association with the events of mitosis, the fluorescence changes are observed only in the MA and not in the non-spindle cytoplasm. It is unknown what these two probes are reporting on in the MA, but they too could be sensitive to changes in Ca^{2+} in the environment of the membranes. Because of their different charges, it seems likely that dipentyldioxacarbocyanine and anilinonaphthalene sulphonate are either partitioning to different compartments within the endomembrane domain or are responding to different ion species. Despite our lack of understanding about dye-binding properties, the fact that the fluorescence changes are coupled tightly both temporally and spatially to the events of mitosis indicates that substantial membrane-associated ion redistributions have occurred that may be important in regulating the metaphase/anaphase transition (Wolniak *et al.*, 1983).

Evidence that free Ca^{2+} increases at the onset of anaphase is provided indirectly by different studies in which the $[Ca^{2+}]$ has been controlled both internally (Izant, 1983) and externally (Hepler, 1985; Wolniak and Bart, 1985a,b). By reducing $[Ca^{2+}]$, the results generally show that the initiation of anaphase is slowed or even blocked. Studies on stamen hair cells of *Tradescantia* indicate that the Ca^{2+} required for the transition from metaphase to anaphase initially comes from the external space, and may be regulated by voltage-dependent Ca^{2+} channels on the PM (Hepler, 1985; Wolniak and Bart, 1985b). Internal membranes may also participate in this process, but the mechanism remains to be elucidated. Realizing that the PM may be a key regulator of Ca^{2+} in mitotic events, one is reminded of its profound structural changes observed in PtK cells (Sanger *et al.*, 1984). Note is simply made that microvillus formation occurs over the chromosome-to-pole region at a time when Ca^{2+} may increase in that portion of the underlying MA.

The most significant results derive from several recent studies in which attempts have been made to record directly and measure fluctuations in free Ca^{2+} in single dividing cells as they progress through mitosis. Owing to the difficulty of measuring changes in $[Ca^{2+}]$ at the submicromolar level in individual cells, it is understandable that this obvious question has resisted clarification. Even now, as the new results emerge, there is a disturbing lack of consistency in the published data that makes it difficult to arrive at firm conclusions. Nevertheless, some interesting and exciting observations have been made that deserve attention.

In sea-urchin embryos that have been loaded with the fluorescent indicator fura-2, Poenie *et al.* (1985) report that in a small percentage of cells there are several peaks of $[Ca^{2+}]$ increase during the first cell cycle. Beyond the large $[Ca^{2+}]$ spike at fertilization, smaller but definite increases occur that corres-

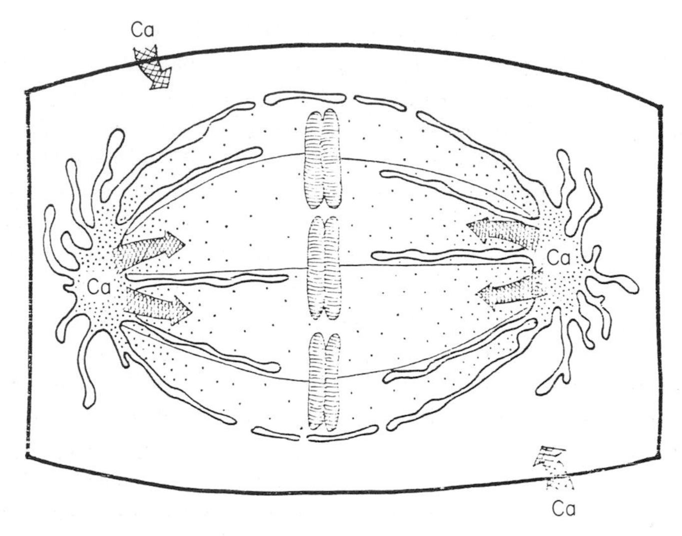

Figure 7.8. A diagrammatic representation of a cell in late metaphase. The suggestion is made that Ca^{2+} initially enters from the outside through voltage-gated channels. This influx may directly stimulate anaphase onset or it may induce a further release of Ca^{2+} from the spindle-associated membrane system, which then initiates anaphase. (From Hepler, 1985.)

pond approximately to pronuclear migration, streak stage, nuclear envelope breakdown, metaphase/anaphase, and cleavage. The problem with these observations is the fact that, with the exception of the fertilization spike, the fluctuations during the cell cycle usually do not occur, yet the cells divide normally. More recently, Steinhardt and Alderton (1988) have analysed the prophase/prometaphase transition more completely and find evidence for a Ca^{2+} increase that precedes nuclear envelope breakdown by 1–3 min. The ability to block this event by buffering the internal calcium and preventing its rise provides further evidence that the ion flux serves a regulatory function (Steinhardt and Alderton, 1988; Twigg *et al.*, 1988).

Quite perplexing, however, are the several published accounts of Ca^{2+} changes in cultured PtK cells. Keith *et al.* (1985a) originally reported that $[Ca^{2+}]$ declined from metaphase to anaphase in quin-2 loaded PtK_2 cells. When the study was redone using fura-2 as the indicator, it was reported that $[Ca^{2+}]$ became elevated at anaphase, although no kinetic data were provided (Ratan *et al.*, 1986). Subsequently, Poenie *et al.* (1986), examining PtK_1 cells loaded with fura-2, noted a brief 20-second spike in the $[Ca^{2+}]$ near the metaphase/anaphase transition. However, in a review on this topic, Tsien and Poenie (1986) showed a $[Ca^{2+}]$ elevation that lasted 2 min or the time of anaphase and was not restricted to a window at the metaphase/anaphase transition. Most recently, Ratan *et al.* (1988) report the occurrence of [Ca^{2+}] spikes except that they are not correlated temporally with the metaphase/anaphase transition. They also show lower-magnitude, long-term increases in [Ca^{2+}] that begin during late metaphase and extend through anaphase. It should be apparent to the reader that there is remarkable disagreement in the published data; among the several reports on PtK cells there are as many different results. Thus, the [Ca^{2+}] declines at anaphase (Keith *et al.*, 1985a), the [Ca^{2+}] elevates at anaphase (Ratan *et al.*, 1986; Tsien and Poenie, 1986), the [Ca^{2+}] spikes at the metaphase/anaphase transition (Poenie *et al.*, 1986), the [Ca^{2+}] spikes but not at the metaphase/anaphase transition (Ratan *et al.*, 1988), and finally the [Ca^{2+}] increases slowly during metaphase (Ratan *et al.*, 1988). Again the problem of dye compartmentation becomes an issue. It is quite possible that the variations observed relate more to capricious dye behaviour than to biological variations in dividing cells.

We have recently attempted to measure free [Ca^{2+}] changes in dividing stamen hair cells of the flowering plant, *Tradescantia*, using the metallochromic absorbance indicator dye arsenazo III (Hepler and Callaham, 1987). The main disadvantage with this indicator is the necessity of injecting relatively high concentrations of the dye (300–500 μM), at which point buffering of the internal Ca^{2+} becomes evident and mitotic progression is slowed relative to controls. Also, we have not been able to detect changes in [Ca^{2+}] below 0.3 μM. Arsenazo III, however, has the favourable property of remaining in the cytoplasm as a reliable Ca^{2+} indicator for hours and not loading into the vacuole as does fura-2. Furthermore, even though the events of mitosis may be slowed, the dye does not stop the process; cells in metaphase invariably progress through to cytokinesis. The results show that the [Ca^{2+}] remains at a basal level during metaphase. However, following the metaphase/anaphase transition, the [Ca^{2+}] begins to increase and ramps upward during the next 10–15 min while the chromosomes move to the poles. Coincident with the arrival of the chromosomes at the poles, the [Ca^{2+}] reaches a high plateau (1 μM) and then declines, returning to the basal level at about the time the cell plate first appears (Hepler and Callaham, 1987). Our observations suggest that the rise in [Ca^{2+}] is the consequence rather than the cause of the metaphase/anaphase transition. The close temporal correlation between the [Ca^{2+}] increase and the movement of chromosomes is consistent with the contention that the ion regulates anaphase rate.

While there is still considerable uncertainty about the timing, magnitude, and location of the Ca^{2+} fluxes during mitosis, it seems likely that they at least occur. Clearly, though, much more work is needed to establish a general pattern, if one exists. Regarding the generation of Ca^{2+} fluxes, it will be important to determine from which membrane compartment they are derived, e.g. the PM or the ER, to decipher the signal that causes them to occur in the first place, and finally to establish the targets for Ca^{2+} action. Answers to these questions may provide important new information about the regulation of mitosis.

B. Membranes as structural components of the MA

In addition to their presumed role in the regulation of Ca^{2+}, the endomembranes, especially the ER, may contribute to the structure of the mitotic cytoskeleton (Hepler and Wolniak, 1984). The presence in many cell types of membranes closely appressed to the spindle microtubules allows for the possibility of direct connections between these two elements (Fig. 7.9). A structural interaction between membranes and microtubules at the very least might impart stability to both elements but in addition could provide the capability for the development of shearing forces (Miller and Lasek, 1985). The idea is explored in this section that membranes in the MA may act as a structural scaffolding to which microtubules bind. A point that is often overlooked in the consideration of the structural properties of the MA is the requirement that the chromosomal microtubules be anchored at the spindle pole with sufficient strength that during anaphase the chromosome moves to the pole rather than the distal end of the microtubule moving to the chromosome.

Considerable attention has been given to the observation that microtubules in the MA may be cross-bridged to one another (Hepler *et al.*, 1970). In the central spindle of diatoms, for example, the packing of microtubules is highly

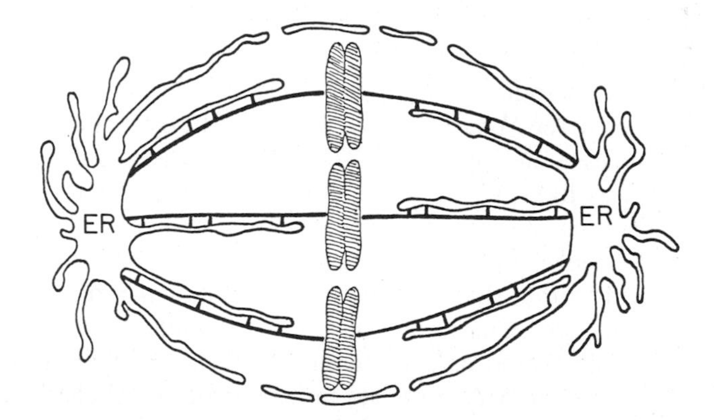

Figure 7.9. A diagrammatic representation of a cell in late metaphase as shown in Fig. 7.8. Here the possible structural association between membranes and kinetochore microtubules is emphasized by showing cross-bridges between these elements. (From Hepler and Wolniak, 1984.)

ordered, presumably owing to the extensive cross-linking that is evident in electron micrographs of these MAs (McDonald *et al.*, 1977; McDonald, this volume). Cross-bridges, similar in structure to those observed between adjacent microtubules, also occur between microtubules and nearby membrane elements in the MA of plant and animal cells (Fux, 1974; Hepler *et al.*, 1970). Much less attention has been given to these microtubule–membrane links, but it seems possible that they too could contribute to stability and force generation in the MA.

Recent biochemical investigations on membranes have provided important and timely information on the presence of cytoskeletal proteins that may help us understand how fibrous structures like microtubules join to membranes. A portion of the cellular tubulin, for example, is now recognized to be an integral membrane protein (Rubin, 1985; Stephens, 1985) and it seems possible that linkages between tubulin dimers in the membranes and dimers on the microtubule lattice could be effected by microtubule-associated proteins (MAPs). An additional noteworthy development has been the discovery that proteins similar to spectrin and ankyrin of the erythrocyte cell membrane occur in a wide variety of cell types (Bennett, 1985). The ability of these proteins to bind microtubules and MAPs also make them prime candidates for linking microtubules to membranes. Ankyrin, for example, has been localized in the MA of HeLa cells by immunofluorescence microscopy and reveals a pattern of staining that closely resembles the pattern of membranes (Bennett and Davis, 1981). There are thus a variety of building blocks that could participate in linking spindle membranes and microtubules. It is premature to specify how the interactions occur; the important point is to realize the likelihood that they do occur. If spectrin/ankyrin are, indeed, components of mitotic membranes they might also impart a structural rigidity to the membranes that could permit the latter to withstand the imposition of forces produced in the MA.

In addition to the stabilizing function that a membrane–microtubule interaction might impart to the MA, the association could be dynamic and contribute to transport processes during mitosis. The cross-bridges might be mechanochemical ATPases like dynein (Pratt, 1984) or kinesin (Vale *et al.*, 1986) capable of generating shearing forces between membranes and microtubules. Considerable evidence supports the view that dynein is present in the MA (Pratt, 1984) and further that it participates in pole separation (anaphase B) (Candé, 1982; see also Chapters 5 and 9). Although the case is less well established, kinesin, based on immunofluorescence labelling, appears to be present in the asters and MA of the sea-urchin embryo (Scholey *et al.*, 1985). Kinesin in axoplasmic preparations has the interesting property of moving particles toward the plus end of microtubules or moving the microtubule itself towards its minus end (Vale *et al.*, 1986). In contrast, the microtubule-associated protein MAP 1C, which shares many properties with dynein, moves microtubules toward their plus ends (Paschal and Vallee, 1987). At the least it seems likely that these motors, and possibly others that might emerge, are responsible for the myriad of directed vesicular and organellar transport (Schliwa, 1984), some of which occurs in the MA (Bajer and Mole-Bajer, 1982). Indeed, it is attractive to imagine that kinesin, attached to kinetochore microtubules, would be ideally poised to draw membrane components into the MA, pulling them towards the plus end of the microtubules.

If a membrane–microtubule shear-force-generating system exists in the MA, it is reasonable to ask whether it could contribute to chromosome-to-pole transport during anaphase. Certainly in those organisms in which the endomembranes interdigitate along the kinetochore fibres the positional relationship of these two components is ideal for force generation. It is conceivable, therefore, that MTs, through the action of mechanochemical cross-bridges, might crawl along membrane elements and thereby pull their attached chromosome to the pole (Fig. 7.10). Such a mechanism would be consistent with the observed poleward migration of UV-microbeam regions of reduced birefringence (Forer, 1966; Chapter 4) and with the increased tension derived from longer spindle fibres (Hays *et al.*, 1982). At the moment, however, the concept of a motor distributed along the length of the kinetochore fibres is inconsistent with the recent results showing that regions of reduced fluorescence following laser-induced photobleaching remain fixed in position during anaphase A (Gorbsky *et al.*, 1987) and further that microtubule subunits are removed at the kinetochore (Koshland *et al.*, 1988). These current views thus place the anaphase A motor in the kinetochore itself. Nevertheless, it may be premature to discount totally the possibility that a membrane–microtubule mechanochemical cross-bridge unit could contribute to the poleward movement of chromosomes during anaphase. But even if membranes do not participate in force generation for anaphase A, it is again important to emphasize the necessity of an anchor for the kinetochore microtubules that is sufficient to hold the end of the microtubule in place while the chromosome

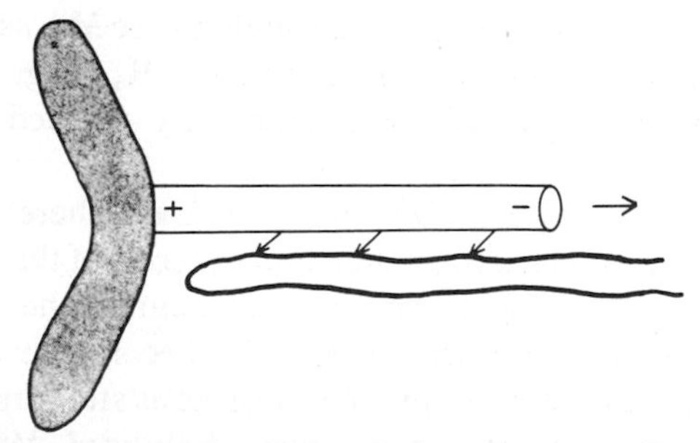

Figure 7.10. Chromosome-to-pole motion during anaphase. The diagram shows a kinetochore microtubule that is cross-bridged to an ER element. Through mechanochemical activity the cross-bridges cause the microtubule to translocate with its (–) end forward. This motion, together with a concurrent depolymerization of the microtubule, carries the attached chromosome to the pole.

moves to the pole. The spindle-associated endomembrane system, at the least, could be this essential anchor.

V. Conclusions

Although they have received little attention, there is an increasing awakening to the existence of membranes as major components of the MA. We are limited in what we see by how we prepare our specimens. The modern concept of the spindle fibre benefited enormously from the introduction of techniques that made it possible to preserve microtubules at the electron-microscopic level, and more recently to observe cytoskeletal spindle components with labelled antibodies. A similar renaissance awaits the membrane component of the MA. It becomes imperative to study the MA without the use of detergents or other agents that destroy or modify the membrane components. That membranes may function as regulators of Ca^{2+} is commonly accepted. In addition, membranes may be structural elements in the MA to which spindle microtubules attach and become stabilized. The association may also participate in different transport processes in the MA. The difficult but important problem of determining the function of these membranes is largely ahead of us. Elucidating their role may help enormously in our quest to decipher the mechanism of mitosis.

VI. Summary

Membranes occur in the MA of most cells. These often consist of smooth, tubular elements of ER, which surround and invade the MA. In some instances the ER establishes close structural association specifically with the kinetochore microtubules. Golgi vesicles also occur throughout the MA usually in a random pattern. During division, membranes including ER, Golgi and PM undergo changes that may be structurally and temporally coupled to the events of mitosis.

Membranes may control the [Ca^{2+}] in the MA and thereby regulate one or more processes involved in the formation and function of the MA. Sequestered Ca^{2+} has been identified in membrane compartments in the MA and release of these stores may cause the internal free [Ca^{2+}] to become elevated at key points during the mitotic cycle. Membranes may also act as structural components to which spindle microtubules bind and become stabilized. Membranes may be the anchor at the spindle pole and thus serve a crucial function in the movement of chromosomes during anaphase.

Acknowledgements

I thank my colleagues at the University of Massachusetts for many helpful criticisms and comments. K. McDonald, Department of Molecular and Developmental Biology, University of Colorado, Boulder, kindly provided Figure 7.4. For assistance in the preparation of the manuscript, I thank P. Bonsignore, S. Lancelle and K. Nelson. This study has been supported by grants from the NSF (DCB 85-02723 and DCB 88-01750).

REFERENCES

Bajer, A. S. and Mole-Bajer, J. (1982). *Cold Spring Harbor Symp. Quant. Biol.* **46**, 263–283.

Bakhuizen, R., van Spronsen, P. C., Sluiman-den Hertog, F. A. J., Venverloo, C. J. and Goosen-de Roo, L. (1985). *Protoplasma* **128**, 43–51.

Bennett, V. (1985). *Annu. Rev. Biochem.* **54**, 273–304.

Bennett, V. and Davis, J. (1981). *Cold Spring Harbor Symp. Quant. Biol.* **46**, 647–657.

Candé, W. Z. (1982). *Nature* **295**, 700–701.

Caswell, A. H. (1979). *Int. Rev. Cytol.* **56**, 145–181.

Caswell, A. H. and Brandt, N. R. (1981). *J. Membrane Biol.* **58**, 21–33.

Chandra, S., Harris, W. C., Jr. and Morrison, G. H. (1984). *J. Histochem. Cytochem.* **32**, 1124–1230.

Church, K. and Lin, H.-P. (1982). *J. Cell Biol.* **93**, 365–373.

Clayton, L., Black, C. M. and Lloyd, C. W. (1985). *J. Cell Biol.* **101**, 319–324.

DeMey, J., Moeremans, M., Geuens, G., Nuydens, R., Van Belle, H. and DeBrabander, M. (1980). In *Microtubules and Microtubule Inhibitors 1980* (ed. M. DeBrabander and J. DeMey), p. 227. Elsevier/North-Holland, Amsterdam.

Dixon, D., Brandt, N. and Haynes, D. H. (1984). *J. Biol. Chem.* **259**, 13 737–13 741.

Forbes, M. S., Plantholt, B. A. and Sperelakis, N. (1977). *J. Ultrastruct. Res.* **60**, 306–327.

Forer, A. (1966). *Chromosoma* **19**, 44–98.

Forer, A. (1985). *Can. J. Biochem. Cell Biol.* **63**, 585–598.

Fuge, H. (1977). *Int. Rev. Cytol. Suppl.* **6**, 1–58.

Fux, T. (1974). *Chromosoma* **49**, 99–112.

Gerace, L. and Blobel, G. (1980). *Cell* **19**, 277–287.

Gorbsky, G. J., Sammak, P. J. and Borisy, G. G. (1987). *J. Cell Biol.* **104**, 9–18.

Guerriero, V., Rowley, D. R. and Means, A. R. (1981). *Cell* **27**, 449–458.

Hafner, M. and Petzelt, C. (1987). *Nature* **330**, 264–266.

Hamaguchi, Y. and Iwasa, F. (1980). *Biomed. Res.* **1**, 502–509.

Harris, P. (1978). In *Monographs on Cell Biology* (ed. J. R. Jeter, I. L. Cameron, G. M. Padilla and A. M. Zimmerman), pp. 75–104. Academic Press, New York.

Hays, T., Wise, D. and Salmon, E. D. (1982). *J. Cell Biol.* **93**, 374–382.

Heath, I. B. (1980). *Int. Rev. Cytol.* **64**, 1–80.

Hepler, P. K. (1976). *J. Cell Sci.* **21**, 361–390.

Hepler, P. K. (1980). *J. Cell Biol.* **86**, 490–499.

Hepler, P. K. (1982). *Protoplasma* **111**, 121–133.

Hepler, P. K. (1985). *J. Cell Biol.* **100**, 1363–1368.

Hepler, P. K. and Callaham, D. A. (1987). *J. Cell Biol.* **105**, 2137–2143.
Hepler, P. K. and Wolniak, S. M. (1983). *Modern Cell Biol.* **2**, 93–112.
Hepler, P. K. and Wolniak, S. M. (1984). *Int. Rev. Cytol.* **90**, 169–238.
Hepler, P. K., McIntosh, J. R. and Cleland, S. (1970). *J. Cell Biol.* **45**, 438–444.
Hepler, P. K., Wick, S. M. and Wolniak, S. M. (1981). In *International Cell Biology 1980–1981* (ed. H. G. Schweiger), pp. 673–686. Springer-Verlag, Berlin and New York.
Inoué, S. and Bajer, A. (1961). *Chromosoma* **12**, 48–63.
Izant, J. G. (1983). *Chromosoma* **88**, 1–10.
Jackson, W. T. and Doyle, B. G. (1982). *J. Cell Biol.* **94**, 637–643.
Jorgensen, A. O., Shen, A. C.-Y. and Campbell, K. P. (1985). *J. Cell Biol.* **101**, 257–268.
Keith, C. H., Maxfield, F. R. and Shelanski, M. L. (1985a). *Proc. Natl. Acad. Sci. USA* **82**, 800–804.
Keith, C. H., Ratan, R., Maxfield, F. R., Bajer, A. and Shelanski, M. L. (1985b). *Nature* **316**, 848–850.
Kiehart, D. P. (1981). *J. Cell Biol.* **88**, 604–617.
Koshland, D. E., Mitchison, T. J. and Kirschner, M. (1988). *Nature* **331**, 499–504.
Kubai, D. F. (1978). In *Nuclear Division in the Fungi* (ed. I. B. Heath), pp. 177–229. Academic Press, New York.
Malgoroli, A., Milani, D., Meldolesi, J. and Pozzan, T. (1987). *J. Cell Biol.* **105**, 2145–2155.
McDonald, K. (1984). *J. Ultrastruct. Res.* **866**, 107–118.
McDonald, K., Pickett-Heaps, J. D., McIntosh, J. R. and Tippit, D. H. (1977). *J. Cell Biol.* **74**, 377–388.
Means, A. R. and Dedman, J. R. (1980). *Nature* **285**, 73–77.
Miller, R. H. and Lasek, R. J. (1985). *J. Cell Biol.* **101**, 2181–2193.
Mitchison, T. J. and Kirschner, M. W. (1985). *J. Cell Biol.* **101**, 766–777.
Moskalewski, S., Thyberg, J., Hinek, A. and Friberg, U. (1977). *Tissue Cell* **9**, 185–196.
Motzko, D. and Ruthmann, A. (1984). *Eur. J. Cell Biol.* **33**, 205–216.
Nadezhdina, E. S., Fais, D. and Chentsov, Y. S. (1979). *Eur. J. Cell Biol.* **19**, 109–115.
Nicklas, R. B., Brinkley, B. R., Pepper, D. A., Kubai, D. F. and Rickards, G. K. (1979). *J. Cell Sci.* **35**, 87–104.
Ottaviano, Y. and Gerace, L. (1985). *J. Biol. Chem.* **260**, 624–632.
Paschal, B. M., and Vallee, R. B. (1987). *Nature* **330**, 181–184.
Paweletz, N. (1974). *Cytobiologie* **9**, 368–390.
Paweletz, N. (1981). *Cell Biol. Int. Rep.* **5**, 323–336.
Paweletz, N. and Fehst, M. (1984). *Cell Biol. Int. Rep.* **8**, 117–125.
Paweletz, N. and Schroeter, D. (1974). *Cytobiologie* **8**, 238–246.
Petzelt, C. (1979). *Int. Rev. Cytol.* **60**, 53–92.
Petzelt, C. (1984). *Eur. J. Cell Biol.* **33**, 55–59.
Petzelt, C. and Hafner, M. (1986). *Proc. Natl. Acad. Sci. USA* **83**, 1719–1722.
Petzelt, C. and Wulfroth, P. (1984). *Cell Biol. Int. Rep.* **8**, 823–840.
Poenie, M., Alderton, J., Tsien, R. Y. and Steinhardt, R. A. (1985). *Nature* **315**, 147–149.
Poenie, M., Alderton, J., Steinhardt, R. A. and Tsien, R. Y. (1986). *Science* **233**, 886–889.
Porter, K. R., Prescott, D. and Frye, J. (1973). *J. Cell Biol.* **57**, 815–836.

Pratt, M. M. (1984). *Int. Rev. Cytol.* **87**, 83–105.
Pratt, M. M., Otter, T. and Salmon, E. D. (1980). *J. Cell Biol.* **86**, 738–745.
Ratan, R. R. and Shelanski, M. L. (1986). *TIBS* **11**, 456–459.
Ratan, R. R., Shelanski, M. L. and Maxfield, F. R. (1986). *Proc. Natl. Acad. Sci. USA* **83**, 5136–5140.
Ratan, R. R., Maxfield, F. R. and Shelanski, M. L. (1988). *J. Cell Biol.*, in press.
Rieder, C. L. and Nowogrodzki, R. (1983). *J. Cell Biol.* **97**, 1144–1155.
Rubin, R. W. (1985). *BioEssays* **1**(4), 157–160.
Ryan, K. G. (1984). *Protoplasma* **122**, 56–67.
Sanger, J. M., Reingold, A. M. and Sanger, J. W. (1984). *Cell Tissue Res.* **237**, 409–417.
Schatten, G., Schatten, H. and Simerly, C. (1982). *Cell Biol. Int. Rep.* **6**, 717–724.
Schliwa, M. (1984). *Cell Muscle Motil.* **5**, 1–82.
Scholey, J. M., Porter, M. E., Grissom, P. M. and McIntosh, J. R. (1985). *Nature* **318**, 483–486.
Schroeder, T. E. (1978). *Develop. Biol.* **64**, 342–346.
Schroeter, D., Ehemann, V. and Paweletz, N. (1985). *Biol. Cell* **53**, 155–164.
Silver, R. B. (1986). *Proc. Natl. Acad. Sci. USA* **83**, 4302–4306.
Silver, R. B., Cole, R. D. and Candé, W. Z. (1980). *Cell* **19**, 505–516.
Sisken, J. E., Awesu, J. E. and Forer, A. (1981). *J. Cell Biol.* **91**, 315a.
Steinhardt, R. A. and Alderton, J. (1988). *Nature* **332**, 364–366.
Stephens, R. E. (1985). *J. Cell Biol.* **100**, 1082–1090.
Stossel, T. P., Chaponnier, C., Ezzell, R. M., Hartwig, J. H., Janmey, P. A., Kwiatkowski, D. J., Lind, S. E., Smith, D. B., Southwick, F. S., Yin, H. L. and Zaner, K. S. (1985). *Annu. Rev. Cell Biol.* **1**, 353–402.
Suprynowicz, F. A. and Mazia, D. (1985). *Proc. Natl. Acad. Sci. USA* **82**, 2389–2393.
Thyberg, J. and Moskalewski, S. (1985). *Exp. Cell Res.* **159**, 1–16.
Tsien, R. Y. and Poenie, M. (1986). *TIBS* **11**, 450–455.
Twigg, J., Patel, R. and Whitaker, M. (1988). *Nature* **332**, 366–369.
Vale, R. D., Scholey, J. M. and Sheetz, M. P. (1986). *TIBS* **11**, 464–468.
Vantard, M., Lambert, A.-M., DeMey, J., Picquot, P. and Van Eldik, L. J. (1985). *J. Cell Biol.* **101**, 488–499.
Weisenberg, R. C. (1972). *J. Cell Biol.* **117**, 1104–1105.
Wick, S. M. and Duniec, J. (1984). *Protoplasma* **122**, 45–55.
Wick, S. M. and Hepler, P. K. (1980). *J. Cell Biol.* **86**, 500–513.
Wick, S. M., Muto, S. and Duniec, J. (1985). *Protoplasma* **126**, 198–206.
Williams, D. A., Fogarty, K. E., Tsien, R. Y. and Fay, F. S. (1985). *Nature* **318**, 558–561.
Willingham, M. C., Wehland, J., Klee, C. B., Richert, N. D., Rutherford, A. V. and Pastan, I. H. (1983). *J. Histochem. Cytochem.* **31**, 445–461.
Wise, D. (1984). *Chromosoma* **90**, 50–56.
Wise, D. and Wolniak, S. M. (1984). *Chromosoma* **90**, 156–161.
Wolniak, S. M. and Bart, K. (1985a). *Eur. J. Cell Biol.* **39**, 33–40.
Wolniak, S. M. and Bart, K. (1985b). *Eur. J. Cell Biol.* **39**, 273–277.
Wolniak, S. M., Hepler, P. K. and Jackson, W. T. (1980). *J. Cell Biol.* **87**, 23–32.
Wolniak, S. M., Hepler, P. K. and Jackson, W. T. (1981). *Eur. J. Cell Biol.* **25**, 171–174.
Wolniak, S. M., Hepler, P. K. and Jackson, W. T. (1983). *J. Cell Biol.* **96**, 598–605.
Zavortink, M., Welsh, M. J. and McIntosh, J. R. (1983). *Exp. Cell Res.* **149**, 375–385.

CHAPTER 8

Genetic Approaches to Spindle Structure and Function

FERNANDO CABRAL

Department of Pharmacology, University of Texas Medical School, Houston, Texas, USA

I. Introduction

Mitosis is a complex process involving a number of coordinated events and requiring many genes and proteins. Although the morphological description of mitosis was reported over 100 years ago (Flemming, 1878, 1880), our understanding of the process is still at a rudimentary stage (Pickett-Heaps *et al.*, 1982). Over the years, a large number of laboratories have amassed a detailed picture of the events associated with mitosis in several organisms (see McIntosh, 1979) but only a few of the proteins and regulatory controls governing the process have been elucidated. I will make no attempt to review this earlier work, since the topics have been addressed in other chapters in this volume and in many excellent reviews (Wilson, 1925; Schrader, 1953; Mazia, 1961; Nicklas, 1971; Dustin, 1978; McIntosh, 1982). Instead, I will restrict myself to several recent genetic studies which have provided new ways of approaching problems in mitosis research and which, when coupled with biochemical and morphological studies, show promise for identifying new genes and proteins involved in the structure and regulation of the mitotic apparatus.

Before describing these genetic studies, however, it will be helpful to emphasize briefly some of the questions these studies are attempting to address. While these may vary somewhat from one organism to the next, it is likely that the basic principles and mechanisms that govern mitosis are similar to all organisms. Indeed, it will be evident later in this chapter that genetic

MITOSIS: Molecules and Mechanisms
ISBN 0-12-363420-2

studies in several different organisms have all come to similar conclusions regarding the role of microtubules in mitosis and cell cycle progression. Morphological differences certainly exist, however. In all organisms, chromosomes must condense and sister chromatids must segregate during mitosis. Centrosomes in mammalian cells or spindle pole bodies in yeast and other lower eukaryotes must duplicate, migrate, and organize spindle microtubules into a functional mitotic apparatus. A mechanism must exist to ensure the equal division of DNA into daughter cells during cell division. On the other hand, mammalian cells lose the continuity of the nuclear envelope during mitosis, while others, like those of yeast and *Aspergillus*, retain their nuclear envelopes throughout. Mammalian cells divide by constriction of a cleavage furrow formed by a ring of actin/myosin filaments attached to the plasma membrane at the cell's equator, while yeasts divide by budding or fission. Although there is no *a priori* reason to expect that all organisms have evolved the same complement of proteins to carry out these processes, this nevertheless is frequently the case. For example, tubulin is the major constituent of all mitotic spindles thus far examined, and recent evidence demonstrates the existence of myosin heavy-chain at the junction between the mother cell and bud during division in *S. cerevisiae* (Watts *et al.*, 1985; 1987). Still, some differences in the proteins involved in cell division are likely to be found among different organisms. One obvious example is that the enzymes involved in chitin digestion and deposition during budding in yeast are not found in mammalian cells, but more subtle examples will probably also be found.

This review will focus on the use of genetics to study mitosis and, especially, to study the role of microtubules during mitosis. Of the many thousands of proteins that exist in the cell, only a small handful have been shown to function in mitosis. One of the most prominent of these is tubulin, the major structural component of the mitotic spindle apparatus (Fuller *et al.*, 1975; Weber *et al.*, 1975). Some of the questions we can ask regarding the function of these microtubules include the following.

1. Are microtubules necessary for mitosis? Their prominence in the spindle structure and the inhibition of mitosis by drugs which interfere with microtubule assembly would certainly suggest that they are, but neither of these arguments is conclusive nor do they tell us how microtubules are involved in the events that occur during mitosis.
2. What proteins other than tubulin are involved in spindle assembly and in the interaction of microtubules with the centrosome and with kinetochores? Microtubule-associated proteins (MAPs) have been well characterized in brain (Vallee, 1984) and several cultured cell lines (Bulinski and Borisy, 1979; Weatherbee *et al.*, 1980; Olmsted and Lyon, 1981), but recent evidence suggests the existence of spindle-specific MAPs (McCarty *et al.*,

1981; Lydersen and Pettijohn, 1980; Izant *et al.*, 1982; Zieve and Solomon, 1982; Pepper *et al.*, 1984; Chapter 5). Are these proteins *bona fide* MAPs and, if so, what is their role in microtubule assembly and function *in vivo*?

3. How is the conversion of the cytoplasmic microtubule complex into the mitotic spindle apparatus carried out? It is well established that spindle microtubules are assembled from the pool of tubulin produced by the dissociation of cytoplasmic microtubules (Bibring and Baxandall, 1977; Fulton and Simpson, 1979). What are the signals that trigger the depolymerization of cytoplasmic microtubules, the migration of the centrosomes, and the reassembly of microtubules from the spindle poles?
4. What force(s) is (are) responsible for anaphase chromosome migration to the spindle poles? Microtubules are clearly involved, but considerable uncertainty exists concerning the nature of this involvement. Do microtubules provide the force for chromosome migration or do they only direct and organize this movement (Inoué, 1981)?
5. Are microtubules involved in cytokinesis and nuclear migration?
6. Are microtubules necessary for cell-cycle progression?

This list of questions is by no means complete, but provides an idea of the kinds of questions that can be approached using genetics.

Many of these questions can also be approached using biochemistry, morphology, drug inhibition, and micromanipulation. However, genetics coupled with these other approaches provides the unique opportunity to identify genes involved in mitosis. Subsequent identification of the mutant gene product and the function altered in the mutant cells can provide an unambiguous demonstration of the involvement of a particular gene and protein product in a given function. Even when the gene product is not identified, complementation analysis of a large number of mutants can give an estimate of the minimum number of genes involved in a process such as cell division; and construction and analysis of suitable double mutants can indicate the order in which the gene products are involved. Examples of these approaches will be given in subsequent sections.

A. Selections for mutant isolation

Two general approaches for the isolation of mutants with alterations in mitosis and cell division have been employed and these are summarized in Fig. 8.1. The first (Fig. 8.1A) makes no *a priori* assumptions about the proteins involved in the process. A selection is devised in which cells that are defective in the process under study are identified and isolated. These mutants may then be arranged into complementation groups to identify how many different genes have been affected; and they may be studied physiologically to determine the

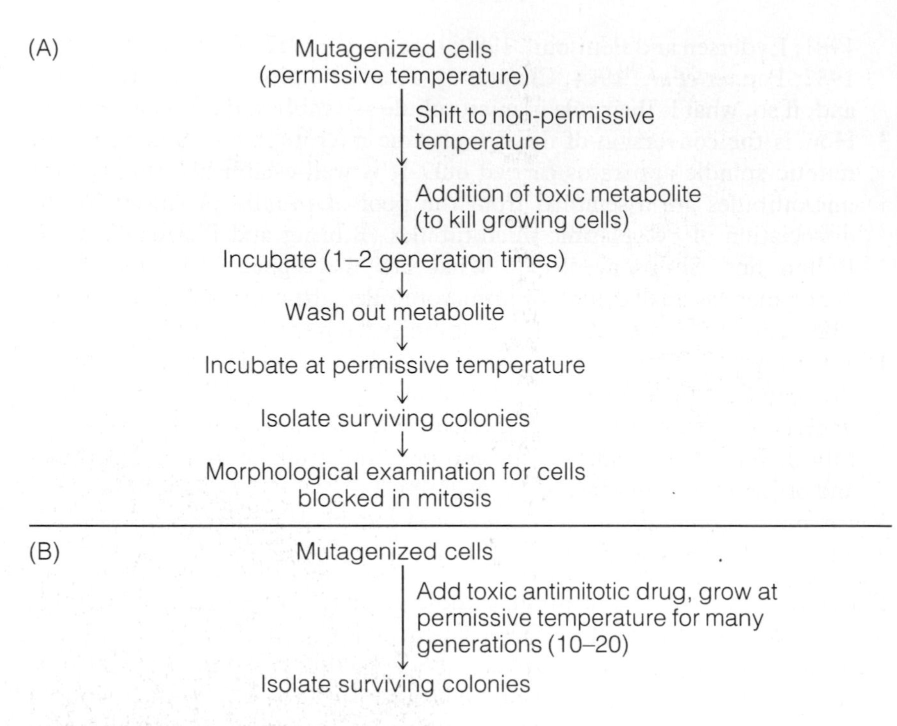

Figure 8.1. Scheme illustrating two commonly used selections for obtaining mutant cell lines with defects in mitosis. (A) Selection for cell division cycle mutants. (B) Selection for drug-resistant mutants. See text for details.

effects of the gene alteration on the process. This kind of approach has been used most effectively by Hartwell and his colleagues to isolate cell division cycle (CDC) mutants in *Saccharomyces cerevisiae* (Hartwell, 1974, 1978). It has the advantage of requiring no prior biochemical knowledge of the process and yet being able to identify genes involved in the process. The main disadvantage in this approach lies in the fact that biochemical identification of the altered gene product and elucidation of its molecular interactions are not trivial problems. Thus, the ability to uncover the molecular details of how a gene and its gene product are involved in a process lags far behind the ability of genetics to identify relevant genes. However, recent advances in molecular biology promise to speed up this connection. In yeast and *Aspergillus*, for example, it is now possible to make plasmid libraries and isolate the wild-type homologue of a mutant gene by *in vitro* complementation (gene transfection). Sequencing the gene followed by chemical synthesis of a corresponding oligopeptide and generation of an antibody can then lead to the isolation and *in situ* localization of the relevant gene product. Some promising steps along this route for yeast

CDC mutants have already been taken (Pringle *et al.*, 1984; Baum *et al.*, 1986; Haarer and Pringle, 1987; Wittenberg *et al.*, 1987; Goebl *et al.*, 1988).

The second approach for isolating mutants (Fig. 8.1B) uses existing biochemical knowledge to identify a protein likely to be involved in a given process and devise a selection that will isolate cells with mutations in the gene(s) for that protein. This approach has been well utilized to isolate tubulin mutants in a variety of organisms. Tubulin has long been known to be a major component of the mitotic spindle apparatus (see Dustin, 1978). Thus, mutations affecting tubulin should alter the ability of cells to progress through mitosis. As will be discussed in detail in subsequent sections, this expectation has been amply demonstrated in several laboratories. The simplest method for generating mutants with altered tubulin has been the use of drug-resistance selections. Many natural plant alkaloids and several synthetic derivatives are known to affect microtubule assembly and are cytotoxic to cells (Ludueña, 1979). One might predict, then, that cells resistant to the cytotoxic effects of the drugs might have an alteration in tubulin; and this has indeed been proved to be true. This approach allows the isolation of many mutants with known biochemical alterations but does not readily identify new genes and proteins involved in a given process. Although some interacting proteins can be identified by analysis of extragenic suppressors of the mutant phenotype, this approach cannot generate the diversity of mutants obtainable by the first approach. None the less, it can provide unambiguous demonstration of the involvement of a known protein in a given process and uncover the role of the protein in that process.

A variation on this second approach has recently been made feasible in yeast by the advent of genetic engineering. Once a gene has been isolated, it can be mutagenized *in vitro* by appropriate chemical treatment and reintroduced into the organism by homologous recombination. In this way, the wild-type gene is disrupted by the mutant gene and the effects on the organism can be studied. This approach has the advantage that mutants with alterations in a given protein can be isolated even when appropriate mutant selections cannot be devised.

II. Genetic Systems Used to Study Mitosis

A. Yeast

1. *Cell cycle mutants*

The pioneering work in the use of genetics to study mitosis was carried out by Hartwell and his collaborators (Hartwell, 1974, 1978; Pringle and Hartwell, 1981) on the organism *Saccharomyces cerevisiae*. These investigators made the assumption that an asynchronous population of yeast, when blocked in mitosis, would all assume a similar and possibly altered morphology, the "terminal phenotype". On this basis, mitotic mutants of cell division cycle (CDC) arrest mutants could be identified among cells that are temperature-sensitive for growth. Of 1500 temperature-sensitive mutants originally isolated, 148 were identified as *cdc* mutants and these fell into 32 complementation groups (Hartwell *et al.*, 1973). Included among these mutants are cell lines defective in spindle pole body duplication, spindle pole body separation, DNA synthesis, DNA synthesis initiation, nuclear division, bud emergence and cytokinesis (Hartwell, 1978). This work has allowed the identification of many genes involved in mitosis; and the construction of appropriate double mutants has allowed the temporal order in which these genes function during the cell cycle to be established. Although the protein products associated with these genes remain largely unidentified, the diversity of mutants that have been isolated make this approach capable of ultimately uncovering many of the steps involved in mitosis. It is equally likely, however, that not all genes involved in mitosis are yet represented among *cdc* mutants in *S. cerevisiae*, since mutants with altered tubulin or actin have not yet been identified. Indeed, 5000 additional temperature-sensitive lethal mutants were later screened and only three new genes were identified (Pringle *et al.*, 1984). These authors suggest that many relevant genes may not readily make temperature-sensitive mutations.

It should be emphasized that failure to identify a relevant gene product does not necessarily make a mutant useless. Although molecular mechanisms are not easily approached, correlations between altered morphology and function can still be made. For example, Byers (1981) has reported studies on mutant *cdc 31* (Baum *et al.*, 1986) that challenge the idea that DNA replication and bud emergence in yeast are dependent upon duplication of the spindle pole body. This mutant, when shifted to the restrictive temperature, produces a single large spindle pole body and yet the cells continue to replicate DNA and exhibit budding. The possibility cannot be excluded, however, that duplication of the material in the spindle body rather than duplication of the spindle body itself is sufficient to allow the other two events. In another study, reciprocal shift experiments were used to order the *cdc* mutants along the pathway of dependent events in the yeast cell cycle (Wood and Hartwell, 1982). The absence of a ring of 10 nm filaments in the neck region of an emerging bud in mutants defective in cytokinesis suggests a role for these filaments in that process (Byers and Goetsch, 1976; Haarer and Pringle, 1987). The studies of Watts *et al.* (1985, 1987) cited earlier localize myosin to this same region, suggesting that the missing filaments might be composed of myosin. Recent work utilizing actin and tubulin immunofluorescence with wild-type and mutant cells is consistent with the involvement of microfilaments and microtubules in many morphogenetic events, including nuclear migration, polarized growth, and secretion (Adams and Pringle, 1984; Kilmartin and Adams, 1984; see Huffaker *et al.*, 1987 for a review). These kinds of correlations are not usually conclusive in the absence of a detailed mechanism but they can corroborate existing evidence from other studies, challenge existing concepts, or even establish new ideas for further investigation.

One *cdc* mutant in *S. cerevisiae*, *ndc1-1*, isolated by Moir *et al.* (1982) behaves in a similar fashion to known β-tubulin mutants. At the non-permissive temperature these cells continue to bud, but DNA division is blocked, resulting in the accumulation of aploid cells, i.e. cells lacking nuclear DNA. Although nuclear division does not occur, DNA replication continues so that the mother cell becomes polyploid. The authors conclude that chromosome segregation is not required for cell-cycle progression. Examination of these cells by tubulin immunofluorescence reveals microtubules that are normal in appearance and number. Nevertheless, the properties of these cells and the conclusions of these authors are similar to previously reported observations made in drug-resistant *Aspergillus nidulans* (Oakley and Morris, 1980, 1981) and in CHO cells (Abraham *et al.*, 1983; Cabral, 1983; Cabral *et al.*, 1980, 1983) in which tubulin or microtubule alterations could be found. It is possible that microtubule alterations are also responsible for the defect in *ndc 1-1*, but the changes are too subtle to detect by immunofluorescence in yeast.

In some very exciting recent studies, the *CDC 28* gene of *Saccharomyces cerevisae* has been shown to encode a protein kinase (Reed *et al.*, 1985; Wittenberg and Reed, 1988). Since mutations in this locus affect the G1 and G2 phases of the cell cycle (Piggott *et al.*, 1982), this raises the intriguing prospect that phosphorylation controls progression at various points in the cell cycle.

cdc mutants have also been isolated in *S. pombe* by Nurse *et al.* (1976) using a similar approach. Starting with 500 temperature-sensitive mutants, these authors identified 27 recessive cell division arrest mutants by microscopic observation; and these were found to represent 14 unlinked genes. The mutants displayed four distinct terminal phenotypes. Two mutants representing one gene were defective in DNA synthesis and as a result displayed no nuclear division. Another nine mutants mapped in five genes and were defective in nuclear division. These mutants eventually stopped synthesizing DNA and failed to form a septum or divide. Defects in early cell plate formation were found in seven mutants representing four genes. No cell plate material was found in these mutants. Although cell division was blocked, no effects on RNA or DNA synthesis were found. Thus, the cells were able to accumulate 4–16 nuclei per cell. Eight mutants in four genes were found to be defective in late cell plate formation. Unlike the previous class of mutants, these cells accumulate cell plate material, but it remains disorganized and septum formation fails to occur. Again, these cells accumulate several nuclei per cell.

As was discussed for the *CDC 28* gene of *S. cerevisiae*, it has recently been found that the *cdc* 2^+ gene of *S. pombe* encodes a 34 kDa protein kinase (Simanis and Nurse, 1986). The deduced protein sequences from both species of yeast indicate significant homology (Hindley and Phear, 1984; Nurse, 1985); and the proteins appear to be functionally interchangeable *in vivo* (Beach *et al.*, 1982; Booher and Beach, 1986). Recently, a human homologue of *cdc* 2^+ has been identified (Lee and Nurse, 1987; Draetta *et al.*, 1987).

Approximately 500 cold-sensitive strains of *S. pombe* were also isolated by Yanagida and collaborators (Toda *et al.*, 1983; see Yanagida, 1987; Hirano and Yanagida, 1988 for reviews) and, of these, 13 strains blocked in nuclear division were chosen for further study. These mutants represent 12 unlinked genes which map to three chromosomes. All the mutant cells elongate in the absence of cell division at the non-permissive temperature and retain a single nucleus. Most cells remain viable at the non-permissive temperature for up to 10 hours and proceed synchronously into the cell cycle (including septum formation and cell division) after being returned to the permissive temperature. The 13 mutant strains can be grouped into at least five classes based on their chromatin morphology at the non-permissive temperature.

Two of the mutant strains are of special interest (Umesono *et al.*, 1983). One of them, *nda 2*, has chromatin that is displaced from the centre of the cell body, an altered positioning of the spindle pole body, and no clear spindle structure at the non-permissive temperature. Furthermore, this strain is supersensitive to benzimidazole carbamates and can be complemented by transformation with two cloned α-tubulin sequences from *S. pombe* (Toda *et al.*, 1984). These results indicate that the *nda 2* mutation resides in an α-tubulin gene and that microtubules are necessary structures in *S. pombe* for spindle pole body placement as well as nuclear placement and division. As with other tubulin mutants described in this chapter, cell division is also blocked. The *nda 3* mutation also appears to affect microtubules, since it maps to the *ben 1* locus which confers resistance to benzimidazole carbamates and is complemented by cloned β-tubulin sequences (Toda *et al.*, 1984). This cell line retains microtubules that nevertheless appear to be non-functional. The *nda 2* mutation appears to resemble *ben A16* in *Aspergillus nidulans* and taxol-dependent mutants in CHO cells both in the supersensitivity to agents that disrupt microtubules and in the effects on nuclear division and cytokinesis (see discussion of *Aspergillus* and CHO below). Mutant *nda 3*, on the other hand, resembles benomyl resistant *Aspergillus* mutants and colcemid resistant CHO mutants with respect to drug resistance and mitotic abnormalities. This is especially evident when comparison is made to *ben A33* which fails to exhibit nuclear movement or division as a result of hyperstable microtubules caused by a mutation in the β-tubulin gene of *A. nidulans* (Oakley and Morris, 1981). These observations support the theme repeated throughout this review that different laboratories working on different organisms and isolating mutants by different procedures are nevertheless arriving at similar conclusions regarding the role of microtubules in the progression of cells through mitosis. It also confirms the prediction that tubulin gene mutations should constitute a subclass of *cdc* mutants.

2. *Drug-resistant mutants*

The earliest attempt to select yeast cells resistant to antimitotic drugs was made by Stetten and Lederberg (1970). The mutant cells, however, were never characterized for mitotic defects or altered gene products. Only very recently has this approach been successfully used to study mitosis in yeast (Thomas *et al.*, 1984). Several recessive lethal cold sensitive (cs) and heat sensitive mutants among benomyl-resistant cells of *S. cerevisiae* were isolated and shown to map to a single locus. This locus could be shown to represent the β-tubulin gene by the failure of these mutants to complement a yeast β-tubulin mutant produced by disruptive integration (see below). Shifting the mutants to the non-permissive temperature results in a *cdc* arrest in mitosis in which the cells accumulate as large budded doublet cells. The nuclear DNA is undivided and fails to migrate to the neck of the bud. Such cells appear similar to wild-type cells arrested with nocodazole or benzimidazole carbamates. These results imply that microtubules are involved in chromosome segregation, nuclear migration and fusion, but not bud emergence (see also Huffaker *et al.*, 1987; Jacobs *et al.*, 1988). These conclusions are further strengthened by immunofluorescence studies showing that microtubules are largely but not totally lost at the non-permissive temperature. Although there are substantial differences in the way yeast and mammalian cells go through mitosis, it is significant that most of the conclusions derived from these studies are virtually identical to conclusions previously reached by studying benomyl-resistant mutants in *Aspergillus* (Oakley and Morris, 1980, 1981) and colcemid-resistant (Cabral *et al.*, 1980; Abraham *et al.*, 1983) and taxol-dependent mutants (Cabral, 1983; Cabral *et al.*, 1983) in CHO cells. This is a dramatic reaffirmation of the principle that basic mechanisms in processes such as cell division are well conserved in biology.

3. *Mutants by disruptive integration*

The ability to produce mutants by disruptive integration of *in vitro* mutagenized cloned genes into wild-type yeast cells is a very recent development and few examples of the approach exist. In the study of mitosis, Neff *et al.*

(1983) have shown that disruption of the *S. cerevisiae* β-tubulin gene by transformation results in a recessive lethal mutation, thereby demonstrating that the β-tubulin gene in yeast is essential for viability. These results have recently been expanded by Huffaker *et al.* (1988). Similarly, disruption of the yeast actin gene (Shortle *et al.*, 1982) is a recessive lethal event. Conditional actin mutants isolated by this method exhibit an altered actin distribution at the non-permissive temperature. At the same time, fewer budded cells are seen and the cells become osmotically sensitive. Finally, there is a delocalization of chitin synthesis. These studies suggest that actin plays a role in bud emergence and in the osmotic regulation of cells (Thomas *et al.*, 1984).

B. *Chlamydomonas*

One of the earliest mutants of *Chlamydomonas reinhardii* with a defect in mitosis was reported by Warr (1968). This mutant, *cyt-1*, was selected based on its small colony size on semisolid medium and was later found to be defective in cytokinesis. Large multinucleated cells accumulate in this cell line suggesting continued nuclear division in the absence of cell division. Interestingly, the number of flagellar pairs in the cells correlated well with the number of nuclei that had accumulated. A later study reported interesting effects of vitamin B_{12}, benzimidazoles and cobalt on the multinucleation, but the biochemical defect in the cells remains unknown (Warr and Durber, 1971).

Four temperature-sensitive mutants in this organism were later isolated by Sato (1976). In all four mutants, cell division, but not flagellar regeneration, was affected by temperature. One of these mutants, TS-60, however, was found to be cold-sensitive as well as heat-sensitive for growth and was also shown to be colchine-resistant, implying an alteration in the organism's microtubules. In this mutant, both cell division and flagellar regeneration are resistant to the effects of colchicine, indicating that the same mutant protein participates in both processes. This conclusion has recently been shown to be true by direct biochemical methods (Brunke *et al.*, 1982; L'Hernault and Rosenbaum, 1983). In all revertants of the temperature-sensitive phenotype, the colchicine-resistance phenotype co-reverted, thus proving that both phenotypes are conferred by mutations in a single genetic locus.

A direct selection for colchicine-resistant *Chlamydomonas* was reported by Adams and Warr (1972), who isolated a total of five mutants. These mutants exhibited abnormalities in cell shape and an approximate doubling of their generation times during vegetative growth. In addition, the cells showed severe defects in zygote germination. It was later reported by Warr and Gibbons (1974) that these cells have normal drug uptake and are cross-resistant to vinblastine, indicating that the defect resides in the microtubules and not in membrane permeability. This conclusion received further support from the interesting observation that most of the mutant strains grow better in the presence of colchicine or vinblastine than they do in the absence of those drugs. Pairwise crosses between the mutants allowed these authors to assign the mutants to two different classes, indicating that at least two genetic loci can be mutated to confer colchicine resistance. The non-viability of zygotes obtained from crosses of non-complementing mutants implies that the mutant gene product (most likely a microtubule protein) participates in meiotic as well as mitotic division.

Colchicine-resistant mutants were also isolated by Flavin and Slaughter (1974). Both cell division and flagellar regeneration are colchine-resistant in these mutants, indicating that the defective gene product participates in both processes as was discussed earlier for the TS-60 mutant of Sato.

C. *Aspergillus nidulans*

Extensive studies on tubulin mutations in *Aspergillus* have been carried out by Morris and his co-workers (see Morris, 1986). Initially, these authors analysed a number of benomyl-resistant mutants (Van Tuyl, 1977) by two-dimensional gel electrophoresis and were able to demonstrate an altered mobility of β-tubulin in a significant number of the mutant cells (Sheir Neiss *et al.*, 1978). This work established *ben A* as a gene coding for β-tubulin. The fact that all benomyl-resistant strains with an observable alteration in tubulin occurred in β-tubulin suggested that β-tubulin contains the binding site for benzimidazole carbamates, but this conclusion remains to be demonstrated directly. In later work, three heat-sensitive mutations in the *ben A* locus were identified and used to isolate revertants (Morris *et al.*, 1979). Of 18 revertants able to grow at the non-permissive temperature, four were back mutations in the same gene while 14 were extragenic suppressors. One of the mutants carrying an extragenic suppressor exhibited an altered α-tubulin. This result provided the first genetic evidence for the interaction of α- and β-tubulin *in vivo*. The other extragenic suppressors map to at least three additional loci and may represent mutations in genes coding for other proteins that function in microtubule assembly (Oakley, 1981; Weil *et al.*, 1986).

These mutants have been used to study the involvement of microtubules in mitosis in *Aspergillus*. Two approaches have been used. In the first, two β-tubulin mutants, *ben A15*, which is resistant to benomyl, and *ben A16*, which is supersensitive to benomyl, were used to study the effects of the drug on nuclear migration. Nuclear migration in *Aspergillus* is sensitive to benomyl, but more drug was required to poison the process in *ben A15* than in the wild-type strain. In *ben A16*, on the other hand, less benomyl was required.

Since the resistance or supersensitivity of the mutant strains is known to result from an alteration in β-tubulin, the authors conclude that nuclear migration in *Aspergillus* is β-tubulin-dependent (Oakley and Morris, 1980).

The second approach uses the conditional phenotype of certain mutants to look at the involvement of microtubules in the cellular process directly. *ben A33*, for example, is a temperature-sensitive β-tubulin mutant that arrests in mitosis at the restrictive temperature (Oakley and Morris, 1981). The arrested cells fail to exhibit nuclear movement or division, indicating the involvement of β-tubulin in those processes. Since the phenotype can be suppressed by *tub A1*, α-tubulin must also be involved. Interestingly, the *ben A33* mutation inhibits movement of the chromosomes to the poles even though these cells have apparently normal spindle microtubules by immunofluorescence (Gambino *et al.*, 1984). Furthermore, *ben A33* confers resistance to a variety of microtubule inhibitors and the temperature-sensitive phenotype is suppressed by these inhibitors. It thus seems likely that *ben A33* causes a hyperstabilization of the microtubules and that it is the inability of microtubules to appropriately depolymerize that leads to the anaphase block (Oakley, 1981; Morris *et al.*, 1984). The suppression of *ben A33* by *tub A1* and *tub A4*, which make microtubules less stable, is also consistent with this interpretation (Morris *et al.*, 1984).

Analysis of benomyl-resistant mutants and their revertants by Morris and his colleagues has also demonstrated the existence of two α and two β tubulin genes in *Aspergillus* (Weatherbee and Morris, 1984; Morris, 1986). The *ben A* gene codes for β1 and β2 tubulins (Sheir Neiss *et al.*, 1978) while the *tub C* gene codes for β3 tubulin (Weatherbee *et al.*, 1985; May *et al.*, 1985). Another gene, *tub A*, identified as an extragenic suppressor of a temperature-sensitive *ben A* mutation (Morris *et al.*, 1979) was found to code for α1 and α3 tubulins. The altered electrophoretic mobility of these mutant proteins revealed the existence of a third α-tubulin species, designated α2, that continued to migrate at the normal position. Recently, it was found that the *ben A* mutation confers resistance of vegetative growth to benomyl, but that conidiation remains benomyl-sensitive (Weatherbee *et al.*, 1985). Since the *ben A* mutation alters the electrophoretic mobility of β1 and β2 tubulin but not that of β3, it was proposed that the β3 tubulin is conidiation-specific. This conclusion was confirmed by the isolation of 20 double mutants, all of which lack β3 tubulin, that can form conidia in the presence of benomyl (Weatherbee *et al.*, 1985). The conclusion was further supported by experiments in which disruptive integration into the gene for β3, in a strain that was benomyl-resistant for vegetative growth, led to the isolation of transformants that were also benomyl-resistant for conidiation (May *et al.*, 1985). These studies demonstrate very convincingly how genetics can be used to identify the number and function of expressed tubulins in an organism.

Cell-cycle mutants have also been isolated in *Aspergillus* (Morris, 1976a,b; Orr and Rosenberger, 1976). As with the yeast CDC mutants, these exhibit a "terminal phenotype" at the non-permissive temperature; also as with the yeast mutants, failure to determine the biochemical lesion in these cells has, thus far, limited their usefulness. Two of the mutants isolated by Morris, however, have been studied extensively by electron microscopy and appear to be microtubule mutants. One of them, *bim C*, has an incompletely formed spindle; while the other, *bim D* appears to lack kinetochore microtubules (Oakley, 1981). The *nim A* gene product is required late in G2 for mitotic initiation. Like the yeast *CDC 28/cdc* 2^+ genes described above, *nim* A^+ also encodes a potential protein kinase (Osmani *et al.*, 1988).

D. *Physarum*

Physarum is a complex organism with respect to tubulin expression but also a very interesting one. The organism can exist in three different forms depending on the growth conditions: a microscopic myxamoeba, a motile flagellate, or a macroscopic syncytium termed the plasmodium. Each of these forms appears to express a distinct complement of tubulin polypeptides (Roobol *et al.*, 1984). The myxamoeba expresses two electrophoretically distinguishable tubulins, $\alpha 1$ and $\beta 1$, while the plasmodium expresses $\alpha 1$, $\beta 1$, $\alpha 2$ and $\beta 2$. In addition, the flagellate expresses a third α-tubulin species, $\alpha 3$, which arises from a post-translational modification of $\alpha 1$ (Green and Dove, 1984) as has also been found in *Chlamydomonas* (L'Hernault and Rosenbaum, 1983) and *Crithidia fasciculata* (Russell and Gull, 1984).

The isolation of mutant myxamoebae resistant to benzimidazole carbamates has indicated that at least four unlinked loci, *ben A*, *ben B*, *ben C* and *ben D*, can be mutated to produce drug resistance (Burland *et al.*, 1984). Drug-resistance tests indicate that *ben B* and *ben D* are expressed in myxamoebae and plasmodia, but that *ben A* and *ben C* are specific for the myxamoebae. Segregation analysis of heterozygous diploid plasmodia coupled with Southern blot analysis suggests that *ben A* and *ben D* represent β-tubulin loci while *ben C* represents an α-tubulin locus (Schedl *et al.*, 1984).

The results show that the myxamoeba expresses at least two β-tubulin polypeptides even though it exhibits only one electrophoretically distinguishable β-tubulin on two-dimensional gels. Further evidence for this interpretation comes from the isolation of mutant *ben 210*. This mutant expresses two β-tubulin species: the $\beta 1$ found in the wild-type strain as well as a new β-tubulin with a different electrophoretic mobility. Thus, genetics may be used to show that a single electrophoretic species may contain the products of more than one gene. Earlier examples of the use of this approach in *Aspergillus* have already been mentioned and similar experiments that have been carried out in CHO cells are discussed below. Although the *ben* mutants in *Physarum* have not yet been used to any great extent to study mitosis, the ability of as many as 10^9 nuclei in the plasmodium to go through mitosis in a synchronous fashion should allow mutant analysis in this organism to produce some very dramatic results in the near future.

Temperature-sensitive mutants isolated in haploid plasmodia fell into nine complementation groups and had a variety of phenotypes (Laffler *et al.*, 1979). One mutant, MA67, was defective in DNA but not protein synthesis, implying that the cells are affected in a step, believed to occur in G2, that is necessary for nuclear replication. The nuclei in the blocked cells exhibit an early prophase morphology and mitosis fails to progress. Another 55 temperature-sensitive mutants of the myxamoeba were later isolated and this selection yielded two cell-cycle mutants, ATS 20 and ATS 22 (Burland and Dee, 1980). ATS 20 has an increased DNA content, accumulates large nuclei and nucleoli, and exhibits a low division index. These observations indicate that DNA synthesis can occur in the absence of nuclear division. ATS 22 also has a low division index in the myxamoeba (not in the plasmodium) but the DNA content of the cells is normal, implying that nuclear division is necessary for DNA synthesis. These seemingly conflicting results point out the difficulty in interpreting mutant properties in the absence of a detailed knowledge of the biochemical defect and the mechanism by which the defect alters the cell's physiology.

E. *Drosophila*

The fruit fly has been a very useful species to the geneticist. Much of what we know in classical genetics has come about through the study of this organism. *Drosophila* also represents one of the favourite systems of the developmental biologist; indeed, it is largely through the use of mutants to study development that our knowledge of the role of microtubules in cell division in that organism has evolved. Kemphues *et al.* (1979, 1980) first showed that *Drosophila* express a testis-specific β-tubulin (β2). This β2-tubulin is not found in other tissues or in the females and is expressed only after mitotic divisions are complete and before the onset of meiotic division in the developing spermatocytes. Since expression of the β2 tubulin is restricted to the testis, it is possible to isolate mutations in the gene for β2 without affecting the viability of the organism. Such mutations are found among male sterile mutants.

A dominant male sterile mutation, β2t, that profoundly affects the course of spermatogenesis has been isolated (Kemphues *et al.*, 1979, 1980). Since β2-tubulin is a major component of the sperm flagellum, it is not surprising that this mutant has severe defects in the structure and organization of the axoneme. What is perhaps more surprising is that this mutant also exhibits abnormal meiotic spindle formation, a failure to undergo cytokinesis, and improper chromosome movement. In addition, an increased frequency of chromosomal non-disjunction is seen, as well as some micronucleation. These results demonstrated convincingly that a single tubulin gene product can participate in the formation of multiple classes of microtubules (i.e. spindle and axoneme).

This conclusion was confirmed by the isolation of recessive mutations in the same locus (Kemphues *et al.*, 1982). These recessive mutants fell into at least two classes. In the first class, the mutant β2-tubulin is synthesized at normal levels but is unstable and rapidly degraded *in vivo*. Interestingly, a commensurate portion of the α-tubulin is also degraded, suggesting that stable α/β heterodimers are resistant to intracellular proteases, but free α- or β-tubulin subunits are not. The class I mutants, then, have very low steady-state levels of tubulin. As a result, these mutants exhibit no meiotic spindle, no chromosome movement, no cytokinesis, no cytoplasmic microtubules, no axoneme, and no nuclear shaping (Kemphues *et al.*, 1982). The class II mutants have a stable β2-tubulin with a variety of alterations and varied phenotypes.

These studies establish the involvement of spindle microtubules in chromosome movement and cytokinesis in *Drosophila*. Similar conclusions have been reached in yeast, *Aspergillus*, and CHO cells. Although other mutations affecting mitosis or meiosis in *Drosophila* have been isolated (Rungger-Brandle, 1977; Ripoll *et al.*, 1985), the defective gene products in those cells have not been identified.

F. Cultured mammalian cells

Attempts to isolate mutants with mitotic defects among cultured mammalian cells that are temperature-sensitive for growth have met with success. The selections that have been used are similar to the selections described for the isolation of *cdc* mutants in yeast: i.e. growing cells are killed at the non-permissive temperature by addition of a toxic metabolite such as cytosine arabinoside or BUdR. The non-cycling cells survive this treatment and may be recovered by shifting the population back to the permissive temperature. Mitotic mutants may then be found among these temperature-sensitive cells by observing their morphology. Another selection makes use of the observation that many mammalian cell lines round up and become less adherent when they enter mitosis. Thus, Thompson and Lindl (1976) isolated a clone of CHO cells, MSI-1, that attaches and grows normally at 34°C but is blocked in mitosis and fails to attach to the culture dish at 38.5°C. This cell line has an elevated incidence of polyploid cells at 34°C that increases greatly when the cells are shifted to the higher temperature. The increased polyploidy appears to result from a defect in cytokinesis, but electron-microscopic observation failed to reveal any alterations in microtubules, microfilaments, or other structures believed to be involved in mitosis.

The isolation of mitotic mutants among cell lines that are temperature-sensitive for growth has been accomplished by a number of laboratories. Smith and Wigglesworth (1972) isolated a mutant of BHK-21 cells that fails to complete mitosis at the non-permissive temperature. This cell line produces binuclear cells, presumably as the result of defective cytokinesis. Unfortunately, the molecular defect in this interesting cell line was not found. Another mutant, *ts111*, defective in cytokinesis, was isolated in Chinese hamster fibroblasts a few years later by Hatzfeld and Buttin (1975). *ts111* forms giant cells that are frequently multinuclear at the restrictive temperature. Even after a prolonged arrest, these cells are able to resume cytokinesis if returned to the permissive temperature. An increased frequency of binucleate and anucleate cells is seen even at the permissive temperature. The authors suggest that the mutant cells resemble normal cells treated with cytochalasin B. In fact, the mutant cells appear to be about twice as sensitive to the drug as are the wild-type cells, suggesting a possible defect in the microfilaments. A Japanese group (Shiomi and Sato, 1976) isolated a mutant, *ts2*, of a murine leukaemic cell line that again is defective in cytokinesis. Cells shifted to the restrictive temperature while they are in G1 (but not G2!) fail to progress normally through the first mitotic phase; they accumulate aberrant mitotic figures and become multinucleated. The cells shifted in G2 pass normally through the first mitotic phase but then show similar defects beginning with their second mitosis. As with the other mutants just described, the biochemical defect in ts2 has not been identified.

Wang and his co-workers have isolated over 50 mitotic mutants in hamster HM-1 cells. One cell line, *ts546*, accumulates aberrant prometaphase-like figures at the non-permissive temperature similar to the mitotic figures induced in wild-type cells by colchicine (Wang, 1974). As with mutants isolated in other laboratories, *ts546* eventually exits the block, fails to undergo cytokinesis, and re-enters interphase as large multinucleated cells (Wang and Yin, 1976). The authors hypothesized that *ts546* has an altered tubulin or microtubule-associated protein, but no direct (or indirect) evidence was presented. Another class of mutants, exemplified by *ts655*, exhibits defective progression through prophase (Wang, 1976). The cells enter a prophase-like state at the non-permissive temperature in which the chromatin condenses into clumps rather than morphologically intact chromosomes. The nuclear membrane is lost during this process but is unable to re-form around the clumped DNA. This cell line should be useful for studying the factors regulating chromosome condensation and nuclear membrane formation in mitosis.

The same laboratory has also produced cell lines defective in anaphase chromosome movement and in centriole separation. Mutant *ts687* (Wissinger and Wang, 1978) retains lagging chromosomal material in the cleavage furrow at the non-permissive temperature that inhibits cell division and leads to the production of binucleated and variably sized multinucleated cells. After several days at the restrictive temperature, the cells pinch off cytoplasts and/or karyoplasts leading to an accumulation of these structures in the medium. Complementation analysis suggests that *ts687* represents a different gene from *ts546*. More recently, this group isolated another mutant, *ts745*, which forms monopolar spindles with an attached shell of chromosomes after only 15 min at the restrictive temperature (Wang *et al.*, 1983). Morphological examination of this mutant indicates that centriole separation fails to take place, there is no chromatid separation, and the cells block in a prometaphase-like state. Spindle pole separation and chromosome segregation will proceed if the cells are returned to the permissive temperature. If the cells remain in the blocked state for more than 3–5 hours, however, the cells revert to interphase and continue through the cell cycle. Because they fail to see interpolar microtubules and the spindle poles do not separate, the authors conclude that interpolar microtubules may be necessary for spindle pole separation. This notion is disproved, however, by the existence of a CHO mutant, described below, which also fails to elaborate interpolar microtubules yet has well separated spindle poles.

It is clear that in mammalian cells a great diversity of mutants exist that should provide a valuable resource for the study of mitosis. Before these mutants will have a sizable impact on our understanding of mitosis, however, the difficult work of defining the biochemical lesion in these cells must begin.

III. The Chinese Hamster Ovary Cell System

A. Drug-resistance selections

The isolation of mammalian cells resistant to drugs that interact with microtubules as a means of obtaining mutants with microtubule alterations has been largely restricted to Chinese hamster ovary (CHO) cells. Drugs such as colchicine, colcemid, griseofulvin, podophyllotoxin, maytansine, benzimidazole carbamates and taxol have been used in several laboratories for the isolation of mutants and this work has been reviewed elsewhere (Cabral, 1984; Schibler and Cabral, 1985). The following discussion will be restricted to those mutants in which a mitotic alteration has been demonstrated or those which serve as an example of the kinds of approaches that genetics can provide to answering biological questions.

The earliest CHO mutants resistant to colchicine were isolated by Ling and Thompson (1974) but these were later shown to be permeability mutants (Bech-Hansen *et al.*, 1976). The first drug-resistant mutants with clearly defined alterations in tubulin were reported by Cabral *et al.* (1980, 1981). Again, the vast majority of cells selected for resistance to colchicine, colcemid and griseofulvin had a permeability alteration but a small subset of these mutants displayed a β-tubulin with an altered electrophoretic mobility on two-dimensional gels. Ling *et al.* (1979) have isolated a CHO cell line with altered colcemid binding. Later studies indicated an altered α-tubulin in this cell line (Keates *et al.*, 1981). However, the mutant was isolated in a multiple-step procedure, and so it is not evident that the α-tubulin alteration is responsible for the decreased drug-binding affinity. Furthermore, cDNA sequencing has thus far failed to confirm the existence of a mutant α-tubulin protein in this strain (Elliott *et al.*, 1986). This mutant is the only one so far reported to have altered drug binding. The majority of mutants isolated in our laboratory and in others (Cabral *et al.*, 1980, 1981; Schibler and Cabral, 1985; Warr *et al.*, 1982; Gupta, 1981, 1983; Gupta *et al.*, 1982) display patterns of drug resistance that are inconsistent with an alteration in the binding site for any particular drug. A discussion of the probable mechanism of resistance in these mutants will be delayed to a later section.

B. Conditional mutants

The isolation of drug-resistant mutants with alterations in tubulin is a good approach for studying mechanisms of drug resistance and the interaction of drugs with microtubules; but unless these mutants have a conditional phenotype, they are of limited utility for the study of microtubule involvement in mitosis. By good fortune, many of the drug-resistant tubulin mutants that have been isolated do, in fact, have a conditional phenotype. For example, Cmd 4 and Grs 2 cell lines resistant to colcemid and griseofulvin respectively, are β-tubulin mutants that grow normally at 37°C but fail to survive at 40.5°C, a temperature at which the wild-type strain continues to grow (Abraham *et al.*, 1983). A taxol-resistant α-tubulin mutant, Tax 1, is also temperature-sensitive for growth (Cabral *et al.*, 1981). This property has allowed the selection of revertants (see below) and an analysis of the effects of tubulin alterations on cell growth (Abraham *et al.*, 1983). At the non-permissive temperature, these cells exhibit an increased mitotic index, which is due to an increase in the length of time the cells spend in mitosis relative to other phases of the cell cycle. Normal cytoplasmic microtubule organization and function is seen at the elevated temperature, but the spindles that form have abnormal morphologies and, after a prolonged exposure to the non-permissive conditions, are frequently multipolar. These aberrant spindles lead to abortive cell division and multinucleated cells accumulate in the culture.

A novel class of conditional mutants has recently been isolated in our laboratory. Tax 18 (Cabral, 1983) is a cell line selected for resistance to taxol, a drug that binds to and promotes microtubule assembly (Schiff *et al.*, 1979; Schiff and Horwitz, 1980). This mutant has the interesting property of growing normally in the presence of the drug, but failing to divide if taxol is omitted from the growth medium. Examination of the microtubules by immunofluorescence demonstrates that, even in the absence of taxol, these cells display a normal cytoplasmic microtubule complex. Mitotic cells, however, appear to be arrested in a prometaphase-like state in which the spindle poles are separated and short radial bundles of microtubules surround each spindle pole (Cabral *et al.*, 1983). Electron-microscopic observation reveals that these radial bundles are composed of kinetochore microtubules. Interpolar microtubules are not seen. The cells fail to display the metaphase or anaphase stages of mitosis but proceed directly into a telophase-like state of intense membrane activity. During this time a cleavage furrow forms but is unable to complete cell division. Instead, the cleavage furrow relaxes and the dividing cell re-enters interphase as a single larger cell. These cells continue through the cell cycle, as evidenced by unabated protein synthesis, DNA replication and centriole duplication. The mutant is able to proceed through these series of events several times before metabolic activity is lost, and in the process forms giant cells with many oddly shaped nuclei and micronuclei (Cabral, 1983). As with mutant ts687 (Wissinger and Wang, 1978) these giant cells pinch off plasma membrane vesicles during the "telophase-like" state and also accumulate a vastly elevated chromosome content. Close examination of these chromosomes in cells deprived of taxol for 2–3 days or more reveals many aberrations, including premature chromosome condensation (PCC)-like events, chromosome non-disjunction, ring chromosomes and precipitated DNA. We believe that it is the inability of the cell to process these vastly increased levels of DNA and maintain genetic balance that leads to the eventual death of the cell. Although the biochemical alteration in Tax 18 is not yet known, it clearly affects spindle microtubule formation. More recent work from our laboratory shows that taxol-dependent mutants with altered α- or β-tubulin have very similar properties (Schibler and Cabral, 1986). The properties of Tax 18 and other taxol-dependent mutants demonstrate that spindle microtubule assembly is necessary for organized chromosome segregation and ultimately for cell division. These microtubules are not essential for continued cell-cycle progression. Interestingly, they are also not necessary for short-term viability; but the defects that accumulate as a result of their loss ultimately prove lethal to the cell.

C. Reversion analysis

The conditional phenotype of these mutants has been useful not only for studying the effects of microtubule disruption on the course of mitosis, but also for the isolation of revertants. Although it is difficult to select drug-sensitive revertants directly, it is a relatively simple matter to select revertants able to grow at the non-permissive temperature (in the case of the temperature-sensitive mutants) or in the absence of taxol (in the case of taxol-dependent mutants). This kind of analysis was initially carried out to establish a causal link between the altered tubulin seen in the mutants and their drug resistance. For example, if the altered β-tubulin in Cmd 4 is responsible for the drug resistance and temperature sensitivity of the strain, then all three properties should co-revert at high frequency. This was indeed found to be the case (Cabral *et al.*, 1982).

In the course of these studies, however, we found that reversion can occur through a variety of mechanisms. Two of the earliest kinds of revertants we isolated include cells that no longer display the mutant tubulin on two-dimensional gels (recent data suggest that these subunits are no longer expressed) and cells that are no longer able to assemble the mutant tubulin (Cabral *et al.*, 1982). These latter revertants probably arise from a second mutation that alters the conformation of the mutant tubulin such that it can no longer assemble. Recently, we explored this phenotype further by isolating a large number of revertants of Cmd 4. In this way we were able to identify eight cell lines with assembly-defective β-tubulin (Boggs and Cabral, 1987). Some of these cell lines arise from mis-sense mutations, others from apparent deletions and insertions. In all cases, the assembly-defective tubulin is unstable and fails to accumulate in the cell. Thus, these cell lines resemble the class I recessive mutations in the β2 locus of *Drosophila* (Kemphues *et al.*, 1982). Unlike the *Drosophila* mutants, however, the CHO revertants have normal growth and appearance. Since the unstable subunit represents about 30% of the total β-tubulin synthesized, this indicates that CHO cells have an excess of tubulin subunits and can survive with a lower abundance of tubulin than is found in the wild-type strain. The isolation of cells with assembly-defective tubulin in CHO and *Drosophila* should open new avenues towards studying subunit–subunit interactions and the control of microtubule assembly *in vivo*.

D. The mechanism of drug resistance

As already mentioned, most drug-resistant CHO cells do not appear to have altered drug binding. Although the pattern of cross-resistance to various drugs is sometimes complex, the following generality seems to hold in the majority of the mutants we and others have isolated. Mutants selected for resistance to a

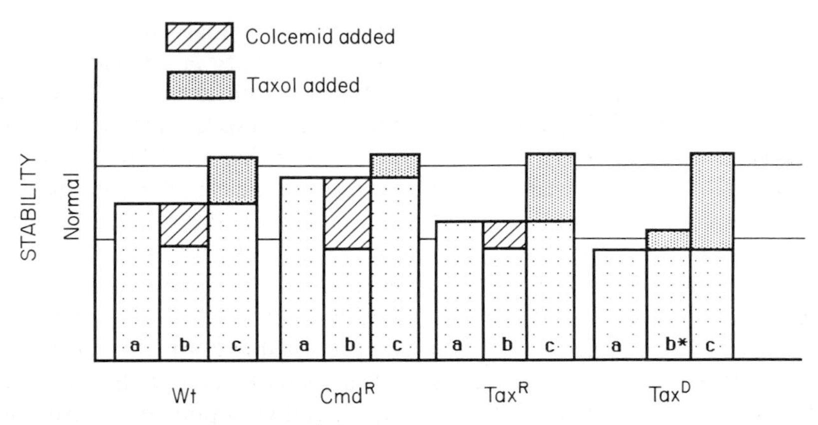

Figure 8.2. Model to explain the mechanism of drug resistance in cell lines with altered microtubules. Wt, wild-type cells; Cmd^R, colcemid-resistant cells; Tax^R, taxol-resistant cells; Tax^D, taxol-resistant and dependent cells. (a) Stability of cellular microtubules in the absence of any exogenously added drug; (b) amount of colcemid (hatched bar) required to reduce microtubule stability below the acceptable limit; (c) amount of taxol (dotted bar) required to raise microtubule stability above the acceptable limit; (b*) a small amount of taxol added to a taxol-dependent cell line restores microtubule stability to the acceptable range.

microtubule-destabilizing drug (e.g. colcemid) are cross-resistant to other destabilizing drugs (e.g. colchicine, podophyllotoxin, vinblastine, maytansine, etc.) but are supersensitive to taxol. Conversely, cells selected for resistance to taxol are usually supersensitive to colcemid and other destabilizing drugs. Other mutant properties are also inconsistent with an altered drug-binding site. For example, we now have good evidence that alterations in α- or β-tubulin can confer resistance to microtubule-active drugs. Although this does not rule out alterations in the drug-binding site, alternative explanations are more likely. Furthermore, the fact that mutants are only 2–3-fold more resistant to the selecting drug than are the wild-type cells, the co-dominant nature of the mutation, the existence of drug-dependent mutants and the recent indication from reversion studies that mutant subunits must assemble in order to confer drug resistance all favour the simple scheme shown in Fig. 8.2 to explain the mechanism of drug resistance in the mutant cells.

According to this scheme, wild-type cells regulate the stability of the microtubules within narrow limits (defined by the thin horizontal lines). The addition of an exogenous destabilizing agent such as colcemid becomes toxic when it lowers the total "stabilizing activity" (exogenous drug plus intrinsic activity) below the normal zone. Similarly, addition of taxol becomes toxic when it raises the total stabilizing activity above the normal zone. A colcemid-resistant mutant is one in which the intrinsic stabilizing activity is higher (owing to increased affinity between tubulin subunits or increased MAP binding, for example) so that more colcemid is required to bring the total activity below acceptable limits. Conversely, a taxol-resistant mutant is one in which the intrinsic activity is lower. This predicts that a colcemid-resistant mutant should be cross-resistant to other microtubule-destabilizing agents but more sensitive to taxol; and a taxol-resistant mutant should be supersensitive to microtubule-destabilizing drugs. This has indeed been found to be the case for the vast majority of mutants that we and others have isolated. When thought of in these terms, a taxol-dependent mutant is one in which the intrinsic stabilizing activity is already below acceptable limits. Such cells can then be rescued by addition of an exogenous stabilizing agent such as taxol. In agreement with this idea, we have found that dimethyl sulphoxide can partially rescue the cells in the absence of taxol (Cabral *et al.*, 1986). The scheme also predicts that high concentrations of taxol will still be toxic even to taxol-dependent cells, as has been found (Cabral, 1983), and predicts the existence of colcemid-dependent mutants. Although these have been more difficult to find, recent work in which multiple copies of the mutant β-tubulin gene of Cmd 4 have been introduced into wild-type cells has resulted in a colcemid-dependent cell line (Whitfield *et al.*, 1986; Cabral *et al.*, 1984).

The study of drug-resistant mutants has also contributed to our knowledge of the mechanism of drug action. The interpretation of *in vivo* effects of a drug are always complicated by the possibility that the drug may exert side-effects independent of its action on microtubules. On the other hand, one can never be certain that effects seen *in vitro*, where side-effects are eliminated, reflects the way the drug acts in living cells. It is perhaps significant that mutant cells under non-permissive conditions frequently resemble drug-treated cells. The mutant cells, for example, generally arrest in prometaphase, fail to segregate chromosomes, fail to divide, and form large multinucleated cells before they die, much as drug-treated cells do. Since evidence suggests that the mutant cells have altered subunit–subunit interactions that affect microtubule stability, it seems possible that at least some drugs also modulate microtubule stability by altering the conformation of the tubulin to which they bind. This mechanism predicts that drug-bound subunits assemble into microtubules thereby affecting their stability, affect microtubule function at substoichiometric levels, and need not act solely by capping the microtubule end.

IV. Conclusions

A. What has been done?

In reviewing the genetic studies over the last ten years, I am struck less by the diversity of mitosis in different organisms than by the common mechanisms that underlie the process. Although different organisms clearly exhibit morphologically distinct events during mitosis (see McIntosh, 1979), the consistency of results from genetic studies in very diverse organisms is significant. Tubulin was found to be essential for mitosis and cell viability in organisms ranging from yeast and fungi to mammalian cells. Microtubules have been implicated in chromosome movement, nuclear migration, and cytokinesis; but microtubule defects that alter those functions have not been found to prevent cell-cycle progression in those organisms for which the appropriate mutants are available. Counterparts for some of the yeast CDC mutants appear also to exist in mammalian cells, indicating that mitosis is carried out using similar steps in these diverse cell types. It should be noted that genetic analysis is not limited to lower eukaryotes or to mammalian cells growing in culture. Mutations affecting progression through mitosis have been isolated in a number of organisms exhibiting developmental defects (Raff *et al.*, 1976; Gounon and Collenot, 1974). In one particularly interesting example (Magnuson and Epstein, 1984), mice homozygous for the oligosyndactyly mutation accumulate cells in mitosis, beginning at the blastocyst stage. When viewed by tubulin immunofluorescence, the mitotic cells are seen to arrest at metaphase with a well-formed mitotic spindle. Thus, the mutation appears to block the transition from metaphase to anaphase and results in the death of the embryo. The defective gene product in this intriguing mutant has not been identified. These and other examples argue that mutations affecting cytoskeletal proteins and mitosis occur at appreciable frequencies in multicellular organisms but are rarely seen outside the laboratory because they are lethal and arrest early development.

The most important aspect of the genetic studies so far is that they have begun to provide links between the genes controlling mitosis, their protein products, and the ultimate functions of those proteins. In addition, the isolation of mutants is providing tools with which to study the number of expressed tubulins in different organisms, the control of microtubule and flagellar assembly (Luck, 1984), the probable mechanism of drug action, and the determination of functional specificity of tubulin. A demonstration of this latter use of the genetic approach has already been described in the studies which demonstrated that β2-tubulin in *Drosophila* is a major component of several microtubule-containing structures, and the studies by May *et al.* (1985) showing that a sporulation-specific β-tubulin gene in *Aspergillus nidulans* can be disrupted but that sporulation is unaffected because it can recruit the cytoplasmic β-tubulin species. In mammalian cells, the CHO mutant, Grs2, was used to demonstrate that mutant β-tubulin is a component of both cytoplasmic and spindle microtubules (Kuriyama *et al.*, 1985).

B. What remains to be done?

Some of the questions posed at the beginning of this chapter have been answered, but many others remain. The spatial and temporal control of microtubule assembly *in vivo* is still largely a mystery, as are the intracellular signals that trigger microtubule breakdown and reassembly into astral fibres in prophase, congregation of chromosomes onto a metaphase plate, and chromosome segregation in anaphase. I believe that mutant analysis will make significant contributions to our understanding of these events in the next several years.

We are on the verge of identifying, by genetic means, what proteins function as MAPs in eukaryotic cells. Extragenic suppressors of tubulin mutations in *Aspergillus* and *Drosophila* have been isolated and should yield interesting results in the near future. It is likely that many of these suppressor genes will prove to be structural genes coding for MAPs. These studies should provide definitive identification of MAPs and proof of their function *in vivo*. Genes coding for MAPs are also likely to be found among the temperature-sensitive mutants in yeast and mammalian cells. Much work remains to be done in characterizing the molecular defects in these cells. The task, however, should be aided by the identification of mutant genes using an *in vitro* complementation assay followed by sequence analysis, synthesis of oligopeptides, and generation of antibodies against the peptides to allow the localization of the corresponding protein in cells as well as its subsequent isolation.

Considerable progress has recently been made in elucidating the mechanism by which tubulin levels are regulated *in vivo*. The initial observation by Ben Ze'ev *et al.* (1979) that microtubule-depolymerizing drugs suppress tubulin synthesis while microtubule-polymerizing or precipitating drugs promote tubulin synthesis suggested that cells could sense the levels of free tubulin in the cytoplasm and adjust the synthesis of new tubulin accordingly. This observation has been elegantly pursued by Cleveland, Kirschner, and collaborators who were able to show that mRNA levels for tubulin are sensitive to the same drugs, indicating that the regulation is pretranslational (Cleveland *et al.*, 1981). Subsequent studies have indicated that transcription is not affected (Cleveland and Havercroft, 1983), that microinjection of tubulin dimers mimics the colchicine effect (Cleveland *et al.*, 1983), that transfected tubulin genes from a heterologous organism are similarly regulated (Lau *et al.*, 1985), and most recently, that regulation is seen in cytoplasts (Pittenger and Cleveland, 1985; Caron *et al.*, 1985). These results argue that the level of free tubulin in the cell affects tubulin mRNA stability by some as yet undetermined mechanism. While this phenomenon represents a novel regulatory mechanism of great interest, care must be taken in assuming that this mechanism operates under normal physiological conditions—i.e. in cells that have not been microinjected or treated with drugs. It is here that appropriate mutants can help us to understand the phenomenon. In CHO revertants with assembly-defective tubulin, for example, one of the normally expressed CHO β-tubulin subunits is unstable and does not accumulate to appreciable steady-state levels. As a result, these cells have 30% less tubulin than wild-type cells, yet they grow and divide normally. These results suggest that CHO cells do not fine-tune their intracellular levels of tubulin as might have been predicted. Furthermore, it appears that normal tubulin levels are in excess of what is required to maintain essential microtubule functions. Thus, the physiological significance of the tubulin autoregulatory control mechanism remains obscure. It is also clear from the class I mutations of the $\beta 2$ locus in *Drosophila* (Kemphues *et al.*, 1982), from the CHO mutants with assembly-defective tubulin, and from CHO cells that have been transfected with multiple copies of β-tubulin genes (Whitfield *et al.*, 1986) that protein degradation is another mechanism by which tubulin levels may be regulated and which ensures coordinate amounts of α- and β-tubulin for assembly.

Finally, it should be pointed out that mapping the site of the mutation in the assembly-defective β-tubulins will help to identify the domains of β-tubulin that participate in microtubule assembly. There is currently much activity in trying to map tubulin domains using biochemical methods (Serrano *et al.*, 1984; Maccioni *et al.*, 1985; Mandelkow *et al.*, 1985) and the complementary genetic information will be most useful. An important first step in this direction has recently appeared (Rudolph *et al.*, 1987).

V. Outlook

The future is bright for mutant approaches to studying microtubule assembly, regulation, and function and to dissecting the mechanisms that operate during mitosis. We have only just begun to appreciate the power of this approach and its ability to provide answers to some of the questions that have confounded us for many years. Results from the study of many diverse organisms are converging to give us confidence that the answers we are finding are universal and true. The marriage of genetics with many of the newer techniques in molecular biology is giving us an unprecedented opportunity at last to dissect out the genes, proteins and regulatory mechanisms involved in mitosis. These approaches in conjunction with the exciting advances that have recently been made in video-enhancement microscopy and biochemical analysis of motility promise to provide us with much excitement for the foreseeable future.

Acknowledgements

I wish to thank the many excellent collaborators with whom I have had the privilege of working over the last several years; and I am grateful to the NIH (NIGMS) and the American Cancer Society for their generous support of my research efforts.

REFERENCES

Abraham, I., Marcus, M., Cabral, F. and Gottesman, M. M. (1983). *J. Cell Biol.* **97**, 1055–1061.

Adams, A. E. M. and Pringle, J. R. (1984). *J. Cell Biol.* **98**, 934–945.

Adams, M. and Warr, J. R. (1972). *Exp. Cell Res.* **71**, 473–475.

Baum, P., Furlong, C. and Byers, B. (1986). *Proc. Natl. Acad. Sci. USA* **83**, 5512–5516.

Beach, D., Durkacz, B. and Nurse, P. (1982). *Nature* **300**, 706–709.

Bech-Hansen, N. T., Till, J. E. and Ling, V. (1976). *J. Cell. Physiol.* **88**, 23–32.

Ben-Ze'ev, A., Farmer, S. R. and Penman, S. (1979). *Cell* **17**, 319–325.

Bibring, T. and Baxandall, J. (1977). *Develop. Biol.* **55**, 191–195.

Boggs, B. and Cabral, F. (1987). *Mol. Cell. Biol.* **7**, 2700–2707.

Booher, R. and Beach, D. (1986). *Mol. Cell Biol.* **6**, 3523–3530.

Brunke, K. J., Collis, P. S. and Weeks, D. P. (1982). *Nature* **297**, 316–318.

Bulinski, J. C. and Borisy, G. G. (1979). *Proc. Natl. Acad. Sci. USA* **76**, 293–297.

Burland, T. G. and Dee, J. (1980). *Mol. Gen. Genet.* **179**, 43–48.

Burland, T. G., Schedl, T., Gull, K. and Dove, W. F. (1984). *Genetics* **103**, 123–141.

Byers, B. (1981). In *Molecular Genetics in Yeast* (ed. D. von Wettstein, J. Friis, M. Kielland-Brandt and A. Stenderup), pp. 119–133. Alfred Benzon Symposium 16.

Byers, B. and Goetsch, L. (1976). *J. Cell Biol.* **70**, 35a.

Cabral, F. (1983). *J. Cell Biol.* **97**, 22–29.

Cabral, F. (1984). In *Cell and Muscle Motility*, Vol. 5 (ed. J. W. Shay), pp. 313–340. Plenum Press, New York.

Cabral, F., Sobel, M. and Gottesman, M. M. (1980). *Cell* **20**, 29–36.

Cabral, F., Abraham, I. and Gottesman, M. M. (1981). *Proc. Natl. Acad. Sci. USA* **78**, 4388–4391.

Cabral, F., Abraham, I. and Gottesman, M. M. (1982). *Mol. Cell. Biol.* **2**, 720–729.

Cabral, F., Wible, L., Brenner, S. and Brinkley, B. R. (1983). *J. Cell Biol.* **97**, 30–39.

Cabral, F., Schibler, M. J., Abraham, I., Whitfield, C., Kuriyama, R., McClurkin, C., Mackensen, S. and Gottesman, M. M. (1984). In *Molecular Biology of the Cytoskeleton* (ed. G. G. Borisy, D. W. Cleveland and D. B. Murphy), pp. 305–317. Cold Spring Harbor Press, New York.

Cabral, F., Brady, R. C. and Schibler, M. J. (1986). In *Dynamic Aspects of Microtubule Biology* (ed. D. Soifer), Vol. 466, pp. 745–756. New York Academy of Science Press, New York.

Caron, J. M., Jones, A. L., Rall, L. B. and Kirschner, M. W. (1985). *Nature* **317**, 648–651.

Cleveland, D. W. and Havercroft, J. C. (1983). *J. Cell Biol.* **97**, 919–924.

Cleveland, D. W., Lopata, M. A., Sherline, P. and Kirschner, M. W. (1981). *Cell* **25**, 537–546.
Cleveland, D. W., Pittenger, M. F. and Feramisco, J. R. (1983). *Nature* **305**, 738–740.
Draetta, G., Brizuela, L., Potashkin, J. and Beach, D. (1987). *Cell* **50**, 319–325.
Dustin, P. (1978). *Microtubules*. Springer-Verlag, New York.
Elliott, E. M., Henderson, G., Sarangi, F. and Ling, V. (1986). *Mol. Cell. Biol.* **6**, 906–913.
Flavin, M. and Slaughter, C. (1974). *J. Bacteriol.* **118**, 56–69.
Flemming, W. (1878). *Arch. Mikrosk. Anat. Entwicklungsmech.* **16**, 302–436.
Flemming, W. (1880). *Arch. Mikrosk. Anat. Entwicklungsmech.* **18**, 151–259.
Fuller, G. M., Brinkley, B. R. and Boughter, M. J. (1975). *Science* **187**, 948–950.
Fulton, C. and Simpson, P. A. (1979). In *Microtubules* (ed. K. Roberts and J. S. Hyams), pp. 117–174. Academic Press, New York.
Gambino, J., Bergen, L. G. and Morris, N. R. (1984). *J. Cell Biol.* **99**, 830–838.
Goebl, M. G., Yochem, J., Jentsch, S., McGrath, J. P., Varshavsky, A. and Byers, B. (1988). *Science* **241**, 1331–1335.
Gounon, P. and Collenot, A. (1974). *J. Microscopie* **20**, 145–150.
Green, L. and Dove, W. F. (1984). *Mol. Cell. Biol.* **4**, 1706–1711.
Gupta, R. S. (1981). *Somat. Cell Genet.* **7**, 59–71.
Gupta, R. S. (1983). *J. Cell. Physiol.* **114**, 137–144.
Gupta, R. S., Ho, T. K. W., Moffat, M. R. K. and Gupta, R. (1982). *J. Biol. Chem.* **257**, 1071–1078.
Haarer, B. K. and Pringle, J. R. (1987). *Mol. Cell. Biol.* **7**, 3678–3687.
Hartwell, L. H. (1974). *Bacteriol. Rev.* **38**, 164–198.
Hartwell, L. H. (1978). *J. Cell Biol.* **77**, 627–637.
Hartwell, L. H., Mortimer, R. K., Culotti, J. and Culotti, M. (1973). *Genetics* **74**, 267–286.
Hatzfeld, J. and Buttin, G. (1975). *Cell* **5**, 123–129.
Hindley, J. and Phear, G. A. (1984). *Gene* **31**, 129–134.
Hirano, T. and Yanagida, M. (1988). In *The Molecular and Cell Biology of Yeasts*. Blackie, Glasgow, in press.
Huffaker, T. C., Hoyt, M. A. and Botstein, D. (1987). *Ann. Rev. Genet.* **21**, 259–284.
Huffaker, T. C., Thomas, J. H. and Botstein, D. (1988). *J. Cell Biol.* **106**, 1947–2010.
Inoué, S. (1981). *J. Cell Biol.* **91**, 131s–147s.
Izant, J. G., Weatherbee, J. A. and McIntosh, J. R. (1982). *Nature* **295**, 248–250.
Jacobs, C. W., Adams, A. E. M., Szaniszlo, P. J. and Pringle, J. R. (1988). *J. Cell Biol.* **107**, 1409–1426.
Keates, R. A. B., Sarangi, F. and Ling, V. (1981). *Proc. Natl. Acad. Sci. USA* **78**, 5638–5642.
Kemphues, K. J., Raff, R. A., Kaufman, T. C. and Raff, E. C. (1979). *Proc. Natl. Acad. Sci. USA* **76**, 3991–3995.
Kemphues, K. J., Raff, E. C., Kaufman, T. C. (1980). *Cell* **21**, 445–451.
Kemphues, K. J., Kaufman, T. C., Raff, R. A. and Raff, E. C. (1982). *Cell* **31**, 655–670.
Kilmartin, J. V. and Adams, A. E. M. (1984). *J. Cell Biol.* **98**, 922–933.
Kuriyama, R., Borisy, G. G., Binder, L. I. and Gottesman, M. M. (1985). *Exp. Cell Res.* **160**, 527–539.
Laffler, T. G., Wilkens, A., Selvig, S., Warren, N., Kleinschmidt, A. and Dove, W. F. (1979). *J. Bacteriol.* **138**, 499–504.

Lau, J. T. Y., Pittenger, M. F. and Cleveland, D. W. (1985). *Mol. Cell. Biol.* **5**, 1611–1620.
Lee, M. G. and Nurse, P. (1987). *Nature* **327**, 31–35.
L'Hernault, S. W. and Rosenbaum, J. L. (1983). *J. Cell Biol.* **97**, 258–263.
Ling, V. and Thompson, L. H. (1974). *J. Cell. Physiol.* **83**, 103–116.
Ling, V., Aubin, J. E., Chase, A. and Sarangi, F. (1979). *Cell* **18**, 423–430.
Luck, D. J. L. (1984). *J. Cell Biol.* **98**, 789–794.
Ludueña, R. F. (1979). In *Microtubules* (ed. K. Roberts and J. S. Hyams), pp. 65–116. Academic Press, New York.
Lydersen, B. K. and Pettijohn, D. E. (1980). *Cell* **22**, 489–499.
Maccioni, R. B., Serrano, L. and Avila, J. (1985). *Bioessays* **2**, 165–169.
Magnuson, T. and Epstein, C. J. (1984). *Cell* **38**, 823–833.
Mandelkow, E.-M., Hermann, M. and Ruhl, V. (1985). *J. Mol. Biol.* **185**, 311–327.
May, G. S., Gambino, J., Weatherbee, J. A. and Morris, N. R. (1985). *J. Cell Biol.* **101**, 712–719.
Mazia, D. (1961). In *The Cell* (ed. J. Brachet and A. E. Mirsky), Vol. 3, pp. 80–412. Academic Press, New York.
McCarty, G. A., Valencia, D. W., Fritzler, M. J. and Barada, F. A. (1981). *New Eng. J. Med.* **305**, 703.
McIntosh, J. R. (1979). In *Microtubules* (ed. K. Roberts and J. S. Hyams), pp. 381–441. Academic Press, New York.
McIntosh, J. R. (1982). In *Developmental Order: Its Origin and Regulation* (ed. S. Subtelny and P. B. Green), pp. 77–115. Alan R. Liss, New York.
Moir, D., Stewart, S. E., Osmond, B. C. and Botstein, D. (1982). *Genetics* **100**, 547–563.
Morris, N. R. (1976a). *Exp. Cell Res.* **98**, 204–210.
Morris, N. R. (1976b). *Genet. Res.* **26**, 237–254.
Morris, N. R. (1986). *Exp. Mycol.* **10**, 77–82.
Morris, N. R., Lai, M. H. and Oakley, C. E. (1979). *Cell* **16**, 437–442.
Morris, N. R., Weatherbee, J. A., Gambino, J. and Bergen, L. G. (1984). In *Molecular Biology of the Cytoskeleton* (ed. G. G. Borisy, D. W. Cleveland and D. B. Murphy), pp. 211–222. Cold Spring Harbor Press, New York.
Neff, N. N., Thomas, J. H., Grisafi, P. and Botstein, D. (1983). *Cell* **33**, 211–219.
Nicklas, R. B. (1971). In *Advances in Cell Biology* (ed. D. M. Prescott, L. Goldstein and E. McConkey), Vol. 2, pp. 225–297. Appleton-Century-Crofts, New York.
Nurse, P. (1985). *Trends Genet.* **1**, 51–55.
Nurse, P., Thuriaux, P. and Nasmyth, K. (1976). *Mol. Gen. Genet.* **146**, 167–178.
Oakley, B. R. (1981). In *Mitosis/Cytokinesis* (ed. A. M. Zimmerman and A. Forer), pp. 181–196. Academic Press, New York.
Oakley, B. R. and Morris, N. R. (1980). *Cell* **19**, 255–262.
Oakley, B. R. and Morris, N. R. (1981). *Cell* **24**, 837–845.
Olmsted, J. B. and Lyon, H. D. (1981). *J. Biol. Chem.* **256**, 3507–3511.
Orr, E. and Rosenberger, R. F. (1976). *J. Bacteriol.* **126**, 895–902.
Osmani, S. A., Pu, R. T. and Morris, N. R. (1988). *Cell* **53**, 237–244.
Pepper, D. A., Kim, H. Y. and Berns, M. W. (1984). *J. Cell Biol.* **99**, 503–511.
Pickett-Heaps, J. D., Tippit, D. H. and Porter, K. R. (1982). *Cell* **29**, 729–744.
Piggott, J. A., Rai, R. and Carter, B. L. A. (1982). *Nature* **298**, 391–394.
Pittenger, M. F. and Cleveland, D. W. (1985). *J. Cell Biol.* **101**, 1941–1952.
Pringle, J. R. and Hartwell, L. H. (1981). In *The Life Cycle of the Yeast*

Saccharomyces: Life Cycle and Inheritance (ed. J. N. Strathern, E. W. Jones and J. R. Broach). Cold Spring Harbor Press, New York.
Pringle, J. R., Coleman, K., Adams, A., Lillie, S., Haarer, B., Jacobs, C., Robinson, J. and Evans, C. (1984). In *Molecular Biology of the Cytoskeleton* (ed. G. G. Borisy, D. W. Cleveland and D. B. Murphy), pp. 193–209. Cold Spring Harbor Press, New York.
Raff, E. C., Brothers, A. J. and Raff, R. A. (1976). *Nature* **260**, 615–617.
Reed, S. I., Hadwiger, J. A. and Lorincz, A. T. (1985). *Proc. Natl. Acad. Sci. USA* **82**, 4055–4059.
Ripoll, P., Pimpinelli, S., Valdivia, M. M. and Avila, J. (1985). *Cell* **41**, 907–912.
Roobol, A., Paul, E. C. A., Birkett, C. R., Foster, K. E., Gull, K., Burland, T. G., Dove, W. F., Green, L., Johnson, L. and Schedl, T. (1984). In *Molecular Biology of the Cytoskeleton* (ed. G. G. Borisy, D. W. Cleveland and D. B. Murphy), pp. 223–234. Cold Spring Harbor Press, New York.
Rudolph, J. E., Kimble, M., Hoyle, H. D., Subler, M. A. and Raff, E. C. (1987). *Mol. Cell. Biol.* **7**, 2231–2242.
Rungger-Brandle, E. (1977). *Exp. Cell Res.* **107**, 313–324.
Russell, D. and Gull, K. (1984). *Mol. Cell. Biol.* **4**, 1182–1185.
Sato, Ch. (1976). *Exp. Cell Res.* **101**, 251–259.
Schedl, T., Owens, J., Dove, W. F. and Burland, T. G. (1984). *Genetics* **108**, 143–164.
Schibler, M. J. and Cabral, F. (1985a). *Can. J. Biochem. Cell Biol.* **63**, 503–510.
Schibler, M. J. and Cabral, F. (1985b). In *Molecular Cell Genetics: The Chinese Hamster Cell* (ed. M. M. Gottesman), pp. 669–710. Wiley, New York.
Schibler, M. J. and Cabral, F. (1986). *J. Cell Biol.* **102**, 1522–1531.
Schiff, P. B. and Horwitz, S. B. (1980). *Proc. Natl. Acad. Sci. USA* **77**, 1561–1565.
Schiff, P. B., Fant, J. and Horwitz, S. B. (1979). *Nature* **277**, 665–667.
Schrader, F. (1953). *Mitosis.* Columbia University Press, New York.
Serrano, L., de la Torre, J., Maccioni, R. B. and Avila, J. (1984). *Proc. Natl. Acad. Sci. USA* **81**, 5989–5993.
Sheir-Neiss, G., Lai, M. H. and Morris, N. R. (1978). *Cell* **15**, 639–647.
Shiomi, T. and Sato, K. (1976). *Exp. Cell Res.* **100**, 297–302.
Shortle, D., Haber, J. and Botstein, D. (1982). *Science* **217**, 371–373.
Simanis, V. and Nurse, P. (1986). *Cell* **45**, 261–268.
Smith, B. J. and Wigglesworth, N. M. (1972). *J. Cell. Physiol.* **80**, 253–260.
Stetten, G. and Lederberg, S. (1970). *Science* **168**, 485–487.
Thomas, J. H., Novick, P. and Botstein, D. (1984). In *Molecular Biology of the Cytoskeleton* (ed. G. G. Borisy, D. W. Cleveland and D. B. Murphy), pp. 153–174. Cold Spring Harbor Press, New York.
Thompson, L. H. and Lindl, P. A. (1976). *Somat. Cell Genet.* **2**, 387–400.
Toda, T., Umesono, K., Hirata, A. and Yanagida, M. (1983). *J. Mol. Biol.* **168**, 251–270.
Toda, T., Adachi, Y., Hiraoka, Y. and Yanagida, M. (1984). *Cell* **37**, 233–242.
Umesono, K., Toda, T., Hayashi, S. and Yanagida, M. (1983). *J. Mol. Biol.* **168**, 271–284.
Vallee, R. B. (1984). In *Cell and Muscle Motility*, Vol. 5 (ed. J. W. Shay), pp. 289–311. Plenum Press, New York.
Van Tuyl, J. M. (1977). Thesis, Agricultural University, Wageningen, The Netherlands.
Wang, R. J. (1974). *Nature* **248**, 76–78.

Wang, R. J. (1976). *Cell* **8**, 257–261.
Wang, R. J. and Yin, L. (1976). *Exp. Cell Res.* **101**, 331–336.
Wang, R. J., Wissinger, W., King, E. J. and Wang, G. (1983). *J. Cell Biol.* **96**, 301–306.
Warr, J. R. (1968). *J. Gen. Microbiol.* **52**, 243–251.
Warr, J. R. and Durber, S. (1971). *Exp. Cell Res.* **64**, 463–469.
Warr, J. R. and Gibbons, D. (1974). *Exp. Cell Res.* **85**, 117–122.
Warr, J. R., Flanagan, D. J. and Anderson, M. (1982). *Cell Biol. Int. Rep.* **6**, 455–460.
Watts, F. Z., Miller, D. M. and Orr, E. (1985). *Nature* **316**, 83–85.
Watts, F. Z., Shiels, G. and Orr, E. (1987). *EMBO J.* **6**, 3499–3505.
Weatherbee, J. A. and Morris, N. R. (1984). *J. Biol. Chem.* **259**, 15 452–15 459.
Weatherbee, J. A., Luftig, R. B. and Weihing, R. R. (1980). *Biochemistry* **19**, 4116–4123.
Weatherbee, J. A., May, G. S., Gambino, J. and Morris, N. R. (1985). *J. Cell Biol.* **101**, 706–711.
Weber, K., Bibring, T. and Osborn, M. (1975). *Exp. Cell Res.* **95**, 111–120.
Weil, C. F., Oakley, C. E. and Oakley, B. R. (1986). *Mol. Cell Biol.* **6**, 2963–2968.
Whitfield, C., Abraham, I., Ascherman, D. and Gottesman, M. M. (1986). *Mol. Cell. Biol.* **6**, 1422–1429.
Wilson, E. B. (1925). *The Cell in Development and Heredity*. MacMillan, New York.
Wissinger, W. and Wang, R. J. (1978). *Exp. Cell Res.* **112**, 89–94.
Wittenberg, C. and Reed, S. I. (1988). *Cell* **54**, 1061–1072.
Wittenberg, C., Richardson, S. L. and Reed, S. I. (1987). *J. Cell Biol.* **105**, 1527–1538.
Wood, J. S. and Hartwell, L. H. (1982). *J. Cell Biol.* **94**, 718–726.
Yanagida, M. (1987). *Microbiol. Sci.* **4**, 115–118.
Zieve, G. and Solomon, F. (1982). *Cell* **28**, 233–242.

CHAPTER 9

Mitosis *In Vitro*

W. ZACHEUS CANDE

Department of Botany, University of California, Berkeley, California, USA

I. Introduction

The logic of what we need to know about the mechanism of chromosome movement is straightforward and has many precedents in molecular biology and in the motility paradigms of muscle and flagella. We need to know the biochemical identity of the structural, mechanochemical and regulatory molecules, their precise localization in the spindle, and how their location changes during anaphase. In order to answer these questions it is essential to have a functional *in vitro* model system that can be used to dissect and then reconstitute those elements of the spindle responsible for chromosome movement. An alternative strategy is to genetically dissect mitosis, generate mutant phenotypes and analyse them with the techniques of molecular biology (Chapter 8). Ultimately, however, these approaches complement each other, since some understanding of function will be required in order to understand the nature of these mutants and the relative roles of the affected gene products during mitosis. This can only come from an *in vitro* analysis of mitotic processes.

Another approach that has recently turned out to be very useful for understanding mitosis has been to introduce proteins or other molecules into dividing cells via microinjection and to use the intact cell as the equivalent of a test tube. McIntosh, Salmon and co-workers (this volume) have microinjected fluorescently labelled tubulin into cells and monitored redistribution of tagged tubulin subunits after photobleaching of the spindles (Salmon *et al.*, 1984; Saxton *et al.*, 1984). Several laboratories have injected calcium, and calcium-

MITOSIS: Molecules and Mechanisms
ISBN 0-12-363420-2

chelating dyes into cells with the stated goal of determining the role of calcium in regulating mitotic events (Poenie *et al.*, 1986; Ratan *et al.*, 1986; Chapter 7). Antibodies against myosin and N-ethylmaleimide-modified myosin have been microinjected into cells to determine whether actomyosin is involved in moving chromosomes during anaphase (Meeusen *et al.*, 1980; Kiehart *et al.*, 1982). All of these studies are attempts to find substitutes for a truly functional *in vitro* system. These approaches to studying function have been successful but are still plagued with all the problems of complexity of interpretation that comes from working with living cells.

One key question that *in vitro* systems could be used to resolve is how the spindle generates the forces that move chromosomes. Two strategies have been used in an attempt to answer this question. On the one hand, spindles have been isolated and purified by a variety of procedures, and attempts were then made to induce them to move chromosomes. Although Mazia and Dan (1952) first isolated mitotic apparatuses from sea-urchin zygotes over 30 years ago, the major questions raised by Mazia in his original study were left unanswered: what exactly is the composition of the spindle and what are the enzymatic motors responsible for chromosome separation? Hoffman-Berling (1954a,b) developed an alternative method for analysing non-muscle cell motility, including mitosis, using gently permeabilized cell models (reviewed in Snyder, 1981). However, lysed cells rival intact cells in complexity and this approach, although useful for describing spindle physiology, has not yet led to an identification of the enzymatic motors responsible for chromosome movement.

In this review of mitosis *in vitro* I have two goals. First, I want to analyse why it has been so difficult to develop model systems for studying mitotic events *in vitro*. I will use this exercise as a framework to describe what types of questions could be answered by an *in vitro* approach, to review briefly some of the literature of spindle isolation, and to analyse what preparations may be good candidates for further experimentation. This will be done in the next section. Second, I want to summarize what has been learned about the mechanism of anaphase chromosome movement *in vitro*, in both a positive and negative way. And in particular I want to focus on my own work and summarize what I have learned about mitosis using permeabilized mammalian tissue culture cells and isolated diatom spindles. This will be done in the last section. Since there have been several good reviews on methods of spindle isolation and spindle biochemistry during the last few years (Snyder, 1981; Zimmerman and Forer, 1981; Pratt, 1984; Petzelt, 1979; Sakai, 1978; McIntosh, 1977), I will not attempt to survey the literature on spindle isolation procedures or spindle composition. There are several articles in this volume that report on the biochemistry of spindle parts, including spindle MAPs (Chapter 5), kinetochores (Chapter 3) and centrosomes (Chapter 2).

II. The Problems Inherent in the *In Vitro* Approach

A. Stabilization

The mitotic spindle has many unique features as an organelle that make it difficult to analyse with biochemical techniques. First of all, the spindle is a very labile structure. Its existence is transitory and, as it moves chromosomes, it self-destructs. A major challenge in isolating mitotic spindles is creating a stable spindle that will survive the rigours of the isolation procedure. This is a difficult task, since the distinction between an unnatural fixation and a physiologically useful stabilization of spindle components may be only a matter of degree. A variety of agents have been used to stabilize spindles and most media have been designed to stabilize microtubules even at the expense of other *bona fide* spindle components such as endoplasmic reticulum (ER)-derived vesicles or chromosomes.

The history of spindle isolation procedures can readily be analysed in terms of the agents used for spindle stabilization (see McIntosh, 1977, for a good review of early efforts at spindle isolation). The first method, developed by Mazia and Dan (1952), relied on cold ethanol to stabilize the spindle and detergents to extract contaminants. This method was quickly acknowledged to be undesirable, because the stabilization was, in essence, irreversible. Mazia and co-workers then found that sucrose and dithiodiglycerol could be used to stabilize spindles (Mazia *et al.*, 1961). Ultimately, Kane found that acidic pH and any of a series of glycols, such as hexylene glycol, would give a useful stabilization of spindle components (Kane, 1962, 1965). These agents were non-specific in nature, and, as Kane has pointed out, there is a good correlation between spindle stabilization and poor protein solubilization.

Spindle isolates prepared by these techniques have been of limited usefulness, since neither do they have the lability properties of *in vivo* spindles nor do they move chromosomes. Their chief value has been to demonstrate that the spindle is a discrete organelle with many fibrous elements, i.e. *microtubules*. The results of structural studies with spindles prepared in this manner have been reviewed in several places (McIntosh, 1977; Zimmerman and Forer, 1981).

The most lifelike spindle isolates have been prepared in media with properties that favour microtubule assembly *in vitro*. These procedures were inspired by the breakthrough discoveries of Weisenberg (1972) for obtaining microtubule polymerization *in vitro*, and by the subsequent development of methods of using cycles of microtubule polymerization/depolymerization to purify tubulin and MAPs (microtubule-associated proteins) from brain homogenates (reviewed in Kirschner, 1978). These media all have in common the following ingredients, including glycerol (and/or DMSO), EGTA, low-salt, slightly acidic pH and buffers like PIPES or MES.† Cytoplasmic contaminants are usually removed from spindles by using non-ionic detergents such as Triton X-100 or mechanical shear and filtration (Silver *et al.*, 1980). The logic of this approach was carried to an extreme by several groups (Rebhun *et al.*, 1974; Inoué *et al.*, 1974) who lysed cells in polymerizable neurotubulin in order to maintain spindle structure. The extent of spindle birefringence was dependent on the tubulin concentration in the homogenization medium, and the augmentation of spindle birefringence by high tubulin concentrations suggests that neurotubulin was incorporated into these spindles during the lysis procedure.

Spindles prepared by homogenizing cells in tubulin or in glycerol/EGTA-based medium are more labile than spindles prepared in hexylene glycol or ethanol. It has been reported that some of these isolates are cold-labile (Forer and Zimmerman, 1974; Rebhun *et al.*, 1974; Inoué *et al.*, 1974), Ca^{2+}-labile (Salmon and Segall, 1980; Dinsmore and Sloboda, 1988), contain vesicles capable of sequestering Ca^{2+} (Silver *et al.*, 1980) and have interesting proteins associated with them such as dynein-like ATPases (Pratt *et al.*, 1980) and creatine kinase (Silver *et al.*, 1983). Sakai *et al.* (1975, 1976, 1979) and Cande and McDonald (1985, 1986) have shown that spindles isolated in glycerol-containing medium will undergo some aspects of chromosome movement *in vitro*. These results will be described in more detail in a later section.

B. Complexity of spindle function

The complexity and multiplicity of the processes that take place during mitosis make it difficult for the investigator to decide what criteria should be used to evaluate the usefulness of the *in vitro* model. The spindle is primarily a sorting device and secondarily a motile machine. In order to ensure proper chromosome segregation, the cell centres must replicate and then during prophase form a bipolar axis. The forces that are required to establish overall spindle architecture and move chromosomes to the metaphase plate are not necessarily the same forces and processes required to move chromosomes polewards during anaphase. Thus, the fully formed metaphase and anaphase spindle may no longer contain elements that are essential for the formation of the spindle. Spindle formation also contains within it a chromosome cycle. During spindle formation the chromatin undergoes structural rearrangements and prepares itself to interact with the spindle. The interactions between the chromosomes and the newly formed spindle may be triggered by nuclear envelope breakdown.

†DMSO, dimethylsulphoxide; EGTA, Ethylene glycol bis(β-aminoethyl ether) N,N,N′-tetra-acetic acid; PIPES, Piperazine-N,N′bis[2-ethanesulfonic acid]; MES, 2-(N,Morpholino)-ethanesulfonic acid.

Although we know very little about spindle formation, several recent developments suggest that early mitotic events involving chromosome condensation and nuclear envelope breakdown may be susceptible to analysis by *in vitro* methods. Using cell-free extracts prepared from *Xenopus* oocytes, Lohka and Maller (1985) and Miake-Lye and Kirschner (1985) were able to induce early mitotic events, including nuclear envelope breakdown, chromosome condensation and spindle formation, *in vitro*. These events can be induced in somatic interphase nuclei from a variety of sources, provided that the nuclei are incubated in extracts prepared from MPF- (maturation promoting factor)-stimulated eggs. Although these extracts are almost as complex as the egg itself, it should be possible to fractionate these homogenates further in order to define what components are required for spindle assembly. The recent papers by Dunphy *et al.* (1988) and Gautier *et al.* (1988) may prove to be an important step in this direction. Second, it has been possible to isolate centrosome complexes from a variety of dividing and interphase cells, to study their ability to nucleate microtubules *in vitro* (Mitchison and Kirschner, 1984; Toriyama *et al.*, 1984; Kuriyama, 1984), and to prepare antibodies against their components (Kuriyama and Borisy, 1985). Eventually it may be possible to combine the above two approaches to understand what processes are involved in regulating centrosome replication and spindle formation.

Most studies of mitosis *in vitro* have focused on the mechanism of anaphase chromosome movement, and attaining chromosome movement *in vitro* has been defined as the chief goal of *in vitro* studies (Zimmerman and Forer, 1981). However, this goal has been difficult to achieve. Anaphase chromosome movement consists of several distinct phases (Inoué, 1981) and the conditions required for the maintenance of one phase may not be compatible with the preservation of other phases *in vitro*. During chromosome-to-pole movement (anaphase A), the kinetochore-attached microtubules must shorten. During spindle elongation (anaphase B), polar-attached microtubules are rearranged as microtubules of one half-spindle slide over microtubules of the other half-spindle in the zone of microtubule overlap (McDonald *et al.*, 1979). These microtubules may also increase in length during this process (McIntosh *et al.*, 1985). Finally, in some cells, there may be forces generated at the spindle poles by interactions between astral microtubules and elements of the surrounding cytoplasm that help to pull the spindle poles further apart at the end of anaphase (see Aist and Berns, 1981).

The conditions required to maintain these three classes of motile events may be mutually exclusive. In order to investigate the relationships between kinetochore microtubule depolymerization and the generation of forces that move chromosomes polewards, kinetochore microtubules must be destabilized *in vitro* in a controlled fashion. However, to understand the mechanisms of microtubule sliding, the overlapping arrays of microtubules in the spindle midzone must be stabilized. Medium designed to stabilize microtubules involved in anaphase B may interfere with anaphase A (Cande *et al.*, 1981). Second, different mechanochemical enzymes may be involved in anaphase A versus B and the conditions required for their reactivation may differ. Finally, attempts to purify a spindle free of cytoplasmic contamination may disrupt the interactions of astral microtubules or polar complexes with cytoplasmic organelles. Only in permeabilized cell models has it been possible to get both anaphase A and B at the same time. In spindle isolates, either anaphase A (Sakai *et al.*, 1976) or anaphase B (Goode and Roth, 1969; Cande and McDonald, 1985) has been favoured.

Another set of problems unique to studying the mechanochemistry of mitosis is the slow speed and irreversible nature of chromosome movements. Chromosomes in a PtK_1 cell move polewards at a rate of 1–2 μm per minute, and spindles elongate at slightly slower rates (see Carlson, 1977, for a review). For comparison, this is two orders of magnitude slower than kinesin-based particle movements *in vitro* (Vale *et al.*, 1985). This suggests either that there are many fewer enzymatic motors involved in moving chromosomes or that non-mechanochemical events regulate or control the rates of spindle elongation and chromosome-to-pole movement. Reactivation and inhibitor studies designed to study the physiology of mitotic chromosome movements are more difficult to complete if the timescale needed to measure a significant translocation of chromosomes and poles must be measured in units of minutes rather than seconds.

C. Appropriateness of starting material

Although mitosis has been described in many different types of cells, only a few classes of dividing cells have been chosen for *in vitro* studies. The favourite starting material has been invertebrate marine eggs, especially those of sea urchins. These cells have been selected because they can be obtained cheaply in large numbers, and because the eggs shed from one female undergo synchronous waves of cell division after fertilization. The spindles isolated from sea-urchin zygotes are very large, on the order of 20–30 μm in length. However, there are several problems associated with using this material. First of all, these spindles are not highly ordered structurally and the large size of the marine egg makes it difficult to develop good fixation regimes and analyse the structure with ultrastructural and immunocytochemical techniques. Second, the marine egg contains within it many components needed for future cell-cycle events, including large amounts of yolk (see also Chapter 5). Since the sea-urchin spindle, like other higher eukaryotic spindles, is embedded in cytoplasm and not set off from the rest of the cell, it is difficult to free these spindles of cytoplasmic contamination except by rigorous detergent-extraction procedures.

Another obvious choice for starting material is mammalian tissue culture cells. Using drug regimes to accumulate cells entering cell division and a mitotic shake-off as an enrichment technique, it is possible to obtain large numbers of dividing cells with synchrony that rivals that of marine eggs after fertilization (Zieve *et al.*, 1980b). However, it is more expensive to grow these cells and more difficult to obtain the bulk amounts required for biochemistry. Cabral *et al.* (1984) have utilized somatic cell genetics to analyse microtubule function in CHO cells, and the ability to use these techniques adds a dimension of analysis for understanding mitosis that is not yet presently available with sea-urchin zygotes (Chapter 8). Although tissue culture cells do not contain yolk, their spindles are surrounded by a cage of intermediate filaments that is difficult to disrupt (Zieve *et al.*, 1980a). These spindles, although less than one-half the size of sea-urchin spindles, are also disordered compared to many lower eukaryote spindles and hence are difficult to analyse at an ultrastructural level.

PtK_1 and PtK_2 cells have been used for many immunofluorescence studies of mitosis because these cells, when compared to other tissue culture cells, have large spindles, and the cells remain attached to coverslips during mitosis. At the onset of mitosis, most tissue culture cells round up; however PtK_1 cells only thicken slightly and pull in their cell margins. Because of these properties, PtK_1 cells have been used for studying chromosome movement and centrosome function *in vitro* by permeabilizing them with non-ionic detergents (reviewed in Snyder, 1981). However, this cell line is not suitable for bulk spindle isolation, because cells grow slowly and mitotic shake-off techniques are ineffective in enriching for dividing cells.

Lower eukaryotic cells have not been used as starting points for *in vitro* studies to any significant degree, yet they offer many obvious advantages. Many fungal spindles, such as those found in yeast, *Aspergillus*, *Physarum* (plasmodial stage) and *Dictyostelium* are intranuclear, hence it may be easier to purify these spindles in a functional form, free of cytoplasmic contamination. The techniques of molecular biology are now being used to analyse mitotic spindle functions in yeast (Broach, 1986; and Chapter 8) and *Aspergillus* (Oakley and Morris, 1981), and significant progress is being made using these organisms. Although attempts have been made to reactivate spindle elongation with isolated yeast spindles (King *et al.*, 1982), these efforts have not yet been successful.

Diatoms and protozoan spindles have many unique structural attributes that make them favourable objects for morphological studies (Pickett-Heaps and Tippit, 1978; Inoué and Ritter, 1975). Diatoms have been important model systems for studying the morphological changes associated with anaphase chromosome movement because the fibrous systems responsible for anaphase A versus B are spatially separated, and the central spindle, which is responsible for anaphase B, is a paracrystalline array of microtubules (McDonald *et al.*,

1977, 1979). Recently, Cande and McDonald (1985, 1986) have succeeded in reactivating anaphase B *in vitro* using isolated diatom central spindles.

From our experience with diatoms, the major difficulty in working with lower eukaryotes is lack of understanding of their natural history. Unlike the case with sea-urchin eggs or tissue culture cells, the methods of cell synchronization and cell culture may not have been well worked out and will require more effort on the part of the investigator before appropriate numbers of dividing cells can be obtained.

III. Studies of Chromosome Movement *In Vitro*

A. Goals

The goal of an *in vitro* approach to mitosis should be to define at a molecular, physiological and structural level those processes that are responsible for the formation of the mitotic spindle and the movement of chromosomes during anaphase. One would like to use a functional model spindle to define and characterize the mechanochemical elements responsible for moving chromosomes and the regulatory machinery associated with it, to make a "parts list" of spindle components that are relevant to these functions, and to know how these parts move and are redistributed during prometaphase and anaphase.

B. Early studies

Although Mazia and Dan (1952) first developed methods of isolating mitotic spindles in bulk over 30 years ago, these preparations were non-functional with respect to chromosome movement. At about the same time, Hoffman-Berling (1954a,b) developed methods of stabilizing fibroblast cytoskeletons using glycerol. Dividing cells in his preparations would undergo anaphase chromosome movements after ATP addition. Movement of chromosomes polewards (i.e. anaphase A) could be obtained using reactivation medium identical to that required for the contraction of the interphase cell cortex. However, lengthening of the spindle (anaphase B) required the addition of a plasticizer such as 0.8 M urea or 40 mM ATP. These differences lead Hoffman-Berling to conclude that the two types of movement had a different mechanochemical basis, and that anaphase A was related to muscle-like contractions. He also suggested that relaxing factors, i.e. Ca^{2+}-sequestering membrane systems, could be responsible for the action of plasticizing agents *in vivo*. Several groups have attempted to repeat these observations using a variety of glycerinated cell types (Cande, McIntosh and others, personal communications) but have met with little success. Since the physiological descriptions of reactivation requirements are not very extensive and no analysis was made of the structural basis of these movements, Hoffman-Berling's observations have not been very useful for understanding mitotic mechanisms.

Using phosphate-buffered salt solutions, Goode and Roth (1969) dissected spindles from the giant amoeba *Chaos carolineus* and were able to remove them as a unit. These spindles are unusual barrel-shaped structures with small chromosomes arrayed as a plate and contain prominent parallel fibres in the spindle midzone. During anaphase, the spindles elongate approximately 10-fold but with little chromosome-to-pole movement. Although Goode could not reactivate spindle elongation after isolation, addition of ATP to the dissection medium lead to doubling of spindle length during the dissection procedure. Because of the difficulty in obtaining material, Goode (personal communication) was unable to characterize this response in any detail. However, this study does demonstrate that anaphase B is an ATP-dependent process.

C. Reactivation of isolated sea-urchin spindles

Sakai and associates (1975) isolated spindles from sea-urchin eggs using a glycerol stabilization medium developed by Sakai and Kuriyama (1974) with the addition of GTP, glutathione, ascorbic acid, cAMP and brain tubulin. These spindles were cold-labile and spindle birefringence was reduced by addition of Ca^{2+} and sulphydryl reagents, but not colchicine. When anaphase spindles were treated with tubulin concentrations calculated to be greater than the subunit pool concentration *in vivo* (greater than 3 mg/ml), tubulin addition occurred and the spindles increased in length as the asters grew. However, chromosome translocation was irregular and in proportion to the general increase in spindle dimensions. In 0.5 mg/ml tubulin, birefringence changes were not as dramatic, but chromosome translocation occurred in a more lifelike manner. Although chromosomes only moved part way to the poles (2–5 μm), chromosome-to-pole distances decreased measurably and reproducibly, and the pole-to-pole distances increased slightly. These movements occurred very slowly: chromosome movements took place over an hour, whereas *in vivo* it would have taken place in less than 5 minutes. These movements only took place in ATP, and other nucleotides would not substitute. Subsequently, Sakai *et al.* (1976) showed that antisera against dynein fragment A inhibited chromosome movement, but antisera against anti-starfish egg myosin had no effect. Sakai *et al.* (1979) were also able to demonstrate that these movements were blocked by vanadate and that vanadate-sensitive ATPase activity was associated with the spindle preparation.

Although these results were potentially very exciting, they have had relatively little impact on our thinking about the mechanism of anaphase chromosome movement. The major arguments made in the literature against accepting these results is that the movements are extremely slow, usually less than 1% of *in vivo* rates (Forer, 1985). Another problem is that it is not possible in many experiments to determine whether the chromosome translocations that occur are the equivalent of anaphase A, anaphase B, or both processes working in concert. The results with the antibodies are the only direct evidence available in the literature that a dynein-like motor may be involved in chromosome movement. Unfortunately, these experiments were not performed with affinity-purified antisera. Recent experiments by Asai (1985) demonstrate that this antiserum reacts not only with a cytoplasmic dynein-like ATPase found in sea-urchin eggs but also with other polypeptides of unknown function. The negative results with antimyosin may be more conclusive, since Mabuchi and Okuno (1977) and Kiehart *et al.* (1982) have injected the same affinity-purified antibodies into dividing sea-urchin and starfish eggs and demonstrated that these immunoglobulins block cleavage but not mitosis.

More recently, using isolated sea-urchin spindles, Salmon (1982) described spindle microtubule rearrangements that can be interpreted as mimicking anaphase A chromosome movements. Salmon isolated calcium-labile spindles in EGTA/glycerol buffers (Salmon and Segall, 1980). In the presence of micromolar ATP and 2 μM Ca^{2+}, as the spindle shrinks and poles move closer together, chromosomal fibres shorten, moving chromosomes polewards (Salmon, 1982). This anaphase-A-like chromosome translocation occurs rapidly with rates approaching 10 μm/min. Similar movements, including rapid inward movement of poles and chromosomes moving polewards, have been observed in permeabilized PtK_1 cells when they were lysed in 1 μM free Ca^{2+} and ATP (Cande, 1981). These movements, although they mimic anaphase A are difficult to interpret because they take place as the spindle collapses. However, given the controversies about the nature of chromosome-to-pole movement and the role of nucleotides during this process, it would be worthwhile to repeat these experiments.

D. Permeabilized PtK_1 cells

Using dividing mammalian tissue culture cells, we have described the requirements for maintenance of anaphase chromosome movement after lysis. This research had two goals. First, we wanted to determine whether the mechanochemistry of chromosome movement was unique to the spindle or whether it was similar to better-described systems such as muscle and cilia. Second, we wanted to compare the physiological requirements for anaphase A versus anaphase B and determine what distinctions could be made between them.

1. *Lysis methods*

We found that our strategy for studying mitosis was limited by the complexity of the lysed cell and the lability of the spindle (Cande, 1982a). Chromosome movement was difficult to maintain after lysis. For example, medium containing 30% glycerol stabilized spindle birefringence at *in vivo* levels but did not stabilize spindle function. Nor would spindle function survive extensive cell extraction. When high concentrations of detergent were used to permeabilize cells, chromosome movement ceased once 30% of the cell's proteins was extracted (Cande, 1982a). As a consequence of these problems, we had to alter our lysis procedure during the course of this research.

At first, dividing cells were permeabilized with Triton X-100 and neurotubulin was included in the medium to stabilize spindle microtubules (Cande *et al.*, 1974). Alternative methods of lysis and spindle stabilization were developed because chromosome movement occurred infrequently under these conditions. In subsequent studies, 2.5% polyethylene glycol (M_r 20 000) was used to partially stabilize spindle structure and a more hydrophobic detergent, Brij 58, was used for cell permeabilization. Under these conditions, spindle birefringence faded slowly for 15 min after lysis and chromosome movements continued for 10 min after lysis at 20–50% of *in vivo* rates (Fig. 9.1). Chromosomes usually moved completely to the poles, and spindle poles moved several micrometres apart. In work reported by Cande and Wolniak (1978), Cande (1979, 1980), and Cande *et al.* (1981) no attempts were made to buffer the free Ca^{2+} levels after lysis; however, in subsequent studies free Ca^{2+} levels

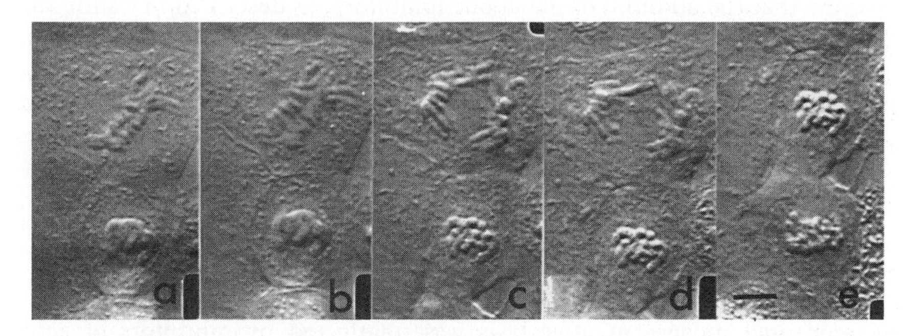

Figure 9.1. Effect of NEM-S_1 on chromosome separation and cleavage in permeabilized PtK_1 cells. (a) 0.1 min before lysis; (b) 1 min after lysis in 7 mg/ml NEM-S_1; (c) 8 min after lysis; (d) 12 min after lysis; (e) adjacent cleaving cell 12 min after lysis. In micrographs (a)–(d) the top half of the cell undergoing cleavage can be seen in the lower half of the frame. Bar = 5 μm. (Reprinted with the permission of the *Journal of Cell Biology*; Cande *et al.*, 1981.)

were buffered to 0.1 μM with a 10 mM EGTA buffer system (Cande, 1982b,c, 1983). This change had important consequences for maintenance of reliable chromosome movement, since spindle stability and function are affected by small changes in the free calcium level. Increasing the free calcium level from 0.1 μM to 0.5 μM accelerated the loss of spindle birefringence and affected the rate of chromosome movement (unpublished data, and Cande, 1981).

These changes in lysis conditions may have affected the interpretation of some of our earlier results. In our original studies, in the absence of a Ca/EGTA buffer system, we showed that metabolic inhibitors and vanadate—an ATPase poison—blocked both anaphase A and B (Cande and Wolniak, 1978; Cande, 1979). In subsequent studies, using a Ca/EGTA buffered medium, I have shown that in the absence of ATP, anaphase A will still continue after lysis, whereas anaphase B will not (Cande, 1982b,c, 1983). Spindle elongation under these conditions is preferentially sensitive to vanadate, sulphydryl reagents (Cande, 1982b), EHNA erythro-9-[3-(2-hydroxynonyl)]adenine (Cande, 1982c) and is preferentially maintained by an endogenous creatinase–kinase system (Cande, 1983). This difference in results suggests that some of the effects of vanadate and metabolic inhibitors on chromosome movement after lysis as described by Cande and Wolniak (1978) may have been indirect, due to the effects on spindle structure by release of sequestered Ca^{2+} from spindle-associated membrane vesicles. This interpretation is supported by our observation (Cande, unpublished results) that cells stained with chlorotetracycline, a dye that fluoresces in an environment of high free Ca^{2+}, lose their stain slowly when lysed in the absence of a Ca/EGTA buffer system but lose their stain rapidly, i.e. in 1–2 min, when lysed in a 10 mM Ca/EGTA buffer. Thus, it is possible that the addition of metabolic inhibitors, as described in Cande and Wolniak (1978), may have caused membrane vesicles in lysed cells to lose sequestered Ca^{2+} and to have altered overall spindle structure and thus affect both anaphase A and B. An alternative explanation suggested by Spurck and Pickett-Heaps (1987) is that lysis in the absence of Ca^{2+} and nucleotides stabilizes kinetochore microtubules and blocks the microtubule depolymerization required for chromosome-to-pole movement.

2. *Role of actomyosin during chromosome movement*

Chromosome movement after lysis was unaffected by inhibitors of actin polymerization/depolymerization and of actomyosin-based contractions, whereas it was affected by agents that altered tubulin stability (Cande *et al.*, 1981). Since cleavage furrow contraction continued after lysis in a medium identical to that which supported anaphase chromosome movement (Cande, 1981), this provided an internal control that showed whether inhibitors of actomyosin-based motility could penetrate and affect contractile processes in a lysed cell. Lysis in NEM-modified S_1, phalloidin and cytochalasin B had no effect on chromosome-to-pole movement or spindle elongation; however, these inhibitors did block cleavage in lysed cells (Fig. 9.1; Cande *et al.*, 1981). In addition, we were able to demonstrate that lysis in neurotubulin augmented spindle birefringence and slowed down chromosome-to-pole movement after lysis, demonstrating that proteins could penetrate the lysed cell and affect spindle function. These studies are consistent with microinjection studies with antibodies against myosin (Mabuchi and Okuno, 1977; Kiehart *et al.*, 1982) and with NEM-modified heavy meromyosin (Meeusen *et al.*, 1980; Zavortink *et al.*, 1981), which demonstrate that mitosis continues in the presence of these proteins while cleavage does not. Although these experiments suggest that the mechanism of anaphase chromosome movement does not involve actomyosin-based mechanochemistry, for a counterview and a critique of these experiments see Forer (1985).

3. *Anaphase A in permeabilized cells*

The physiological requirements for maintenance of chromosome movement after lysis suggest that the mechanochemical systems responsible for anaphase A and B are different. After lysis in 0.1 μM free Ca^{2+}, chromosomes continue to move polewards at 30–50% of *in vivo* rates even in the absence of ATP (Cande, 1982b). This movement is unresponsive to levels of inhibitors, such as EHNA (Cande, 1982b,c), vanadate, sulphydryl reagents and EDTA, that block anaphase B in the same cells. The only way that it is possible to manipulate the rate of chromosome-to-pole movement *in vitro* is by stabilizing or destabilizing spindle microtubules after lysis. Addition of taxol and tubulin preferentially block anaphase A; increasing the free Ca^{2+} levels to 0.5 μM causes an overall spindle collapse, and in a few cases a transient chromosome translocation that may be equivalent to an acceleration of anaphase A (Cande, 1981; Cande *et al.*, 1981) (Fig. 9.2). These results suggest that ATP is not required for anaphase A, provided that the process of cell permeabilization generates an appropriate depolymerization of kinetochore microtubules. Using digitonin-permeabilized tissue culture cells, Spurck and Pickett-Heaps (1987) obtained similar results. After permeabilization, chromosome movement ceased. Chromosome-to-pole movement could be restored by addition of ATP, ATPγS or by addition of a microtubule-destabilization medium containing calcium. They suggested that the role of ATP during anaphase A was to regulate microtubule depolymerization.

Recent studies using isolated chromosomes complexed with microtubules are consistent with the studies of anaphase A in permeabilized cells. Mitchison and Kirschner (1985a,b) demonstrated that kinetochores *in vitro* can capture microtubules and in the presence of ATP incorporate new tubulin onto

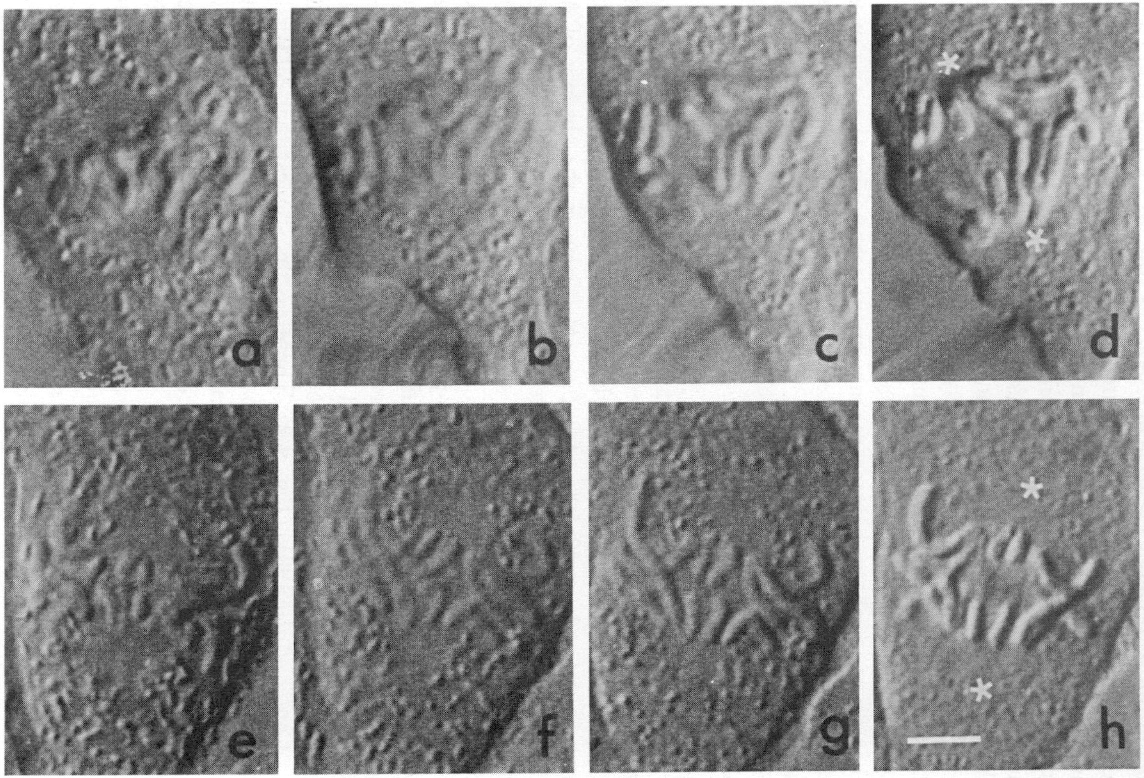

a b c d
e f g h

Figure 9.2.

pre-existing microtubules at the kinetochore. In microtubule-destabilizing medium, microtubules attached to the kinetochore shrink and lose subunits at the kinetochore (Koshland *et al.*, 1988). This latter movement is equivalent to a movement of the chromosome towards the "minus end" of the microtubule and is analogous to the movement of chromosomes polewards during anaphase A. It does not require ATP *in vitro*. Kirschner and co-workers (Mitchison and Kirschner, 1985a,b; Koshland *et al.*, 1988) suggest that regulation of microtubule lengths at the kinetochore is an ATP-dependent process, but that anaphase A does not require ATP.

There are several possible mechanisms of force generation that are consistent with the observations on permeabilized cells. First, as predicted by Inoué and Sato (1967), kinetochore microtubule depolymerization may be the only step required for moving a chromosome poleward. Alternatively, the energy input for anaphase A may occur during spindle assembly. Luby and Porter (1980) and Pickett-Heaps *et al.* (1982) have suggested that kinetic energy is stored in a cytomatrix and that this matrix would be stretched out over the kinetochore microtubules during prometaphase. Then, as the kinetochore microtubules shorten, the collapse of the cytomatrix could generate the forces necessary for anaphase A without a further energy input. However, the observations of the behaviour of kinetochores *in vitro* are inconsistent with this hypothesis (Koshland *et al.*, 1988).

4. *Anaphase B*

The requirements for maintenance of anaphase B after cell permeabilization are consistent with the involvement of a dynein-like ATPase for generating the forces responsible for moving the spindle poles apart. Spindle elongation requires ATP hydrolysis and the nucleotide specificity of this process is similar to that required for maintenance of flagellar beat in demembraned models (Gibbons, 1981). In contrast, both myosin- and kinesin-based motility have less-stringent nucleotide specificity (Pollard and Weihing, 1974; Vale *et al.*, 1985). Spindle elongation in permeabilized PtK_1 cells is blocked by vanadate and sulphydryl reagents and by EHNA (Cande, 1982b,c, 1983). Although these inhibitors are relatively non-specific, they all affect dynein ATPase

---

Figure 9.2. Effect of taxol on chromosome movement in permeabilized PtK_1 cells. (a)–(d) Chromosome movement in the absence of taxol: (a) 0.1 min before lysis; (b) 2 min after lysis; (c) 4 min after lysis; (d) 10 min after lysis. Chromosomes approach the poles (asterisks). (e)–(h) Chromosome movement after lysis in 0.75 μM taxol: (e) 0.1 min before lysis; (f) 2 min after lysis; (g) 4 min after lysis; (h) 9 min after lysis; the chromosomes do not approach the poles (asterisks); however, the spindle does elongate. Bar = 5 μm.

activity, whereas kinesin is not affected by vanadate or sulphydryl reagents except at very high concentrations (Vale *et al.*, 1985). However, unlike the case for dynein-based movements, AMPPNP (adenylylimidodiphosphate) even at equimolar ratios with ATP will inhibit spindle elongation (Cande, 1982b).

Given the morphology of the mammalian spindle and the complexity of lysed cells, it is not possible to study the structural rearrangements of the microtubules that occur during spindle elongation. However, it is unlikely that tubulin polymerization contributes to the extent of anaphase B observed *in vitro*. Anaphase B movements are not stimulated by addition of taxol or tubulin in the lysis medium, although under these same conditions there is an increase in astral microtubule birefringence after neurotubule addition (Cande *et al.*, 1981). If microtubule polymerization was the primary force-generating event after lysis, GTP addition should have been a requirement for spindle elongation. However, *in vitro* GTP could not substitute for ATP. These results are consistent with models of force generation that predict that microtubule–microtubule interactions are the primary mechanochemical event during anaphase B (Margolis *et al.*, 1978; McIntosh *et al.*, 1969). But they do not eliminate the possibility that astral microtubules interact with cytoplasmic components to help pull the spindle poles apart (Aist and Berns, 1981).

In summary, permeabilized mammalian tissue cells have been a useful model system for describing the physiological requirements for anaphase A and B and for demonstrating that these two processes generate mechanochemical forces by different mechanisms, neither of which involves actin or myosin. Unfortunately, given the inability to extract spindle components selectively and perform reconstitution experiments, it has not been possible to identify the enzyme systems responsible for these movements. And given the complexity of the mammalian spindle, it has not been possible to correlate structural rearrangements of spindle microtubules with chromosome translocation.

E. Reactivation of the diatom central spindle

The diatom central spindle is uniquely suited for the study of the mechanism of anaphase B, because the fibrous systems responsible for anaphase A and B are spatially separated (Pickett-Heaps and Tippit, 1978) and the central spindle is a paracrystalline array of microtubules with a prominent, well-defined zone of microtubule overlap (McDonald *et al.*, 1977, 1979). This overlap zone is visible even by light microscopy and decreases as the spindle elongates (Pickett-Heaps *et al.*, 1980). We have recently described a simple procedure for isolating spindles from dividing cells of the centric diatom *Stephanopyxis turris* using glycerol buffers to stabilize microtubules, and filtration and differential centrifugation to purify spindles (Cande and

McDonald, 1985, 1986). The isolated spindles are similar to *in vivo* spindles with respect to size (approximately 8 μm at metaphase/early anaphase), number of microtubules per half-spindle (about 300), molecular order of the central spindle (hexagonal packing) and extent of microtubule overlap (15–25% original spindle length).

The response of the isolated diatom spindle to ATP is complex and mimics many aspects of *in vivo* anaphase B. Time-lapse video studies (Cande and McDonald, 1985) and studies of populations of spindles using immunocytochemical techniques (Cande and McDonald, 1986; McDonald *et al.*, 1986) show that spindle elongation *in vitro* includes at least three classes of ATP-dependent events:

1. initiation of spindle elongation;
2. separation of half-spindles owing to sliding;
3. selective microtubule depolymerization, or tubule addition starting in the zone of microtubule overlap.

As shown by polarization optics and by indirect immunofluorescence, almost all the isolated spindles in these preparations show this behaviour. The development of a preparation that uniformly and reproducibly undergoes spindle elongation *in vitro* should allow us to analyse this process with some assurance that the mechanochemistry is still present in an intact form. The structural regularity of the isolated central spindle will allow us to define precisely the rearrangement of spindle components that occurs during reactivation of elongation *in vitro*.

1. *The mechanism of force generation*

As shown by video microscopy (Cande and McDonald, 1985; McDonald *et al.*, 1986) and in Fig. 9.3, spindle elongation is due to the sliding apart of the two half-spindles with a concomitant decrease in the zone of microtubule overlap. This process is consistent with models of force generation that postulate that mechanochemical interactions in the zone of microtubule overlap generated by cross-bridges between antiparallel microtubules are responsible for moving the half-spindles apart (Margolis *et al.*, 1978; McDonald *et al.*, 1979; McIntosh *et al.*, 1969). Spindle elongation cannot be due to the autonomous swimming apart of the half-spindles, as the two half-spindles always remain associated with each other and the extent of elongation, in the absence of exogenous tubulin, is limited to the extent of microtubule overlap. The forces responsible for spindle elongation are autonomous to the spindle, since no cytoplasmic structures are attached to the spindle and spindle elongation can occur even when all chromatin is removed by digestion with DNAse I (McDonald *et al.*, 1986). These results eliminated the possibility that anaphase B is primarily due

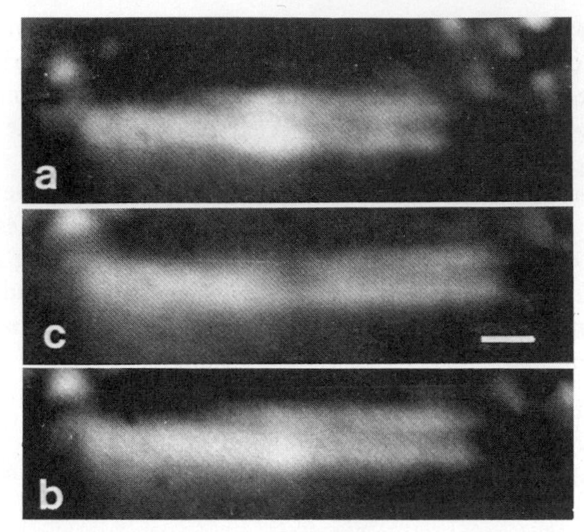

Figure 9.3. The change in diatom central spindle structure after reactivation as monitored by polarization optics and time-lapse video: (a) the central spindle just after addition of 1 mM ATP; (b) after 5 min in ATP; (c) after 13 min in ATP. Notice the decrease in spindle birefringence in the zone of microtubule overlap (spindle midzone) as the spindle increases in length. Bar = 1 μm. (Reprinted with the permission of the *Journal of Cell Science*; McDonald *et al.*, 1986.)

to pulling forces applied at the poles (Aist and Berns, 1981) or to tension generated during spindle formation and stored in the kinetochore fibres (Snyder *et al.*, 1984). Since spindle elongation can occur in the absence of exogenous tubulin pools, contrary to the views first expressed by Inoué and Sato (1967), microtubule polymerization is not a requirement for force generation.

At present it has not been possible to identify the enzymes responsible for spindle elongation owing to difficulties in obtaining sufficient quantities of spindles for conventional biochemistry. However, we do know that microtubule sliding requires ATP hydrolysis, is highly specific for this nucleotide, and is inhibited by vanadate and sulphydryl reagents (Cande and McDonald, 1986). The physiological requirements of reactivation are similar to those required for maintenance of flagellar beat in demembranated axonemes (Gibbons, 1981), suggesting that a dynein-like ATPase with some unique properties may be involved. It is not consistent with the involvement of a kinesin-like motor, since kinesin is unaffected by low concentration of vanadate and sulphydryl reagents (Vale *et al.*, 1985). Since DNAse I binds to actin with high affinity and depolymerizes microfilaments in extracts (Hitchcock *et al.*, 1976), our observation that spindles will elongate after chromatin digestion with

DNAse I reinforces the view that actomyosin is not involved in spindle elongation. Except for a few minor details, such as lack of sensitivity to EHNA, the physiology of central spindle elongation is identical to that required for spindle elongation in permeabilized mammalian cells (Cande *et al.*, 1981; Cande, 1982b,c), suggesting that an analogous mechanism may be important for anaphase B in many cell types.

2. *Role of microtubule polymerization during spindle elongation*

Microtubule polymerization must contribute to the increase in central spindle length *in vivo*, since in *Stephanopyxis turris* the central spindle elongates by at least 2.5-fold during anaphase B (Wordeman *et al.*, 1986; McDonald *et al.*, 1986). In the absence of tubulin, *in vitro* spindle elongation is limited to the extent of the overlap zone. As shown in Fig. 9.4, when isolated spindles are

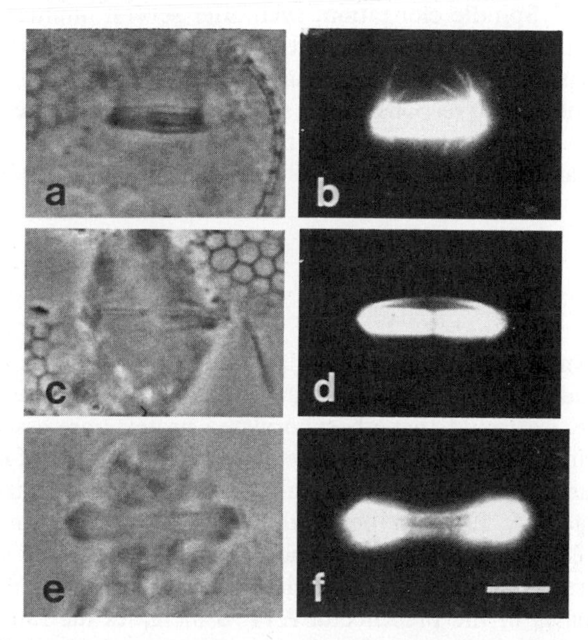

Figure 9.4. Phase-contrast and fluorescence (antitubulin) micrographs of isolated spindles incubated in 20 μM taxol and 25 μM neurotubulin during reactivation. (a, b) A spindle incubated for 10 min in taxol but no tubulin or ATP. (c, d) A spindle incubated for 10 min in taxol but no tubulin or ATP. (c, d) A spindle incubated for 10 min in taxol and ATP. (e, f) A spindle incubated in taxol, tubulin and ATP for 10 min. After incubation in tubulin and ATP the two half-spindles have moved further apart than the original zone of microtubule overlap. Bar = 5 μm. (Reprinted with the permission of the *Journal of Cell Science*; McDonald *et al.*, 1986.)

reactivated in neurotubulin, the extent of spindle elongation increases and is often 2 or 3 times the original overlap zone size (McDonald *et al.*, 1986; Masuda and Cande, 1987). Using biotinylated tubulin as a marker, we find that there are two major sites of tubulin incorporation *in vitro*. In the absence of ATP, labelled—i.e. biotinylated—tubulin is incorporated around the spindle pole complex as astral-like arrays; however, relatively little tubulin is incorporated elsewhere. During spindle reactivation in the presence of ATP, tubulin is also incorporated into the zone of microtubule overlap (Masuda and Cande, 1987). At first, tubulin is incorporated at the edges of the overlap zone where the microtubule plus ends are distributed (McDonald *et al.*, 1979), but subsequently, as the spindle elongates, biotinylated tubulin is distributed throughout the overlap zone. This change in pattern of tubulin incorporation over time suggests that the tubulin addition relevant for spindle elongation occurs only in the zone of microtubule overlap and that as new microtubules are polymerized onto the plus ends of microtubules in each half-spindle, they slide over each other. Spindle elongation, even after several minutes in tubulin, requires ATP hydrolysis and is inhibited by vanadate and AMPPNP. These results are consistent with the view that microtubule sliding is the primary mechanochemical event responsible for spindle elongation, while microtubule polymerization is an auxiliary process that limits the extent of spindle elongation *in vitro* and presumably *in vivo* as well (Margolis *et al.*, 1979; McIntosh *et al.*, 1969).

3. *Regulation of spindle elongation*

The spindle contains many phosphorylated structures (Vandre *et al.*, 1984; Wordeman and Cande, 1987). Using antibodies that recognize phosphorylated epitopes, we found that phosphorylated proteins are associated with the diatom spindle pole complexes, kinetochores and the zone of microtubule overlap. Diatom spindles isolated in media that contain inhibitors of phosphatase activity function better than spindles isolated in more conventional media (Wordeman and Cande, 1987). Preliminary experiments suggest that irreversible thiophosphorylation of 205 kDa protein by spindle-associated kinases in the presence of ATPγS increases the rate and extent of spindle elongation *in vitro*. Although the function of this phosphorylated protein is not known, the diatom central spindle may be similar to other motility systems in that phosphorylation of spindle-associated proteins appears to be essential for maintenance and regulation of spindle elongation *in vitro*.

IV. Concluding Remarks

One of the major challenges facing the cell biologist is unravelling the complexities of mitosis. There has been a renaissance in studying mitosis, and this is due to the willingness of contemporary biologists to narrow the scope of their analysis to just one aspect of mitosis and to use new techniques for studying it. Intellectually, this may seem unsatisfying, but it is an illusion to think that a sudden stroke of genius or a novel insight will suddenly illuminate all aspects of this problem and "solve mitosis". Rather, as demonstrated by the history of *in vitro* analysis of chromosome movement, what is required is the careful development of new experimental approaches. As we begin to understand one aspect of spindle structure and function, this in turn will allow us to analyse the rest of the process.

A variety of lines of evidence from both *in vivo* and *in vitro* studies suggest that several force-generating systems are responsible for anaphase chromosome movements. The spindle is a dynamic structure; as chromosomes move polewards kinetochore microtubules disassemble and as the poles move apart the polar microtubules may increase in length. Although genetical approaches combined with the tools of molecular biology are being used to study mitosis, *in vitro* approaches are essential for defining spindle function.

The mechanochemical enzymes responsible for force generation in the spindle have not yet been identified for any of the processes involved in moving chromosomes. However, studies of spindle elongation using isolated diatom central spindles demonstrate that the key mechanochemical event *in vitro* during anaphase B is microtubule sliding (Cande and McDonald, 1985, 1986). Spindles elongate because the two half-spindles slide apart. Microtubule polymerization is an auxiliary process that may limit the extent of spindle elongation *in vitro*, but it does not in and of itself generate the forces required for spindle elongation (Masuda and Cande, 1987). Given the discrete morphology of the diatom central spindle and the possibilities of growing these organisms in bulk amounts, it should eventually be possible to identify those elements of the spindle midzone that are responsible for spindle elongation.

In order to generate equivalent progress in understanding the mechanism of anaphase A movement, it is necessary to develop analogous well-defined model systems that undergo chromosome-to-pole movement *in vitro*. Although permeabilized cells display this behaviour after lysis, these cell models are too complex for meaningful biochemical and structural analysis. An alternative approach is suggested by the studies of Mitchison and Kirschner (1985a,b) and Koshland *et al.* (1988) on kinetochore function *in vitro*. Eventually it may be possible to reconstitute anaphase A *in vitro* using properly prepared kinetochores, centrosomes, microtubules and other spindle components.

REFERENCES

Aist, J. R. and Berns, M. W. (1981). *J. Cell Biol.* **91**, 446–458.
Asai, D. (1985). *Cell Motility* **5**, 172–173.
Broach, J. R. (1986). *Cell* **44**, 3–4.
Cabral, F., Schibler, M., Abraham, I., Whitfield, C., Kuriyama, R., McCluskin, C., Mackensen, S. and Gottesman, M. M. (1984). In *Molecular Biology of the Cytoskeleton* (ed. G. G. Borisy, D. W. Cleveland and D. B. Murphy), pp. 305–317. Cold Spring Harbor Laboratory, Cold Spring Harbor.
Cande, W. Z. (1979). In *Cell Motility: Molecules and Organization* (ed. S. Hatano, H. Ishikawa and H. Sato), pp. 593–608. University of Tokyo Press, Tokyo.
Cande, W. Z. (1980). *J. Cell Biol.* **87**, 326–335.
Cande, W. Z. (1981). In *International Cell Biology 1980–1981* (ed. H. G. Schweiger), pp. 382–391. Springer-Verlag, Berlin.
Cande, W. Z. (1982a). *Methods Cell Biol.* **25B**, 57–68.
Cande, W. Z. (1982b). *Cell* **28**, 15–22.
Cande, W. Z. (1982c). *Nature* **295**, 700–701.
Cande, W. Z. (1983). *Nature* **304**, 557–558.
Cande, W. Z. and McDonald, K. L. (1985). *Nature* **316**, 168–170.
Cande, W. Z. and McDonald, K. L. (1986). *J. Cell Biol.* **103**, 593–604.
Cande, W. Z. and Wolniak, S. M. (1978). *J. Cell Biol.* **79**, 573–580.
Cande, W. Z., Snyder, J., Smith, D., Summers, K. and McIntosh, J. R. (1974). *Proc. Natl. Acad. Sci. USA* **71**, 1559–1563.
Cande, W. Z., McDonald, K. L. and Meeusen, R. L. (1981). *J. Cell Biol.* **88**, 618–629.
Carlson, J. G. (1977). *Chromosoma* **64**, 191–206.
Dinsmore, J. H. and Sloboda, R. D. (1988). *Cell* **53**, in press.
Forer, A. (1985). *Can. J. Biochem. Cell Biol.* **63**, 585–598.
Forer, A. and Zimmerman, A. M. (1974). *J. Cell Sci.* **16**, 481–497.
Gautier, J., Norbury, C., Lohka, M., Nurse, P. and Maller, J. (1988). *Cell* **54**, 433–439.
Gibbons, I. R. (1981). *J. Cell Biol.* **91** (3, part 2), 107s–124s.
Goode, D. and Roth, L. E. (1969). *Exp. Cell Res.* **58**, 343–352.
Hitchcock, S. E., Carlsson, L. and Lindberg, U. (1976). *Cell* **7**, 531–542.
Hoffman-Berling, H. (1954a). *Biochim. Biophys. Acta* **15**, 332–339.
Hoffman-Berling, H. (1954b). *Biochim. Biophys. Acta* **15**, 226–236.
Inoué, S. (1981). *J. Cell Biol.* **91** (3, part 2), 131s–147s.
Inoué, S. and Ritter, H. (1975). In *Molecules and Cell Movement* (ed. S. Inoué and R. Stephens), pp. 3–30. Raven Press, New York.
Inoué, S. and Sato, H. (1967). *J. Gen. Physiol.* **50** (Suppl.), 259–292.
Inoué, S., Borisy, G. and Kiehart, D. P. (1974). *J. Cell Biol.* **62**, 175–184.
Kane, R. E. (1962). *J. Cell Biol.* **12**, 47–55.
Kane, R. E. (1965). *J. Cell Biol.* **25**, 137–144.
Kiehart, D. P., Mabuchi, I. and Inoué, S. (1982). *J. Cell Biol.* **94**, 165–178.
King, S. M., Hyams, J. S. and Luba, A. (1982). *J. Cell Biol.* **94**, 341–349.
Kirschner, M. W. (1978). *Int. Rev. Cytol.* **54**, 1–71.
Koshland, D., Mitchison, T. J. and Kirschner, M. W. (1988). *Nature* **331**, 499–504.
Kuriyama, R. (1984). *J. Cell Sci.* **66**, 277–295.
Kuriyama, R. and Borisy, G. G. (1985). *J. Cell Biol.* **101**, 524–530.
Lohka, M. J. and Maller, J. L. (1985). *J. Cell Biol.* **101**, 518–523.
Luby, K. J. and Porter, K. R. (1980). *Cell* **21**, 13–23.

Mabuchi, I. and Okuno, M. (1977). *J. Cell Biol.* **74**, 251–263.
Margolis, R. L., Wilson, L. and Kiefer, B. I. (1978). *Nature* **272**, 450–452.
Masuda, H. and Cande, W. Z. (1987). *Cell* **49**, 193–202.
Mazia, D. and Dan, K. (1952). *Proc. Natl. Acad. Sci. USA* **38**, 826–838.
Mazia, D., Mitchison, J. M., Medina, H. and Harris, P. (1961). *J. Biophys. Biochem. Cytol.* **10**, 467–474.
McDonald, K. L., Pickett-Heaps, J. D., McIntosh, J. R. and Tippit, D. H. (1977). *J. Cell Biol.* **74**, 377–388.
McDonald, K. L., Edwards, M. K. and McIntosh, J. R. (1979). *J. Cell Biol.* **83**, 443–461.
McDonald, K. L., Pfister, K., Masuda, H., Wordeman, L., Staiger, C. and Cande, W. Z. (1986). *J. Cell Sci.* **5** (*Suppl.*), 205–227.
McIntosh, J. R. (1977). In *Mitosis, Facts and Questions* (ed. M. Little *et al.*), pp. 167–184. Springer-Verlag, Berlin.
McIntosh, J. R., Hepler, P. K. and van Wie, D. G. (1969). *Nature* **224**, 659–663.
McIntosh, J. R., Roos, U.-P., Neighbors, B. and McDonald, K. L. (1985). *J. Cell Sci.* **75**, 93–129.
Meeusen, R. L., Bennett, J. and Cande, W. Z. (1980). *J. Cell Biol.* **86**, 858–865.
Miake-Lye, R. and Kirschner, M. W. (1985). *Cell* **41**, 167–175.
Mitchison, T. J. and Kirschner, M. W. (1984). *Nature* **312**, 232–237.
Mitchison, T. J. and Kirschner, M. W. (1985a). *J. Cell Biol.* **101**, 755–765.
Mitchison, T. J. and Kirschner, M. W. (1985b). *J. Cell Biol.* **101**, 766–777.
Oakley, B. R. and Morris, N. R. (1981). *Cell* **24**, 837–845.
Petzelt, C. (1979). *Int. Rev. Cytol.* **60**, 53–92.
Pickett-Heaps, J. D. and Tippit, D. H. (1978). *Cell* **14**, 455–467.
Pickett-Heaps, J. D., Tippit, D. H. and Leslie, R. (1980). *Eur. J. Cell Biol.* **21**, 1–11.
Pickett-Heaps, J. D., Tippit, D. H. and Porter, K. R. (1982). *Cell* **29**, 729–744.
Poenie, M., Alderton, J., Steinhardt, R. and Tsien, R. (1986). *Science* **233**, 886–888.
Pollard, T. D. and Weihing, R. R. (1974). *CRC Crit. Rev. Biochem.* **2**, 1–65.
Pratt, M. M. (1984). *Int. Rev. Cytol.* **87**, 83–105.
Pratt, M. M., Otter, T. and Salmon, E. D. (1980). *J. Cell Biol.* **86**, 738–745.
Ratan, R. R., Shelanski, M. L. and Maxfield, F. R. (1986). *Proc. Natl. Acad. Sci. USA* **83**, 5136–5140.
Rebhun, L. I., Rosenbaum, J., Lefebvre, P. and Smith, G. (1974). *Nature* **249**, 113–115.
Sakai, H. (1978). *Int. Rev. Cytol.* **55**, 23–48.
Sakai, H. and Kuriyama, R. (1974). *Develop. Growth and Differ.* **16**, 123–134.
Sakai, H., Hiramoto, Y. and Kuriyama, R. (1975). *Develop. Growth and Differ.* **17**, 265–274.
Sakai, H., Mabuchi, I., Shimoda, S., Kuriyama, R., Ogawa, K. and Mohri, H. (1976). *Develop. Growth and Differ.* **18**, 211–219.
Sakai, H., Hamaguchi, H., Kimura, I. and Hiramoto, Y. (1979). In *Cell Motility: Molecules and Organization* (ed. S. Hatano, H. Ishikawa and H. Sato), pp. 609–619. University of Tokyo Press, Tokyo.
Salmon, E. D. (1982). *Methods Cell Biol.* **25B**, 70–105.
Salmon, E. D. and Segall, R. R. (1980). *J. Cell Biol.* **86**, 355–365.
Salmon, E. D., Leslie, R. J., Saxton, W. M., Karow, J. L. and McIntosh, J. R. (1984). *J. Cell Biol.* **99**, 2165–2174.
Saxton, W. M., Sample, D. L., Leslie, R. J., Salmon, E. D., Zavortink, M. and McIntosh, J. R. (1984). *J. Cell Biol.* **99**, 2175–2186.

Silver, R. B., Cole, R. D. and Cande, W. Z. (1980). *Cell* **19**, 505–516.
Silver, R. B., Saft, M. S., Taylor, A. R. and Cole, R. D. (1983). *J. Biol. Chem.* **258**, 13 287–13 291.
Snyder, J. A. (1981). In *Mitosis/Cytokinesis* (ed. A. Forer and A. Zimmerman), pp. 301–325. Academic Press, New York.
Snyder, J. A., Golub, R. and Berg, S. P. (1984). *Eur. J. Cell Biol.* **35**, 62–69.
Spurck, T. P. and Pickett-Heaps, J. D. (1987). *J. Cell Biol.* **105**, 1691–1705.
Toriyama, M., Endo, S. and Sakai, H. (1984). *Cell Struct. Funct.* **9**, 213–224.
Vale, R. D., Schnapp, B. J., Mitchison, T., Steuer, E., Reese, T. S. and Sheetz, M. P. (1985). *Cell* **43**, 623–632.
Vandré, D. D., Davis, F. M., Rao, P. N. and Borisy, G. G. (1984). *Proc. Natl. Acad. Sci. USA* **81**, 4439–4443.
Weisenberg, R. C. (1972). *Science* **177**, 1104–1105.
Wordeman, L. and Cande, W. Z. (1987). *Cell* **50**, 535–543.
Zavortink, M., Cande, W. Z. and McIntosh, J. R. (1981). *J. Cell Biol.* **91**, 318 (abstr.).
Zieve, G. W., Heidemann, S. R. and McIntosh, J. R. (1980a). *J. Cell Biol.* **87**, 160–169.
Zieve, G. W., Turnbull, D., Mullins, J. M. and McIntosh, J. R. (1980b). *Exp. Cell Res.* **126**, 397–405.
Zimmerman, A. M. and Forer, A. (1981). In *Mitosis/Cytokinesis* (ed. A. M. Zimmerman and A. Forer), pp. 327–336. Academic Press, New York.

CHAPTER 10

Mitosis in Perspective

J. RICHARD McINTOSH

Department of Molecular, Cellular and Developmental Biology,
University of Colorado, Boulder, Colorado, USA

I. Introduction

The study of mitosis has challenged, delighted and frustrated biologists for more than one hundred years. Much progress has been made, but there is every likelihood that the subject will continue to challenge, delight and frustrate for a while longer. There are numerous technical reasons why chromosome movement is difficult to study. Spindles are small, so their visualization requires microscopes. The spindle is a transient structure, assembled for only a brief time during the cell cycle, so it is difficult to obtain large numbers of mitotic cells for analysis. Because of its transience, the spindle is labile. It usually falls apart when the cell is disturbed, so the mitotic machinery is difficult to isolate. Spindles are chemically complex and there is no cell that is committed simply to mitosis, so the essential mitotic components are always mixed with biostuff that is important for other things. Since there is no prokaryotic cell that goes through mitosis, there has been no simple and elegant biochemical genetics to help sort out the situation. It is thus no wonder that the mechanisms of chromosome movement have remained mysterious over the years since 1950 during which so many problems of biological mechanism have been solved.

In the face of these difficulties, it is impressive how much progress has been made in understanding some of the mechanisms of mitosis. The contents of this volume are a testimony to the insight, energy and skill of many investigators. Each chapter shows how much has been learned about some aspect of mitosis. The reader will see that in most cases we now know quite a bit about

MITOSIS: Molecules and Mechanisms
ISBN 0-12-363420-2

the cell biology of chromosome movement. Our understanding of the molecules involved in these cell biological events, on the other hand, is still unsatisfactory. For example, the chapter on spindle structure describes the impressive detail with which we understand the distributions and time-dependent rearrangements of the microtubules in some small spindles. In contrast, the chapters that deal with the molecules by which chromosome motion is effected and regulated are necessarily vague on certain key questions. The dichotomy between cellular knowledge and molecular ignorance is accurately and readably reflected throughout this volume.

Given a mixed report of progress and frustration, how does current knowledge of mitosis stand, compared with that in 1950 when Schrader published the second edition of his book on mitosis, with 1961 when Mazia produced his extraordinary chapter for *The Cell*, edited by Brachet and Mirsky, or with 1981 when Zimmerman and Forer published their volume *Mitosis–Cytokinesis*. This chapter is a brief, historically oriented overview of mitosis with particular emphasis on the topics that are currently "hot"; it also includes an effort to draw attention to those topics likely to heat up soon.

II. Spindle Function

A. A summary of some current ideas about mitosis

The spindle is required for organized chromosome motion, and microtubules are an essential component of the spindle. Proteins that associate with microtubules act in some way to generate a force on each kinetochore and pull it toward the pole it faces. The directions in which these forces pull are defined by the orientations of the kinetochore-attached fibres. Other forces act to push the chromosomes away from the poles. Still others function to keep the poles from collapsing in on the chromosomes. The pole-directed forces found during anaphase can be strong relative to the forces necessary to overcome the viscous drag on chromosomes moving at mitotic velocities.

During prometaphase, the position of a chromosome develops as a result of a balance of forces whose strengths appear to vary with the distance between a kinetochore and the pole it faces. The net force on each metaphase chromosome is approximately zero, but with the separation of the chromatids at anaphase onset, the force balance changes. The forces of prometaphase probably persist through anaphase and help to pull the chromosomes polewards. Meanwhile, the poles are pushed farther apart. The motors that develop mitotic forces are still unidentified, and a characterization of the mechanochemical transducers for mitosis is one of the most important unsolved problems of spindle research. Late in anaphase, the chromosomes contract and the nuclear envelope re-forms around the tightly condensed chromatin. Thereafter the cell is in telophase, and the reconstruction of an interphase nucleus begins.

B. The action of microtubules during spindle formation

Two questions are key for developing an understanding of the prometaphase organization of chromosomes: (1) what do microtubules do, and (2) how do they do it? Microtubules clearly grow out from the duplicate centrosomes at the onset of prometaphase. The position and orientation of these microtubules are defined by the polymer-initiating qualities of the centrosomes. As they form, the two microtubule arrays interact both with the chromosomes and with each other to form the normal bipolar spindle of metaphase. These statements are, however, just a paraphrase of concepts described in the 1925 book by E. B. Wilson, *The Cell in Development and Heredity*. Some of the newly discovered properties of the mitotic poles are summarized in the chapter by Vandré and Borisy. They describe the process of "maturation" by which interphase centrosomes develop into spindle poles capable of initiating 5–10 times as many microtubules as in interphase. They describe evidence that the centrosomes help to define the geometry of the polymers they initiate. They review the relationships between the cycles of the centrioles, the centrosomes, and the cell itself. They summarize evidence about the importance of protein phosphorylation for the mitotic transitions in centrosome behaviour.

The molecules involved in the initiation events of prometaphase are still unknown, and here is a key area for future work. Progress has been slow partly because there are only one or two centrosomes per cell, so descriptive chemistry of the principal actors in the play is difficult. Monoclonal antibodies and sera from humans with autoimmune diseases have contributed something to our knowledge of centrosome mechanism, but the problem has not yet received the full-scale assault that will probably be required to make significant progress. We do not yet know the mechanisms by which a centrosome counts the number of microtubules it initiates. The factors that influence the directionality of the resulting microtubules are still matters for conjecture. Preliminary information suggests that numerous protein kinases act during the transition from interphase to mitosis, affecting not only the components of the spindle but also chromatin condensation and the proteins of the nuclear lamina. These issues are treated below in the section on the control of mitosis.

Kirschner and Mitchison (1986) have discussed the impact of the mitotic change in centrosome activity on the cell's arrangement of microtubules. As reviewed in the chapter by Salmon, these workers have developed a model, termed "dynamic instability", that is based on the observation that microtubules are intrinsically unstable. Tubulin polymers switch back and forth between an assembly state dominated by slow but consistent growth, and a disassembly state in which the polymer falls rapidly apart. A steady-state population of microtubules, like that found during interphase or metaphase, is envisioned as a balance between many slowly growing polymers and a few polymers that rapidly disassemble, thereby maintaining a constant pool of soluble subunits. This model for microtubule behaviour can account for the prophase disassembly of the interphase microtubule array and the formation of the structures characteristic of prometaphase simply on the basis of a change in the number of microtubule-initiating sites at the centrosome (Kirschner and Mitchison, 1986). The model can also account for the extraordinary dynamism of spindle microtubules, as reviewed by Salmon. The model is therefore currently dominating the minds of researchers who are interested in coming to understand the relationship between microtubule behaviour and chromosome movement.

It seems likely, however, that simply the properties of tubulin and of the structures that initiate microtubules are not going to explain mitotic mechanism. Evidence reviewed in the chapter by Bloom and Vallee suggests that the spindle contains microtubule-associated proteins that probably have an impact on microtubule function during cell division. The information currently available about these molecules is insufficient to allow us to know what they do *in vivo*, but the work summarized by Salmon leaves no doubt as to the extent of the difference in the behaviours of microtubules in interphase and mitotic cells. While the dynamic instability model of Mitchison and Kirschner will probably provide the intellectual underpinnings by which we may come to understand microtubule assembly, microtubule function probably depends on the microtubule-associated proteins that copolymerize with tubulin in each of the fibrous systems of the spindle. They are likely to modulate the assembly and disassembly rate constants of particular groups of spindle microtubules. They probably bind particular classes of microtubules together. They may help to couple microtubules to other structures, such as poles, kinetochores, endoplasmic reticulum and fibrous matrices. Finally, these proteins may include some of the motors important for chromosome movements. Here again is an area where much additional work will be required to identify and characterize the important functional components of the spindle.

Kinetochores have long been recognized as the most important part of the chromosome for mitosis, but recent research has shown that kinetochores do far more than was previously suspected. As summarized by Brinkley *et al.*, kinetochores are capable of initiating microtubules both *in vivo* and *in vitro*. Both structural and biochemical arguments suggest, however, that the normal pathway of spindle formation involves the capture by kinetochores of microtubules initiated from the pole. Kinetochores appear to be organized by specific sequences of DNA interacting with numerous proteins that ultimately connect them with microtubules. The specifics in this chain of interaction remain to be determined, but the pattern is now established. Centromeric DNA will serve to bind kinetochore proteins, which in turn attach to the pole-distal ends of microtubules. This is an area in which recent progress has been impressive, and, as reviewed by Cabral, it seems likely that the tools of genetics and molecular biology will lead to further rapid advances in the immediate future.

The concept that kinetochores bind to microtubules initiated at the poles and thereby establish significant mechanical connections between a chromosome and a pole has been picked up by Kirschner and Mitchison (1986) and put together with their concept of dynamically unstable microtubules to form a "selective" model for the interaction between chromosomes and spindle fibres. These workers propose that microtubules are continuously initiating from the spindle pole, growing forth at a slow rate, but always living with a constant probability of converting to the disassembly state and disappearing. Occasionally such a microtubule may encounter some other object, for example a kinetochore, which (hypothetically) confers upon it a stability appropriate to a functional connection and thereby withdraws it from the pool of polymers that is dynamically unstable. This highly attractive idea may well form the rationale for dynamic instability of microtubules and explain important aspects of the interactions between microtubules and chromosomes. There are, however, some spindle geometries and some physiological observations that are inconsistent with the model. Here again is an area where further work will be essential for our developing an understanding of a key part of mitosis.

C. The chromosome movements of prometaphase

The complex chromosome movements of prometaphase have received considerable attention in recent years. This is an important step forward for mitotic research. The study of prometaphase is daunting because the movement of an individual chromosome is unpredictable. None the less, the patterns of the movements are clear and generalizations should be possible. The work from Nicklas' laboratory during the 1960s and 1970s did much to clarify our notions about the pole-directed forces acting at kinetochores (reviewed in Nicklas, 1985). More recently, Nicklas and his co-workers have succeeded in relating spindle structure to the physiology of chromosome attachment and have demonstrated that microtubules entering the vicinity of a kinetochore are a universal part of the initial, functional attachment between a chromosome and the spindle (Nicklas, 1985, 1987). It seems clear from this work that chromosomes are pulled polewards during prometaphase by fibres that include microtubules.

Since Ostergren's publications of the 1950s, many students of mitosis have favoured the idea of a balance of forces acting to pull prometaphase chromosomes to the spindle equator. This balance has been thought to result from interactions that couple the two kinetochores on a single chromosome to opposite poles, pulling on the kinetochores with a force that is proportional to the distance between each kinetochore and pole it faces. In this book Salmon summarizes recent evidence in favour of this hypothesis. He also treats the evidence that chromosomes are pushed away from the poles by forces that act over the entire, pole-facing surface of the chromosome. Salmon discusses the hypothesis that normal metaphase chromosome positions result from competition between the forces pushing away from the pole and the pole-directed forces acting at the kinetochore, as described by Nicklas. Experimental data to test this model are currently being obtained by microsurgical work with laser microbeams and micromanipulators.

The mechanisms for prometaphase movements are not yet fully understood, but it seems likely that they are even more complex than is currently envisioned. For one thing, Pickett-Heaps has described active movements of kinetochores away from the poles along what appear to be bundles of microtubules (reviewed in Pickett-Heaps *et al.*, 1982). Such descriptive work has led this group to propose that spindle fibres can generate forces that act in two directions on a single kinetochore: an AP force that pushes the chromosomes away from the pole and a P force that pulls them in. These workers propose that both kinds of force come from interactions between the kinetochores and the walls of microtubules, rather than from events occurring at the butt end of microtubules, as is commonly expected from the structure of microtubule attachments to the chromosomes of higher plants and animals. The ideas of this group have been strongly influenced by Keith Porter and the studies of his group on granule movements in chromatophores.

Support for these ideas about kinetochore–microtubule interaction has recently come from a study on the influence of ATP on microtubule–kinetochore interactions *in vitro* (Mitchison and Kirschner, 1985, as reviewed here by Brinkley *et al.*). This work suggests that the corona of fibres emanating from a kinetochore can interact with microtubules in any orientation, but that in the presence of ATP a rather disorganized set of interactions is altered so that only the pole-distal microtubule ends are left associated with the kinetochores. While this phenomenon has not yet been described in sufficient detail to allow us to conclude that kinetochores can walk over a microtubule's surface, this is a likely interpretation of the data. The kinetochore is thus coming to be seen as more than simply a microtubule-capturing organelle. It appears to have the kinetic properties of a motor.

A motile kinetochore is consistent with some of Pickett-Heaps' observations on kinetochore behaviour *in vivo*, but it does not account for the observations summarized by Salmon that a monopolar spindle can interact with chromosomes so as to set up a stable configuration with the chromosomes held at about a half-spindle's length from the pole. Since the AP force should cease when the chromosome reaches the pole-distal end of its microtubule bundle, there would be no force acting from the monopolar spindle to counteract the pole-directed forces normally seen. Salmon interprets stable monopolar arrangements of chromosomes with the idea that there is a pushing force acting between the pole and the whole, pole-facing surface of the chromosome, as cited above. An alternative would be to postulate an additional feature of the kinetochore fibre itself, such as a stability that allows it to resist the action of the pole-directed mitotic motor, so long as the motor never becomes too strong. Nicklas has suggested that a stable interaction between the chromosome and spindle depends on the existence of strong, opposing forces acting to pull the chromosome toward both poles at once. Perhaps a monopolar spindle never develops such forces and hence never acts strongly enough on the kinetochores to pull them poleward against the structural stability of the kinetochore fibre.

D. Spindle structure changes and the events of anaphase

The largest single body of data relevant to understanding the mechanisms of spindle action is probably that found in the fine structure studies of numerous mitotic cells. This work from the 1960s through the early 1980s has helped to define the complexity of the changes that underlie the stately organization and segregation of the chromosomes. As reviewed in the chapters by McDonald and by Hepler, the work expands and clarifies both the results from light microscopy of fixed specimens and the elegant studies of living cells using polarization microscopy. It shows how some microtubules elongate while others shorten during anaphase. It demonstrates that the mitotic rearrangements of spindle structure cannot be understood simply as an assembly and disassembly of microtubules; there is also a repositioning of existing microtubules as the spindle poles move apart. Such observations have defined a new body of mitotic phenomena that is an integral part of chromosome movement and must be accounted for by the ideas put forward to explain spindle action.

E. The chromosome movements of anaphase

The motion of chromosomes to poles (anaphase A) has always been seen as one of the most significant and beautiful aspects of mitosis. Its mechanism, however, is still elusive. The motion probably results from the concerted function of the pole-directed forces that initially acted on the two oppositely directed kinetochores of each chromosome during prometaphase, but which now act on the chromatids that have just separated from their sisters. (The process by which sisters separate at anaphase onset is another important unsolved problem. Some recent work suggests there are proteins that glue sister chromatids together through metaphase but disperse at anaphase onset (Cooke *et al.*, 1987).)

There is no question that during anaphase A the microtubules running between chromosomes and poles must shorten, and there is strong current interest in the pathway by which kinetochore microtubules disassemble. Evidence from polarization optical studies suggests that spindle fibres are at rest during anaphase A and that the chromosomes move up their fibres as they approach the poles (Schaap and Forer, 1984). This idea has recently received support from microinjection studies that used labelled tubulin. Mitchison *et al.* (1986) injected biotinylated tubulin during metaphase to label the kinetochore-proximal ends of kinetochore microtubules. When injected cells were allowed to enter anaphase before fixation and processing, the biotinylated label was gone from the kinetochore-proximal region of the microtubule bundle, suggesting a kinetochore-proximal loss of tubulin subunits during anaphase A. A similar conclusion has been reached by Borisy's group on the basis of fluorescence bleaching studies using cells injected with DTAF tubulin (Gorbsky *et al.*, 1987). The latter two studies provide strong evidence that microtubule subunits are lost at the kinetochores during anaphase A, but they do not exclude the possibility that kinetochore microtubule subunits are also lost at the poles or from the microtubule surface, as suggested by Inoué and Sato (1967). None the less, the idea that the pole might reel in the kinetochore fibres to achieve anaphase A can no longer be taken as a sufficient explanation for chromosome-to-pole motion. This is an important advance in our understanding of mitotic movements.

Some investigators have proposed that kinetochores act to disassemble the microtubules to which they are attached, doing work on the kinetochore microtubules and the chromosomes at the same time. Such a view ascribes to the kinetochore two of the most important kinetic roles in mitosis. An alternative that fits the data is that the kinetochore holds onto the kinetochore microtubules as they disassemble, and that it is the disassembly of microtubules which pulls the chromosomes to the poles. It is also possible that a contractile matrix couples the chromosome to the poles, and microtubule disassembly is simply permissive for chromosome-to-pole movement. This view, sketched by McIntosh (1981) and developed by Pickett-Heaps (1986), is based on the surmise that the necessary contractile machinery exists in parallel with the kinetochore microtubules. Pickett-Heaps' laboratory has recently obtained some evidence from PtK cells that is consistent with the view, but it is difficult to see how such a contractile matrix could be significant in acentric spindles, where anaphase A is accomplished without any polar structure to which the contraction would be attached. On current evidence, it seems likely that kinetochores do possess kinetic properties that are important for chromosome motions during both prometaphase and anaphase A. This focus on the kinetochore as a motile organelle is new.

F. Anaphase elongation of the spindle

Spindle elongation (anaphase B) is a common component of chromosome movement. As reviewed in the chapters by Cande and by McDonald, the structure most likely to promote this movement is the set of interdigitating microtubules that connects the two poles. Most spindles contain prominent bundles of microtubules that run from pole to pole, and these probably provide the mechanical forces that hold the poles apart during prometaphase and metaphase when the still-duplicate chromosomes are being pulled poleward and there is a resulting compression acting on the spindle ends. The structural studies reviewed by McDonald have provided the comparatively new insight that after prometaphase there are only a few microtubules running from pole to pole, so the interpolar support framework is probably provided by interactions between the microtubules that emanate from both poles, pass the chromosomes, and interdigitate near the spindle equator. This situation is most easily seen in the spindles of some lower eukaryotes, but studies of spindle microtubule polarity reviewed by McDonald demonstrate that an analogous design is found in both mammals and higher plants.

The only well-studied system for spindle elongation is that found in diatoms. The relevant structural and experimental work *in vivo* (e.g. the chapter by McDonald and the work by Leslie and Pickett-Heaps, 1983) can now be combined with the results from a reconstitution of anaphase B *in vitro*, as reviewed in the chapter by Cande. The data demonstrate that diatom spindles develop sliding forces to move the two interdigitating families of microtubules apart. Spindle elongation beyond that expected from the extent of microtubule interdigitation at metaphase is achieved by the elongation of the interdigitating microtubules through subunit addition at the pole-distal ends of the microtubules (Masuda and Cande, 1987). This mechanism lays additional "track" along which the sliding motors for anaphase B can run in order to effect a greater spindle elongation.

A persistent question for students of mitosis is the extent to which a mechanism found in one organism may be used to interpret mitosis in other species that are not so amenable to experimental study. In particular, is the diatom mechanism for anaphase B applicable to all eukaryotes? Recent work by Saxton and McIntosh (1987) suggests that a system similar to that in diatoms operates in the spindles of mammalian cells, strain PtK. Experiments based on the injection of fluorescent tubulin into living cells suggest that the anaphase elongation of interdigitating microtubules occurs by the addition of tubulin subunits to their pole-distal ends. Fluorescence bleaching experiments demonstrate that the two half-spindles move apart as the PtK spindle elongates. While this study was limited to the late stages of spindle elongation by the fast turnover of mammalian spindle microtubules during early anaphase, the data suggest that the mechanism for anaphase B may be quite well conserved across a broad range of phylogeny.

In the fungus *Fusarium*, on the other hand, the severing of interzone microtubules during anaphase promotes a speeding up of pole separation, rather than a blocking of the process (Aist and Berns, 1981). This result suggests that in some organisms elongation of the interzone spindle is merely permissive for the increase in interpolar distance, not causal for it. These workers have proposed a "front-wheel drive", whereby interactions between the spindle poles and the cytoplasm outside the spindle would pull the poles apart. This idea has not yet been tested directly by experiment, but one can hope that future studies will evaluate the possibility. Certainly one must conclude from the work reviewed here by Cande that front-wheel drive is not important in diatoms, so the universality of mechanisms for anaphase B is still very much in question. Perhaps one will need to think about anaphases A, B and C in order to put all these ideas together into a single framework.

Most workers would agree that spindle elongation involves some sort of microtubule-dependent motor by which chemical energy can be converted into forces that slide microtubules through their surroundings. Two motors that might serve such a function are now known; dynein and kinesin. Immunocytochemistry has provided evidence that both of these molecules reside in sea-urchin spindles (Scholey *et al.*, 1985 for kinesin; Hisanaga *et al.*, 1987 for dynein). The significance of these localizations for mitotic mechanism is, however, still unclear. One is naturally concerned by the possibility that microtubule-binding motors are present in an egg simply because they are part of the maternal dowry to the embryo. Motor molecules in eggs may be destined for specialized cell functions, such as axonal transport and axonemal beating, in which their importance is well documented. Perhaps they are found in the spindle simply because they bind to microtubules and there is nothing to prevent them from becoming localized there. Certainly studies on the immunolocalization of dynein have provided conflicting evidence over the years, and a recent study localizing kinesin in cultured cells suggests that while the molecule is present in the spindles of cells from vertebrates, it is localized at the spindle poles, not at the zone of interdigitation where it might contribute to microtubule sliding during anaphase B (Neighbors *et al.*, 1988). Kinesin at the spindle poles would seem to have no function if all subunit loss from kinetochore microtubules during anaphase A is at the kinetochores themselves.

The identification of spindle motors and the roles that they may play during chromosome movement are still enigmas requiring further work.

III. Control of Spindle Function

An essential feature of mitosis is the fidelity with which the chromosomes are segregated. Certainly the speed and power demonstrated by a spindle are of low order when compared with those of other biological machines like muscle, cilia and the spasmoneme. It is more important that the spindle makes few mistakes. Estimates for mitotic error rates in yeasts suggest that chromosomes are misplaced with a frequency of only about one in one hundred thousand (Hartwell *et al.*, 1982). Control, in its broadest sense, would therefore seem to be one of the most significant aspects of spindle mechanism and *ergo* of spindle research. In fact, little research has thus far been done on the control of spindle action. This situation is not an evaluation of the importance of the problem, it reflects the difficulty of studying the control of a process whose mechanisms are not yet understood.

None the less, the question of how the spindle manages to be so accurate has received some attention. Nicklas and his co-workers have shown that incorrect chromosome arrangements on the meiotic spindle (those that would result in chromosomal non-disjunction) are intrinsically unstable. These workers have proposed that the attachment of chromosomes to the spindle is stabilized by the tension generated by the mitotic motors when a chromosome is in the correct, bipolar orientation (reviewed in Nicklas, 1985). Here is a beautiful example of control at a high level of structural organization. Although we do not yet understand how mitotic stability is developed by tension, the phenomena described have given us a direction for future work. Such work will be important because chromosome segregation errors that occur during gametogenesis can lead to congenital malformations that have profound medical and personal repercussions.

The controls of individual parts of mitosis are also important subjects for future work. We know that there is a change in the control of microtubule assembly at the onset of spindle formation. As described above, the centrosome goes through a transition that increases the number of microtubules initiated. This change occurs at about the time of nuclear envelope breakdown. The protein phosphorylations discussed by Vandré and Borisy are likely to be important here, but the aspect of cell cycle progression that turns on the relevant protein kinases is yet to be determined. The genetic approaches reviewed by Cabral should provide powerful tools for identifying the controls for entry into mitosis. Indeed, some recent work by Russell and Nurse (1987) on the yeast *Schisosaccharomyces pombe* has revealed a cascade of genes whose products are essential for setting up mitosis in a vegetatively growing cell. The biology of "maturation promoting factor" is revealing some of the pleotropic effects of one particularly important protein kinase in the cell cycles of embryos and probably of all cells (Gerhart *et al.*, 1985; Karsenti *et al.*, 1987; Hunter, 1987). Investigations of this kind are likely to be key for improved understanding of the problem.

As a further example of mitotic control, some cells in both animals and higher plants contain a mechanism that controls the orientation of the spindle axis and thus the plane of the ensuing cytokinesis. This aspect of mitotic regulation is, at the moment, more like a mystery than a defined set of scientific questions. So little is known about such mechanisms that one can hardly pose an experiment. None the less, careful descriptive work on such behaviour during embryogenesis in the nematode worm *Caenorhabditis elegans* has begun (Hyman and White, 1987) and this problem is likely to be among the most important for mitosis research during the not too far distant future.

At the moment, much of the interest in mitotic control is focused on the potential roles of calcium ions for regulating microtubule assembly, microtubule movement and the onset of anaphase. Some of the key facts related to this problem are reviewed in the chapters by Hepler and by Welsh and Sweet. There is evidence that many spindles contain a system of vesicles that can sequester calcium ions and release them at critical times. Elevated levels of calcium in the physiological range can induce spindle microtubule disassembly, and may contribute to the signals that start anaphase. The ubiquitous calcium-binding protein calmodulin is clearly a spindle protein, and it may serve to transduce transient changes in calcium concentration into biologically significant signals. What the signals are, however, and how they might work remain to be determined.

There are probably many issues pertinent to the regulation of spindle motors and microtubule polymerization that we are still too ignorant to appreciate. Moreover, at the end of anaphase, the chromosome arms usually contract in toward their centromeres and the nuclear envelope re-forms. While biochemical studies have strongly implicated histone phosphorylation in the regulation of chromatin folding and lamin phosphorylation in the control of nuclear envelope assembly (Burke and Gerace, 1986; Newport and Forbes, 1987), the relevant mechanisms for cellular regulation of the important protein kinases are still essentially unknown. One must conclude that spindle regulation is a fascinating subject for future work, but that at the moment our ignorance is too pervasive to permit rapid progress.

IV. From Where Will New Knowledge About Mitosis Come?

The principal questions concerning mitotic mechanism are now matters of molecular biology, broadly defined. The focus for the coming research will most likely be on the identity and character of the macromolecules that play key roles in the cellular events that have already been well described. In this context, what seems likely to be the most productive avenues for future research?

Spindle chemistry has proved to be extremely difficult as a result of

1. the small size of the mitotic spindle;
2. its lability and resistance to isolation in a pure form;
3. the lack of fully functional isolated spindles;
4. the tiny amounts of material available for work under most circumstances; and
5. the complexity of the isolates that are so far available.

None the less, some progress in spindle chemistry has been made. A few spindle microtubule-associated proteins have been identified, as reviewed in the chapter by Bloom and Vallee. Spindle tubulin has been characterized and compared with the subunits of other microtubules. Direct description of spindle chemistry is, however, hard enough that workers concerned have sought alternative methods for the identification and characterization of relevant proteins. The use of taxol as a way to prepare stable microtubules that can be isolated from mitotic cells, complete with associated proteins, has led Vallee and his co-workers to be able to identify a significant number of microtubule-associated proteins in the mitotic spindle of sea urchins. Their review in this book provides a description of this method. The biochemistry of microtubule-associated motors is also a promising avenue for mitotic research.

Descriptive biochemistry, in combination with cytochemical methods, is likely to identify a significant number of components from the mitotic apparatus, but one needs a functional approach to assess the roles of these putative spindle components in the mitotic process. The microinjection of antibodies or of proteins modified to disrupt their normal function provides one route for obtaining evidence on the function of particular molecules within the context of the living cell. Construction of vectors that carry the antisense gene for a hypothetically important mitotic protein offers another. All the techniques necessary for these investigations are in hand, but progress with the methods has so far been slow.

The analysis of protein function *in vivo* is naturally the realm of genetics. As summarized by Cabral, a significant amount of genetic research is underway on mitosis. It is disappointing that so little has so far been learned from a genetic approach, but as the biology of fungi and other microorganisms comes under ever closer scrutiny, the analysis of mitosis in these favourable organisms should improve our knowledge of the process. Certainly an organism like *Saccharomyces cerevisiae*, in which one can go back and forth between genetics, molecular biology, protein chemistry and immunology, offers great possibilities for taking a lead from one promising result and pushing it forward to yield a rather complete understanding of a molecular process.

Mitosis is unquestionably a tough nut to crack. The people lucky enough to have found an interest in this subject will not soon be out of work, and they are likely to be faced with enjoyable challenges as long as they can endure.

REFERENCES

Aist, J. R. and Berns, M. S. (1981). *J. Cell Biol.* **91**, 446–458.
Burke, B. and Gerace, L. (1986). *Cell* **44**, 639–652.
Cooke, C. A., Heck, M. M. S., Earnshaw, W. C. (1987). *J. Cell Biol.*, **105**, 2053–2067.
Gerhart, J., Cyert, M. and Kirschner, M. (1985). *Cytobios* **43**, 335–347.
Gorbsky, G. J., Sammak, P. J. and Borisy, G. G. (1987). *J. Cell Biol.* **104**, 9–18.
Hartwell, L. H., Dutcher, S. K., Wood, J. S. and Garvick, B. (1982). In *Advances in Yeast Molecular Biology: Recombinant DNA*, Vol. 1, pp. 28–38. Lawrence Berkely Laboratories Press, Berkeley.
Hisanaga, S., Tanaka, T., Mosaki, T., Sakai, H., Mabuchi, I. and Hiramoto, Y. (1987). *Cell Motil. Cytoskel.* **7**, 97–109.
Hunter, T. (1987). *Cell* **50**, 823–829.
Hyman, A. A. and White, J. G. (1987). *J. Cell Biol.*, **105**, 2123–2135.
Inoué, S. and Sato, H. (1967). *J. Gen. Physiol.* **50**, 259–284.
Karsenti, E., Bravo, R. and Kirschner, M. (1987). *Develop. Biol.* **119**, 442–453.
Kirschner, M. and Mitchison, T. (1986). *Cell* **45**, 329–342.
Leslie, R. J. and Pickett-Heaps, J. D. (1983). *J. Cell Biol.* **96**, 548–561.
Masuda, H. and Cande, W. Z. (1987). *Cell* **49**, 193–202.
McIntosh, J. R. (1981). In *International Cell Biology 1980–1981* (ed. H. G. Schweiger), pp. 359–368. Springer-Verlag, Berlin.
Mitchison, T. J. and Kirschner, M. W. (1985). *J. Cell Biol.* **101**, 766–777.
Mitchison, T. J., Evans, L., Schulze, E. and Kirschner, M. (1986). *Cell* **45**, 515–527.
Neighbors, B. W., Williams, R. C. and McIntosh, J. R. (1988). *J. Cell Biol.* **106**, 1193-1204.
Newport, J. and Forbes, D. J. (1987). *Annu. Rev. Biochem.* **56**, 535–565.
Nicklas, R. B. (1985). In *Aneuploidy* (ed. V. L. Dellarco, P. E. Voytek and A. Hollaender), pp. 183–195. Plenum Press, New York.
Nicklas, R. B. (1987). In *Chromosome Structure and Function: The Impact of New Concepts*, 18th Stadler Genetics Symposium (ed. J. P. Gustafson, R. Appels, and R. J. Kaufman). Plenum Press, New York, in press.
Pickett-Heaps, J. D., Tippit, D. H. and Porter, K. R. (1982). *Cell* **29**, 729–744.
Pickett-Heaps, J. D. (1986). *Trends Biochem. Sci.* **11**, 504–507.
Russell, R. and Nurse, P. (1987). *Cell* **49**, 559–567, 568–576.
Saxton, W. M. and McIntosh, J. R. (1987). *J. Cell Biol.* **105**, 875–886.
Schaap, C. J. and Forer, A. (1984). *J. Cell Sci.* **65**, 21–40.
Scholey, J. M., Porter, M. E., Grissom, P. M. and McIntosh, J. R. (1985). *Nature* **318**, 483–486.

Index